CRC HANDBOOK OF ELECTROPHORESIS

Volume I
Lipoproteins: Basic Principles and Concepts
Lena A. Lewis and Jan J. Opplt, Editors

Volume II
Lipoproteins and Disease
Lena A. Lewis and Jan J. Opplt, Editors

Volume III
Lipoprotein Methodology and Human Studies
Lena A. Lewis, Editor

Volume IV
Lipoprotein Studies of Nonhuman Species
Lena A. Lewis and Herbert K. Naito, Editors

CRC Handbook of Electrophoresis

Volume IV

Lipoprotein Studies of Nonhuman Species

Editors

Lena A. Lewis
Emeritus Staff
Divisions of Research and Laboratory Medicine
Cleveland Clinic Foundation
Clinical Professor of Chemistry
Cleveland State University
Cleveland, Ohio

Herbert K. Naito
Head, Lipid, Protein, Metabolic Disease Section
Department of Biochemistry
Division of Laboratory Medicine
Cleveland Clinic Foundation
Clinical Professor of Chemistry
Cleveland State University
Cleveland, Ohio

CRC Press Inc.
Boca Raton, Florida

Library of Congress Cataloging in Publication Data (Revised)
Main entry under title:

CRC handbook of electrophoresis.

  Bibliography: p.
  Includes index.
  CONTENTS: —v. 2. Lipoproteins in
disease.— —v. 4. Lipoprotein studies of nonhuman
species/editors, Lena A. Lewis, Herbert K. Naito. · ·
  1. Electrophoresis. I. Lewis, Lena Armstrong,
1910   . II. Opplt, Jan J., joint author. [DNLM:
1. Electrophoresis. 2. Lipoproteins. QU 85 C106 1980]
QD117.E45C18   543'.0871   78-10651
ISBN 0-8493-0571-3 (v. 1)

    Direct all inquiries to CRC Press, Inc., 2000 Corporate Blvd., N.W., Boca Raton, Florida, 33431.

© 1983 by CRC Press, Inc.

International Standard Book Number 0-8493-0571-3 (Volume I)
International Standard Book Number 0-8493-0572-1 (Volume II)
International Standard Book Number 0-8493-0573-X (Volume III)
International Standard Book Number 0-8493-0574-8 (Volume IV)

Library of Congress Card 78-10651
Printed in the United States

PREFACE

When the concept of a *CRC Handbook of Electrophoresis* series was first suggested, it was an exciting idea. Upon consideration, it became obvious that the field was very vast, and it was decided to concentrate on the use of electrophoresis in the study of lipoproteins. *CRC Handbook of Electrophoresis,* Volumes I and II, which were published in 1980, were devoted, as their titles show, respectively, to *Lipoproteins: Basic Principles and Concepts,* and *Lipoproteins in Disease.*

Volume III, which is to be published concurrently with this Volume IV, is subtitled *Lipoprotein Methodology and Human Studies.*

Volume IV is devoted to a coverage of lipoproteins of nonhuman species. Each chapter summarizes the results of investigations of lipoproteins by authors who have devoted much time to a complete understanding of the species or strain of species under consideration. This volume is designed to bring together such material and make available to the reader a vast amount of information which was previously available only in many different journals and books which were difficult if not impossible to obtain. Some of the information is entirely new. This book, which deals with lipoproteins of species from snakes and lizards to nonhuman primates, will fill a great need for the scientist, physician, or educated layman who is interested in lipid and lipoprotein metabolism in general but especially as a background for possible further elucidation of their role in the etiology of atherosclerosis.

EDITORS

Dr. Lena A. Lewis is Emeritus Staff at The Cleveland Clinic Foundation in the Division of Laboratory Medicine and in the Division of Research. She is a Clinical Professor of Chemistry at Cleveland State University. Dr. Lewis has an A.B. from Lindenwood College, St. Charles, Missouri (1931), an M.A. from Ohio State University (1938), and the Ph.D. from Ohio State University. She served for 10 years on the editorial board of *Clinical Chemistry,* and is author of *Electrophoresis in Physiology* (1950; 2nd edition, 1960, Charles C Thomas, Springfield, Illinois). Dr. Lewis received an honorary L.L.D. degree in 1952 from Lindenwood College, was elected a fellow in the New York Academy of Science in 1977, and received the Boehringer-Mannheim Award from the American Association of Clinical Chemists for outstanding contributions to clinical chemistry in the field of lipids and lipoproteins.

Dr. Lewis was co-editor with Jan J. Opplt of Volumes I and II of the *CRC Handbook of Electrophoresis.* She is a contributor to *Endocrinology, Clinical Endocrinology and Metabolism,* the *American Journal of Medicine, Clinical Chemistry*, and the *American Journal of Physiology, Science.*

Dr. Lewis was the President (1970—1971) of the Northeast Ohio chapter of the American Association for Clinical Chemists. She also is a member of the American Physiological Society, the Endocrine Society, and the American Association for the Advancement of Science AAAS; Fellow of the Atherosclerosis Council of the American Heart Association, fellow of the New York Academy of Science, and member of the Quota Club (International Women's Service Club).

She is listed in *American Men of Science, Who's Who in America* (Midwest section), *Who's Who in American Education, American Women in Science, Personalities of the West & Midwest,* and *Who's Who in the World of Women.*

Dr. Herbert K. Naito is Head of the Lipid, Protein, and Metabolic Disease Section in the Biochemistry Department of the Division of Laboratory Medicine and Division of Research at The Cleveland Clinic Foundation. He is also Clinical Professor of Chemistry in the Department of Chemistry at Cleveland State University. He received his A.B. (1963) and M.A. (1965) degrees from the University of Northern Colorado (Greeley, Colorado) and his Ph.D. (1971) at Iowa State University of Science and Technology (Ames, Iowa).

He serves at present on the Editorial Boards of the *Journal of the American College of Nutrition, Atherogenesis,* and *Clinical Physiology and Biochemistry.* He has edited two books, *Nutritional Elements and Clinical Biochemistry,* Plenum Press, New York (1980) and *Nutrition and Heart Disease,* Spectrum Publications, Inc., New York (1982). Dr. Naito serves on the Board of Directors for the National Academy for Clinical Biochemistry, and the American College of Nutrition, and is Past-President of the American Association for Clinical Chemistry, Northeast Ohio Section. He is a Fellow of the Council on Atheriosclerosis of the American Heart Association, Council on Cardiovascular Disease of the American College of Nutrition, Member of the American Association of Pathologists, the Society for Experimental Biology and Medicine, the American Institute of Nutrition, and the Endocrine Society.

In 1977 he was awarded the American Association for Clinical Chemistry Young Investigator Award and, in 1981, the National Academy of Clinical Biochemistry George Grannis Award. He was recently listed in *Who's Who in America* (Midwest section). His major research interest has centered on the study of lipoproteins and their role in atherosclerosis.

CONTRIBUTORS

Petar Alaupovic, Ph.D.
Head, Laboratory of Lipid and
 Lipoprotein Studies
Oklahoma Medical Research Foundation
Oklahoma City, Oklahoma

M. John Chapman, Ph.D.
Maître de Recherches
Laboratoire de Recherche sur le
 Métabolisme des Lipides, Unité 35
Institut National de la Santé et de la
 Recherche Médicale
Hôpital Henri Mondor
Créteil, France

Nassrin Dashti, Ph.D.
Staff Scientist
Laboratory of Lipid and Lipoprotein
 Studies
Oklahoma Medical Research Foundation
Oklahoma City, Oklahoma

Mark Fitch
Research Associate
Department of Nutritional Sciences
University of California
Berkeley, California

Gunther M. Fless, Ph.D.
Research Associate (Assistant Professor)
Department of Medicine
University of Chicago
Chicago, Illinois

Kenneth C. Hayes, D.V.M., Ph.D.
Associate Professor of Nutrition
Harvard School of Public Health
Department of Nutrition
Boston, Massachusetts

Kang-Jey Ho, M.D., Ph.D.
Professor of Pathology
University of Alabama in Birmingham
Medical Center
Birmingham, Alabama

Richard L. Jackson, Ph.D.
Professor of Pharmacology and Cell
 Biophysics, Biological Chemistry, and
 Medicine and
Head, Division of Lipoprotein Research
University of Cincinnati College of
 Medicine
Cincinnati, Ohio

Arthur W. Kruski, Ph.D.
Assistant Professor of Pathology and
 Biochemistry
Department of Pathology
University of Texas Health Science
 Center
San Antonio, Texas

Lena A. Lewis, Ph.D.
Emeritus Staff
Divisions of Research and of Laboratory
 Medicine
Cleveland Clinic Foundation, and
Clinical Professor of Chemistry
Cleveland State University
Cleveland, Ohio

Robert W. Mahley, M.D., Ph.D.
Director
Gladstone Foundation Laboratories for
 Cardiovascular Disease
Cardiovascular Research Institute
Departments of Pathology and Medicine
University of California, San Francisco
San Francisco, California

Herbert K. Naito, Ph.D.
Head, Lipid, Protein, Metabolic Disease
 Section
Department of Biochemistry
 Division of Laboratory Medicine
Cleveland Clinic Foundation, and
Clinical Professor of Chemistry
Cleveland State University
Cleveland, Ohio

Robert J. Nicolosi, Ph.D.
Associate Professor of Comparative
  Pathology
Harvard Medical School, and
Chairman, Division of Nutrition
New England Regional Primate Research
  Center
Southborough, Massachusetts

Rosemarie Ostwald, Ph.D.
Professor of Nutrition
Department of Nutritional Sciences
University of California
Berkeley, California

Irvine H. Page, M.D.
Emeritus Director of Research
Cleveland Clinic
Cleveland, Ohio

Shi-Kaung Peng, M.D., Ph.D.
Associate Professor of Pathology
Albany Medical College of Union
  University, and
Laboratory Service
Veterans Administration Medical Center
Albany, New York

Donald L. Puppione, Ph.D.
Associate Research Molecular Biologist
University of California
Molecular Biology Institute
Los Angeles, California

David C. K. Roberts, Ph.D.
Lecturer
Human Nutrition Unit
Department of Biochemistry
University of Sydney
Sydney, New South Wales
Australia

Catherine A. Rolih
Research Assistant
Department of Medicine
University of Chicago
Chicago, Illinois

Angelo M. Scanu, M.D.
Professor
Departments of Medicine and
  Biochemistry
University of Chicago
Chicago, Illinois

C. Bruce Taylor, M.D.
Research Investigator
Research Service
Veterans Administration Medical Center,
  and
Research Professor of Pathology
Albany Medical College of Union
  University
Albany, New York

Karl H. Weisgraber, Ph.D.
Associate Director
Gladstone Foundation Laboratories for
  Cardiovascular Disease
Cardiovascular Research Institute, and
Associate Professor of Pathology
Department of Pathology
University of California, San Francisco
San Francisco, California

NOMENCLATURE OF LIPOPROTEINS

While the nomenclature of lipoproteins has developed over the years, and was especially adapted to description of the lipoproteins of the human being and of the primates, the terms are for most animal lipoproteins entirely adequate. In any of the chapters covering different species in this book, if special consideration of nomenclature is involved, the terms used are clearly defined. The nomenclature used in this volume is that used in *CRC Handbook of Electrophoresis, Volumes I, II,* and *III,* and is summarized in the following table (Table I), and in the figure (Figure 1).

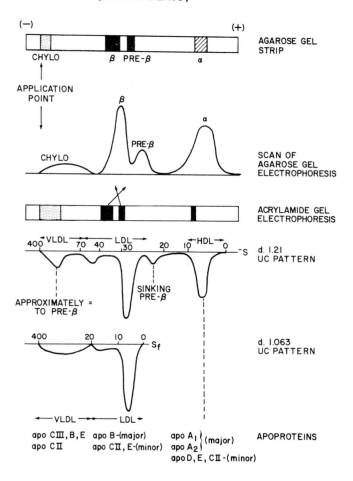

FIGURE 1. A diagrammatic relationship between electrophoretic and ultracentrifugal patterns of serum lipoproteins. Electrophoretic mobilities of fractions in different media are purely diagrammatic and have been drawn to emphasize the relation to fractions resolved by ultracentrifugation.

Table 1
VARIOUS NOMENCLATURES OF LIPOPROTEINS

Method of study on which nomenclature based	Name of fraction
α-Lipoproteins	
Electrophoresis	α-lipoproteins, α-Lp, lipoproteins with electrophoretic mobility of $α_1$-globulins; they may be resolved as single or multiple bands depending on type of support media
Ultracentrifugation	HDL, high-density lipoprotein (d 1.063—1.21 g/mℓ); HDL_2, subclass of HDL (d 1.063—1.125 g/mℓ); HDL_3, subclass of HDL (d 1.125—1.21 g/mℓ); flotation rate at d 1.21 of $-SO$—10.
Apolipoprotein composition — chemical, immunologic	Apo A, apolipoprotein A consisting of two nonidentical polypeptides, A-I and A-II: A-I contains glutamic acid as C-terminal and aspartic acid as *N*-terminal amino acid A-II contains glutamic acid as C-terminal and pyrrolidine carboxylic acid as *N*-terminal amino acid apo D, apolipoprotein D present in HDL_3; apo E_{1-3}, apolipoprotein E_{1-3}, present in HDL; apo C-II, apolipoprotein C-II present in HDL
Pre-β-Lipoproteins	
Electrophoresis	Pre-β-lipoproteins, pre-β-Lp, lipoproteins with electrophoretic mobility of $α_2$-globulins; they have this mobility when agarose, paper, or starch powder is used as support medium; they migrate slower than β-lipoprotein when gels with sieving effect, such as acrylamide or starch gel, are used; pre-β-Lp may be resolved as single or multiple bands
Ultracentrifugation	VLDL, very low-density lipoproteins (d < 1.006 g/mℓ); isolated from serum after previous removal of chylomicron; flotation rate at d 1.21 of $-S70$—400
Apolipoprotein composition — chemical, immunologic	Apolipoprotein of VLDL contains apo C-II, B, and E_{1-3} polypeptides; apo C is an apolipoprotein consisting of 3 nonidentical polypeptides: C-I is characterized by *N*-threonine and *C*-serine, C-II by *N*-threonine and *C*-glutamic acid, C-III by *N*-serine and *C*-alanine
β-Lipoproteins	
Electrophoresis	β-lipoproteins, β-Lp, lipoproteins with electrophoretic mobility of β-globulins; β-Lp may be resolved as single or double bands, depending on buffer and support medium used
Ultracentrifugation	LDL, low-density lipoproteins (d 1.006—1.063 g/mℓ); LDL_1, subclass of LDL (d 1.006—1.019 g/mℓ); LDL_2, subclass of LDL (d 1.019—1.063 g/mℓ); flotation rate at d 1.21 of $-S$ 25—40; at d 1.063, S_f 0—20 may be divided into $-S$ 40—70, i.e., S_f 12—20, intermediate density fraction and $-S$ 25—40, S_f 0—12 LDL
Apolipoprotein composition — chemical, immunologic	Apo B, apolipoprotein B is major apoprotein of β-lipoprotein; β-Lp also contains C-II and E_{1-3}
Other fractions	
Electrophoresis on agar	Lp-X, lipoprotein X, a lipoprotein of β- or slow β-globulin mobility and low density, characteristically found in obstructive jaundice patients' sera; best identified by electrophoresis on agar, where it has unusual property of migrating to γ-globulin position
Immunologic and genetic studies	Lp(a) is a polymorphic form of β-Lp which is of importance in genetic studies, and in those patients who have received multiple transfusions

TABLE OF CONTENTS

Volume IV

Introduction and Background

INTRODUCTION AND BACKGROUND

Lena A. Lewis and Herbert K. Naito

Appreciation of the importance of lipoproteins of blood, body fluids, and tissues has increased so rapidly since 1950 that it is at the present, 1983, a major undertaking to cover the literature dealing with a single species. For any consideration of a comparison between species it is almost overwhelming. This volume is designed to make information and references dealing with the lipoproteins of many species more accessible. Chapters have been prepared by people who have worked with the species under consideration and can give critical evaluation to the materials included.

When this volume was designed, some earlier studies had already stimulated interest in the use of a given species for lipoprotein investigation. In 1952 Lewis[1] pointed out the difference in the lipoprotein patterns of different species of laboratory animals. Electrophoretic patterns obtained in more recent studies are shown in Figures 1 and 2. Thus dogs and rats, species in which atherosclerosis was very difficult to produce experimentally and in which it practically was unknown to occur spontaneously, had low levels of serum β-lipoprotein (β-Lp). Their total cholesterol levels were lower than those of the human being and most of the cholesterol was in the α-lipoprotein (α-Lp) fraction. In contrast, the pig had a serum lipoprotein pattern very much more like that of the human being.[2] They are a species in which atherosclerosis can be induced experimentally with ease. It was also noted that two strains of miniature swine, the one short and fat, the other long and lean, had very different serum lipoprotein patterns when fed the same diet. This was probably the earliest observation of differences in serum lipoprotein patterns in different strains of a laboratory animal.[2] Variations in lipid and lipoprotein of serum or body fluid of additional strains within a species have been reported since these early studies on swine. Species thus studied which showed differences between strains include rats,[3] mink,[4] ponies,[5] and pigeons.[6] Much more precise identification of apoproteins and of lipoprotein families by expanding and rapidly developing techniques has aided these investigations.

With increased appreciation of the importance of lipoproteins as an etiologic agent in atherosclerosis and the need for an animal species in which the disease can be produced and treated, more species have been evaluated, including many mammals, reptiles, amphibians, fish, insects, and birds[1-21] (Tables 1, 2).

With the evolution of time, many diverse animals have evolved to their present form and behavior as a result of gradual change over an immense period of geologic time (Table 3). The exact relationships of the members of the animal kingdom are often vague according to our present knowledge. Most existing organisms are related indirectly to each other through common ancestors which are now extinct (see Figure 3). Most common ancestors were sufficiently generalized in structure to give rise to many divergent groups, but such ancestors have either undergone evolution or else became extinct because of their inability to adapt to a changing environment. If similarity of structure and function mean anything in an evolutionary interpretation, then it is obvious that certain groups of animals are closely connected because the evidence stands out clearly. It is a generalization widely accepted in biology that if two different organisms share many common traits it is logical to assume that there is a relationship basis for this similarity and that it has not been due to convergent or coincidental evolution. Any attempts to integrate the lipoprotein patterns of various animal models to their significance to lipid transport and to the phylogenetic origin of lipoproteins still remains speculative. An ambitious attempt by Chapman[22] to review the animal lipoprotein chemistry, structure, function, and comparative aspects reveals the complexity of the evolution of serum lipoproteins as a function of the specialization of the parent organism.

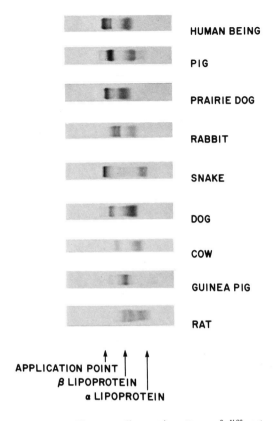

HUMAN BEING

PIG

PRAIRIE DOG

RABBIT

SNAKE

DOG

COW

GUINEA PIG

RAT

APPLICATION POINT
β LIPOPROTEIN
α LIPOPROTEIN

FIGURE 1.   The serum lipoprotein patterns of different species of animals are shown. Agarose gel electrophoresis, barbital buffer 0.05 *M*, pH 8.6 was used. The patterns were stained with methyl alcoholic fat red-7B stain.

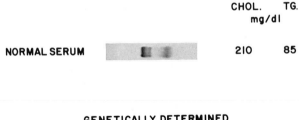

	CHOL.	TG.
	mg/dl	
NORMAL SERUM	210	85

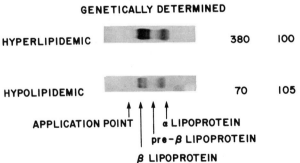

GENETICALLY DETERMINED

HYPERLIPIDEMIC	380	100
HYPOLIPIDEMIC	70	105

APPLICATION POINT   α LIPOPROTEIN
pre-β LIPOPROTEIN
β LIPOPROTEIN

FIGURE 2.   The serum lipoprotein patterns of genetic variants of the human being are shown. Agarose gel electrophoresis, barbital buffer 0.05 *M*, pH 8.6 was used. The patterns were stained with methyl alcoholic fat red-7B stain.

Table 1
REPRESENTATIVE EARLY STUDIES OF SERUM LIPOPROTEINS OF DIFFERENT ANIMAL SPECIES

Author(s)	Year — Ref. no.	Material studied	Evaluation	Characteristics	Methods and electrophoretic conditions
Lewis, L. A.	1952—1	Material studied sera of eight different species of laboratory animals	The species in which a low concentration of low density, i.e., β-Lp and a high proportion of α-Lp was present were resistant to development of atherosclerosis, while those with higher β-Lp and less α-Lp were prone to atherosclerosis	Each species had a typical lipoprotein pattern; guinea pig was unique in showing no high-density α-Lp	Ultracentrifugation at d 1.21 and 1.063. Free-moving boundary electrophoresis pH 7.8
Lewis, L. A. and Page, I. H.	1956—2	Two strains of miniature swine, one long-lean, the other short-fat	The serum lipoprotein pattern of the two strains was markedly different even when all animals were on the same controlled food intake and diet	The lipoprotein patterns of the swine was somewhat like that of the human being in having a greater proportion of lipoprotein as β-Lp, than is found in most laboratory animals	Free-moving boundary and paper electrophoresis and ultracentrifugation d 1.21
Belyaev, D. K., Baranov, O. K. et al.	1974—4c	Sera of adult mink *Mustela vison* of both sexes	These studies permitted identification of 5 Lpm allotypes in minks; Lpm 1,2,3,4,5	Allotypic proteins are those which have antigenic determinants or variants in which individuals of one species differ from one another; these proteins are called *allotypes*; one of the mink allotypes of lipoproteins in α_2-globulin position, similar to that of α_2-esterases; another lipoprotein allotype has α-mobility; it also has strong esterase activity	High-voltage microimmuno electrophoresis; Sudan black stain for lipids; agar or agarose in varying concentrations used in double diffusion studies; electrophoresis using 160 V for 20—30 min; buffer; pH 8.6, 0.025 *M* veronal-medinal; precipitin lines in double diffusion did not form in agarose, in some instances when well-resolved bands formed in agar gel
Baranov, O. K., Belyaev, D. K.	1975—4b	Male and female mink; same methods as above	Precipitin and staining and enzymatic reactivity as in previous studies were the basis of evaluation of sera of 342 mink studied; all 8 allotypes show unusual property of reacting differently when studied by double diffusion technique in agar or in agarose	New allotypes 6,7,8 had similar electrophoretic properties to earlier described allotypes, i.e., α_2-globulin mobility, lipid stainable and esterase activity. All allotypes were already present in sera of newborn mink	Methods same as for 1974 (mink) except besides a mixture of α and β naphthylacetate, each substrate was also used individually to reveal esterase activity

Table 1 (continued)
REPRESENTATIVE EARLY STUDIES OF SERUM LIPOPROTEINS OF DIFFERENT ANIMAL SPECIES

Author(s)	Year — Ref. no.	Material studied	Evaluation	Characteristics	Methods and electrophoretic conditions
Baranov, O. K., Savina, M. A., Belyaev, D. K.,	1976—4e	Mink sera from mink bred in Experimental Farm, Experimental Sciences, U.S.S.R., Siberian Branch	Precipitation patterns obtained using agar and agarose for gel varied greatly; two precipitin lines were resolved on agar, when only one was visible on agarose; using lipid, protein, and esterase stains it was concluded that all five allotypes should be assigned to common protein molecule of α_2-globulin mobility; this may likely be a VHDL	Characteristic mink very low-density (VLDL) low-density (LDL) and high-density (HDL) lipoproteins lacked Lpm 1, Lpm 2, Lpm 3, Lpm 4 and Lpm 5; they were all present in very high-density (VHDL) lipoprotein. These Lpm specificities are present in newborn mink sera, and levels appear stable throughout life; five allotypes Lpm 1, Lpm 2, Lpm 3, Lpm 4 and Lpm 5 have been identified in the lipoprotein of α_2-globulin mobility in mink sera, and their usefulness in genetic studies established; more recently Lpm 6, Lpm 7 and Lpm 8 have been characterized	Antisera to whole mink sera or fractions with Freund's adjuvant were developed in mink and in rabbits; lipoprotein fractions were separated by ultracentrifugation using Havel's technique;[4f] gels for immunoelectrophoresis and immunodiffusion were prepared on barbiturate buffer pH 8.6 0.025 M; agar or agarose gels were used
Baranov, O. K. and Ermolaev, V. I.,	1977—4a	Serum of mink (*Mustela vison*) after 16-hr fast	The lipoprotein fractions obtained by UC included fraction (3) HDL-higher density d<1.10-1.21 (1) VLDL d 1.006 (2) LDL d 1.006—1.063 HDL lower density d 1.063—1.10 (4 and 5) VHDL d >1.21 The immunologic evaluation by double diffusion and electrophoresis makes the reviewer feel that this species is especially suitable for the "family" concept of lipoprotein classification proposed by Alaupovic[4e]	Mink sera contains in addition to chylomicron at least five lipoproteins; two of these identified as #2 and #3 contain most of the lipid; #2 was of d between 1.006—1.10, while #3 had broad spectrum of d 1.006—1.21 but judging by antigenic determinants was a distinct fraction; #4 and #5 were VHDL; fractions 2 and 4 were unique in having genetic alloantigenic determinants, detection of Lpm in the VHDL fraction makes mink alloantigenic determinants unique as in most species they are chiefly in LDL fractions, or some in HDL but never in VHDL	Chylomicron removed after UC at 20,000 g for 40 min; remaining serum separated in the UC at varying density; 24-hr spin at 105,000 g at 5°C; fractions obtained were dialyzed against 0.025 veronal® medinal® buffer and studied by immunoelectrophoresis in agar gel and dried plates stained for lipid with Sudan black; antisera to whole mink sera developed in rabbits was used; antisera developed in rabbits to mink Lpm determinants was also used, as was antisera to allotype 6

Delcourt, R.	1969—8	Material studied 300 vertebrate species: crocodiles, llamas, alpaca, to humans		Results obtained by electrophoresis and by chromatography were in good agreement and led to the conclusion that in general most vertebrate species had lipoproteins of two distinct electrophoretic mobilities and molecular sizes and contained lipids and apo-proteins of characteristic properties; in general, large molecular weight lipoproteins were rich in glyceride; lower molecular weight lipoproteins contained relatively more PL	Studied by paper electrophoresis, acrylamide gel electrophoresis, chromatography on agarose and high-molecular-weight dextrans
Alexander, C. and Day, C. E.	1973—10	36 mammals, reptiles and amphibians salamander, frog, turtle, garter snake, red barred garter snake, fish, trout, carp, dogfish, birds (quail, pigeon, pheasant, chicken, goose, turkey, duck)	Quantitative distribution of lipoprotein classes measured after electrophoresis on agarose gel	Amphibians and reptiles had simplest patterns, the frog and red barred garter snake having only one lipid-stained band which was of slow mobility; the electrophoretic mobility of the fish lipoproteins varied greatly, that of the trout having several bands of high electrophoretic mobility in addition to a very slow migrating band; pig, oppossum and garter snake serum contained relatively large amounts of LDL and in this respect resembles the lipid distribution of human serum; most species had at least two clearly resolved lipid-stained bands after electrophoresis	Agarose gel electrophoresis using Bio-Rad Laboratory's system including staining of electrophoresed gels overnight at 37° with oil red O stain. Electrophoretic mobility was gauged by tracking dye, bromophenol blue and albumin
Goldstein, S. and Chapman, M. J.,	1976—11	A survey of many species including: (1) mammals, humans, nonhuman primates pigs, guinea pigs; (2) birds, chickens; (3) reptiles, snakes, *Pryas mucosus*, *Bungaris fascicaleus*; (4) fish (Bonnaterre, Hagfish)	Visual inspection and electrophoretic mobility measurements made on lipoprotein patterns after staining with Sudan black for lipids	Study of serum LDL lipoproteins of selected numbers from different species of mammals, birds, reptiles and fish indicate a relatively high preservation of LDL structure through evolution; this is based on the degree of cross immunoreactivity and the presence of lipoproteins of LDL characteristics in the materials studied; these observations suggest the continued important transport and metabolic function of these lipoproteins during developmental stages	Immunoelectrophoresis;[13] Electrophoresis in agarose;[12] 0.025 M Veronal® buffer pH 8.6; water cooling of electrophoresis cell during electrophoresis at 4 V/cm; antisera against whole serum were prepared and also against lipoproteins prepared by ultracentrifugation

Table 1 (continued)
REPRESENTATIVE EARLY STUDIES OF SERUM LIPOPROTEINS OF DIFFERENT ANIMAL SPECIES

Author(s)	Year — Ref. no.	Material studied	Evaluation	Characteristics	Methods and electrophoretic conditions
Zilversmit, D. B., Clarkson, T. B., and Hughes, L. B.	1977—17	Mink (*Mustela vison*) male and female	Electrophoretic patterns were examined for relative distribution of lipid stainable material; arteries were examined for plaque formation grossly and microscopically	Plasma cholesterol levels of mink whether on usual type of diet or a cholesterol-free diet were comparable to those of the human being; the distribution of the cholesterol in mink lipoproteins was very different from the human, approximately 80% of cholesterol being in high-density α-Lp in young mink and 70% HDL in older mink; mink as old as 8 years had no atherosclerotic plaques on aorta or coronary arteries; the high % of α-Lp may have had protective effect; the lipoproteins were resolved by electrophoresis into α and β fractions; the α-Lp appeared to probably be a double fraction with greater concentration in the fast moving part	Plasma was separated from blood drawn from anesthetized animals; ultracentrifugal fractionation by method of Hatch and Lees; agarose electrophoresis for plasma lipoprotein by method of Noble
Thomas, K. K. and Gilbert, L. I.,	1968—18,19	Hemolymph of American silk moth, *Hyalophora cecropia*; pupae of the giant silk moth may yield several milliliters of hemolymph; the moths were stored several months at 6° before sample collection; adult development had not been started	Chemical, disc electrophoresis, analytical UC; molecular weight by Archibald method; while all three lipoprotein classes floated as single peaks in analytical UC disc electrophoresis usually resolved LDL into two or three bands; HDL into three or four bands; VHDL two bands; faster migrating sharp narrow band, slower more diffuse	Dialysis against pH 6.7 buffer for longer than 10—15 hr led to breakdown of purified lipoprotein fraction; lipids pres- tained with lipid crimson before electro- phoresis (disc); LDL contained relatively higher concentration of lipid stainable material than HDL and VHDL; VHDL had highest % of PL, the LDL the least; diglyceride is more than 66% of hemolymph total lipid; the understanding of hemolymph lipopro- tein and lipid chemistry gives a firm ba- sis for understanding their hormonal transport	5 μmol glutathione added per mℓ hemolymph; to the hemolymph after removal of any cells EDTA was added; lipoproteins sepa- rated ultracentrifugally at appro- priate densities for hemolymph lipoproteins LDL d 1.063, HDL d 1.158—1.17; VHDL d 1.26; special techniques involved in VHD (see original article); frac- tions evaluated by disc electrophoresis

Table 2
LIPOPROTEINS OF HEMOLYMPH OF AMERICAN SILK MOTH, *HYALOPHORA CECROPIA* CLASS OF LIPOPROTEIN ISOLATED BY ULTRACENTRIFUGATION[18]

	LDL	HDL	VHDL
Density	1.046—1.063	1.156—1.17	1.26

Special note of lipid characteristics.
Phosphatidyl choline major PL of all fractions.
About 75% of lipid in HDL is diglyceride.
Electrophoretic properties (disc electrophoresis) after
 prestaining with lipid crimson.
Each lipoprotein fraction can be resolved electrophoret-
 ically into several bands.

Recognition of spontaneously occurring dyslipoproteinemia within a species has also provided additional information.[15]

As has been clearly demonstrated in human beings, rabbits and pigs,[16] mink (4 abc) have been found to have in their sera genetically determined antigenic determinants called allotypes. The mink Lpm allotype differs from that of most other species where the Lp allotypes belong with classes with low and very low density and electrophoretic mobility less than that of high-density lipoprotein (HDL). The Lpm allotypes of mink have the lowest electrophoretic mobility of any lipoprotein fraction when studied by immunoelectrophoresis, suggesting that it probably has a low density also. The precipitin band of the allo-lipoprotein vs. antisera showed esterase activity. The results indicate the existence of eight Lpm allotypes in mink, but animals with only one Lpm allotype have been demonstrated also. The major part of the lipoproteins of the mink like that of the dog is found in the α-lipoprotein (HDL) fraction.

Recognition of the need to know the lipoprotein pattern of a species to understand the metabolism and development of the animal has resulted in further broadening of the groups studied.[22] In all but three species thus far evaluated, at least two distinct types or classes of lipoproteins have been demonstrated when their body fluids (serum or hemolymph) was studied. Guinea pig, frog, and salamander sera showed only one lipid-stained band by agarose electrophoresis. Differences in lipoprotein characteristics between species were shown by differences in electrophoretic mobility and in many studies by differences in ultracentrifugal flotation rate, in immunologic, chromatographic, and precipitation properties.

From these preliminary studies no clear cut conclusions can be drawn concerning the evolutionary processes involved in the development of the serum lipoproteins.

This present volume, *Handbook of Electrophoresis,* Volume IV, collates the latest, comprehensive review of a wide range of common and not-so-common vertebrate animals which should provide an immense wealth of information for all those interested in the field of lipoprotein metabolism.

Table 3
GEOLOGICAL TIME SCALE[23]

Era	Period	Epoch	Time at beginning of each period (millions of years)	Geological events and climate	Biological characteristics
Cenzoic (age of mammals)	Quaternary	Recent	0.025	End of fourth ice age; climate warmer	Dominance of modern man; modern species of animals and plants
		Pleistocene	0.06—1.0	Four ice ages with valley and sheet glaciers covering much of North America and Eurasia; continents in high relief; cold and mild climates	Modern species; extinction of giant mammals and many plants; development of man
	Tertiary	Pliocene	12	Continental elevation; volcanic activity; dry and cool climate	Modern genera of mammals; emergence of man from man-apes; peak of mammals; invertebrates similar to modern kinds
		Miocene	25	Development of plains and grasslands; moderate climates; Sierra mountains renewed	Modern subfamilies rise; development of grazing mammals, first man-apes; temperate kind of plants; saber-toothed cat
		Oligocene	34	Mountain building; mild climates	Primitive apes and monkeys; whales; rise of most mammal families; temperate kind of plants; archaic mammals extinct
		Eocene	55	Land connection between North America and Europe during part of epoch; mountain erosion; heavy rainfall	Modern orders of mammals; adaptive radiation of placental mammals; subtropical forests; first horses
		Paleocene	75	Mountain building; temperate to subtropical climates	Dominance of archaic mammals; modern birds; dinosaurs all extinct; placental mammals; subtropical plants; first tarsiers and lemurs
Mesozoic (age of reptiles)	Cretaceous		130	Spread of inland seas and swamps; mountains (Andes, Himalayas, Rocky, etc.) formed; mild to cool climate	Extinction of giant land and marine reptiles; pouched and placental mammals rise; flowering plants; gymnosperms decline
	Jurassic		180	Continents with shallow seas; Sierra Nevada Mountains	Giant dinosaurs; reptiles dominant; first mammals; first toothed birds
	Triassic		230	Continents elevated; widespread deserts; red beds	First dinosaurs; marine reptiles; mammal-like reptiles; conifers dominant

Era	Period	Time (millions of years)	Physical conditions	Biological events
Paleozoic	Permian	260	Rise of continents; widespread mountains; Appalachians formed; cold, dry, and moist climate; glaciation; red beds	Adaptive radiation of reptiles which displace amphibians; many marine invertebrates extinct; modern insects; evergreens appear
(Age of amphibians)	Pennsylvanian	310	Shallow inland seas; glaciation in Southern Hemisphere; warm, moist climate, coal swamp-forests	Origin of reptiles; diversification in amphibians; gigantic insects
	Mississippian	350	Inland seas; mountain formation; warm climates; hot swamp lands	Amphibian radiation; insects with wings; sharks and bony fish; crinoids
	Devonian	400	Small inland seas; mountain formation; arid land; heavy rainfall	First amphibians; mostly freshwater fish; lungfish and sharks; forests and land plants; brachiopods; wingless insects; bryozoans and corals
(Age of fishes)	Silurian	425—430	Continental seas; relatively flat continents; mild climates; land rising, mountains in Europe	Eurypterids; fish with lower jaws; brachiopods; graptolites; invasion of land by arthropods and plants
(Age of invertebrates)	Ordovician	475	Oceans greatly enlarge; submergence of land; warm mild climates into highest latitudes	Ostracoderms (first vertebrates); brachiopods; cephalopods; trilobites abundant; land plants; graptolites
	Cambrian	550	Lowlands; mild climates	Marine invertebrates and algae; all invertebrate phyla and many classes; abundant fossils; trilobites dominant
Proterozoic (Precambrian)		2000	Volcanic activity; very old sedimentary rocks; mountain building; glaciations; erosions; climate warm moist to dry cold	Fossil algae 2.6 billion years old; sponge spicules; worm burrows; soft-bodied animals; autotrophism established
Archeozoic (Precambrian)		4000—4500	Lava flows, granite formation; sedimentary deposition; erosion	Origin of life; heterotrophism established

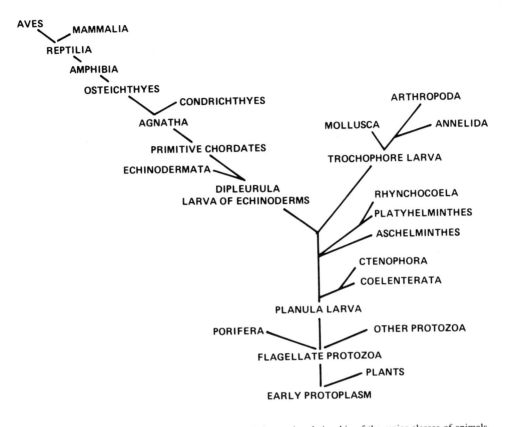

FIGURE 3.   A diagrammatic representation of the phylogenetic relationship of the major classes of animals.

REFERENCES

1. **Lewis, L. A.,** Electrophoretic and Ultracentrifugal Analysis of Lipoproteins, *Proc. Council for High Blood Pressure Res.,* 1, 19, 1952.
2. **Lewis, L. A. and Page, I. H.,** Hereditary obesity: relation to serum lipoprotein concentration in swine, *Circulation,* 14, 55, 1956.
3a. **Yamori, Y., Ohta, K., Morie, R., Ohtaka, M., Nara, Y., and Coshima, A.,** A new model for cerebral thrombosis and its pathogenesis, *Jpn. Heart J.,* 20 (Suppl. 1), 343, 1979.
3b. **Yamori, Y., Horie, R., Akiguchi, I., Ohtaka, M., Nara, Y., and Fukase, M.,** New models of SHR for studies of stroke and atherogenesis, *Clin. Exp. Pharmacol. Physiol. Suppl.,* 3, 199, 1976.
3c. **Yamuri, Y.,** A selection of arteriolipidosis-prone rats (ALR), *Jpn. Heart J.,* 18, 602, 1977.
4a. **Baranov, O. K. and Ermolov, V. I.,** Immunoelectrophoretic analysis of mink serum lipoproteins fractionated by the method of preparative ultracentrifugation (transl.) *Biokimiya,* 42, 60, 1977.
4b. **Baranov, O. K. and Belyaev, D. K.,** Three new allotypes of mink serum α_2-lipoprotein Lpm 6, Lpm 7 Lpm 8 (transl.) *Genetika,* 11, 32, 1975.
4c. **Belyaev, D. K., Baranov, O. K., Savina, M. A., Yuriahima, N. A., Titonlu, N. V., and Evsikov, V. I.,** Identification of five allotypes of serum α_2-lipoprotein esterases in American mink (*Mustela vison*) (transl.) *Genetika,* 10, 62, 1974.
4d. **Baranov, O. K., Savina, M. A., and Balyaev, D. K.,** Immunogenetic study on the polymorphism of serum alpha$_2$-lipoproteins in mink. I. Identification and genetic control of five Lpm allotypes, *Biochem. Genet.,* 14, 3, 1976.
4e. **Alaupovic, P.,** *Handbook of Electrophoresis,* Vol. 1, Lewis, L. A. and Opplt, J. J., Eds., CRC Press, Boca Raton, Fla., 1980, 27—46.

4f. **Havel, R. J., Eder, H. A., and Bragdon, J. H.,** The distribution and chemical composition of ultracentrifugally separated lipoproteins in human serum, *J. Clin. Invest.,* 34, 345, 1955.

5a. **Robie, S. M., Janson, C. H., Smith, C., and O'Connor, J. T., Jr.,** Equine serum lipids: lipid composition and electrophoretic mobility of equine serum lipoprotein fractions, *Am. J. Vet. Res.,* 36, 1715, 1975.

5b. **Wensing, T., Van Gent, C. M., Schotman, A. J. H., and Kronman, J.,** Hyperlipoproteinemia in ponies: mechanism and response to therapy, *Clin. Chim. Acta,* 58, 1, 1975.

6. **Lewis, L. A.,** Lipids and lipoproteins of pigeons, *CRC Handbook of Electrophoresis,* Vol. 4, Lewis, L. A. and Naito, H. K., Ed., CRC Press, Boca Raton, Fla., 1984, 00.

7. **Naito, H. K. and Gerrity, R. G.,** Unusual resistance of the ground squirrel to the development of dietary induced hypercholesteremia and to atherosclerosis, *Exp. Mol. Pathol.,* 31, 128, 1979.

8. **Delcourt, R.,** Serum lipoproteins of vertebrates, electrophoretic and immunologic study, *Acta Zool. Pathol. Antwerp,* 48, 197, 1969.

9. **Mills, G. L. and Taylaur, C. E.,** The distribution and composition of serum lipoproteins of eighteen animals, *Comp. Biochem. Physiol.,* 401, 689, 1971.

10. **Alexander, C. and Day, C. E.,** Distribution of serum lipoproteins of selected vertebrates, *Comp. Biochem. Physiol.,* 46B, 195, 1973.

11. **Goldstein, S. and Chapman, M. J.,** Comparative immunochemical studies of the serum low-density lipoprotein in several animal species, *Biochem. Genet.,* 14, 883, 1976.

12. **Noble, R. P.,** Electrophoretic separation of plasma lipoproteins in agarose gel, *J. Lipid Res.,* 9, 693, 1968.

13. **Scheidegger, J. J.,** Une micromethode de l'immunoelectrophorese, *Int. Arch. Allerg. Appl. Immunol.,* 7, 103, 1955.

14. **Mahley, R. W.,** Alterations in plasma lipoproteins induced by cholesterol feeding in animals including man, in Physiology of Lipids and Lipoproteins in Health and Disease, Dietschy, J. M., Ed., Williams & Wilkens, Baltimore, 1978.

15. **Lee, J. A. and Morris, M. D.,** Characterization of serum low density lipoproteins of normal and two rhesus monkeys with spontaneous hyperbeta lipoproteinemia, *Biochem. Med.,* 10, 245, 1974.

16. **Rapasz, J.,** Lipoprotein, immunogenetics and atherosclerosis, *Am. J. Med. Genet.,* 1, 377, 1978.

17. **Zilversmit, D. B., Clarkson, T. B., and Hughes, L. B.,** High plasma cholesterol in mink (*Mustela vison*) without atherosclerosis, *Atherosclerosis,* 1, 97, 1977.

18. **Thomas, K. K. and Gilbert, L. J.,** Isolation and characterization of the hemolymph lipoproteins of the American Silkmoth, *Hyalophora cecropia, Arch. Biochem. Biophys.,* 127, 512, 1968.

19. **Chino, H. and Gilbert, L. J.,** Lipid release and transport in insects, *Biochem. Biophys. Acta,* 98, 94, 1965.

20. **Swaney, J. B., Reese, H., and Eder, H. A.,** Polypeptide composition of rat high density lipoprotein: characterization by SDS-gel electrophoresis, *Biochem. Biophys. Res. Commun.,* 59, 513, 1974.

21. **Swaney, J. B. and Kuehl, K. S.,** Separation of apolipoproteins by an acrylamide-gradient sodium dodecyl sulfate gel electrophoresis system, *Biochem. Biophys. Acta,* 446, 561, 1976.

22. **Chapman, M. J.,** Animal lipoproteins: chemistry, structure and comparative aspects, *J. Lipid Res.,* 21, 789—853, 1980.

23. **Hickman, C. P., Ed.,** *Integrated Principles of Zoology,* 2nd ed., C. V. Mosby, St. Louis, Mo., 1961, 31.

Lipoproteins of Nonhuman Primates

APPLICATION OF GRADIENT GEL ELECTROPHORESIS TO THE STUDY OF SERUM LIPOPROTEINS AND APOLIPOPROTEINS OF RHESUS MONKEYS

G. M. Fless, C. A. Rolih, and A. M. Scanu

INTRODUCTION

Electrophoretic techniques have played an important role in the fractionation and characterization of both human and non-human serum lipoproteins.[1,2] By conducting electrophoresis in media containing detergents such as sodium dodecyl sulfate (SDS) it has been possible to determine the apolipoprotein composition of the parent lipoproteins. Because electrophoretic procedures are generally easy to carry out, rapid, and need only small amounts of sample while yielding precise information, they have become one of the most popular tools available to the researcher and the clinician in studying serum lipoproteins in both normal and disease states. A technique which recently has been shown to have a high power of resolution is gradient gel electrophoresis. The principle of this method, its detailed description, and its potential use in the study of rhesus lipoproteins and apolipoproteins are the subject of this chapter.

GRADIENT GEL ELECTROPHORESIS

Principles and Application

Gradient gel electrophoresis (GGE) has been developed for the simultaneous analysis of macromolecules in multicomponent mixtures in a gel of increasing acrylamide concentration.[3-8] During electrophoresis particles migrate into a gel of increasing density which progressively decreases their mobility as pore size decreases. Since the leading edge of a zone encounters more resistance than its trailing edge, the macromolecules become concentrated into sharp bands, making possible a high degree of resolution. If electrophoresis is continued long enough, the migration of most proteins is drastically curtailed, allowing those with a lower charge to catch up to the faster, more highly charged molecules. Molecular weights can thus be determined by comparing the penetration distance of unknowns with that of standard proteins because charge differences do not affect their position in the gel.

The above principles apply also to GGE of water-insoluble proteins and peptides from membranes and lipoproteins performed in the presence of SDS.[9-12]

GGE has been used to study plasma lipoproteins of patients with disorders of lipoprotein metabolism.[13-15] Recently GGE of human HDL revealed the presence of three particle-size ranges within this lipoprotein class.[16] Furthermore, when HDL was subfractionated by density gradient centrifugation, GGE established an inverse relationship between particle size and particle hydrated density which was corroborated by electron microscopy and analytical ultracentrifugation. In a followup paper from the same laboratory, GGE was applied in conjunction with automated densitometry to the identification and estimation of three additional subpopulations in HDL_3 from the d <1.20 g/mℓ fraction of human plasma.[17] It was found that GGE is more sensitive in identifying HDL heterogeneity than ultracentrifugal flotation analysis using the schlieren optical system. Recently multiple subclasses of human plasma low density lipoproteins (LDL) have been identified with the aid of GGE.[18] It was found that LDL can be grouped into seven size intervals, with the major peaks of each interval exhibiting an inverse size-density correlation. GGE has also been applied to the study of apo LDL in solutions containing SDS and urea.[19] Depending on the mode of preparation of apo LDL it was possible to resolve numerous oligomeric peptides which differed in size by as little as 6400 daltons.

As yet, GGE has not been exploited in the study of either rhesus lipoproteins or their apoproteins. Furthermore, in comparison to their human counterparts, rhesus lipoproteins have received little attention by any method. Before describing the use of GGE in their analysis, we will provide a brief literature review of the lipoproteins and apolipoproteins in the rhesus monkey.

Rhesus Monkey Serum Lipoproteins and Apolipoproteins

The rhesus monkey (*Macacca mulatta*) has proven to be a popular animal for studying experimental atherosclerosis, particularly in relation to the influence of diet on serum lipids and lipoproteins. Some of the reasons for this popularity are their close phylogenetic relationship to humans and their susceptibility to diet-induced atherosclerosis. The serum of rhesus monkeys contains all the major lipoprotein classes, i.e., chylomicrons, VLDL, LDL, and HDL, which are rather similar in structure and function to those found in humans. However, there are major differences in the distribution of these lipoproteins. In monkeys, HDL, LDL, and VLDL account for about 70%, 22%, and 7%, respectively, of the total lipoproteins whereas humans have more LDL (46%) and less HDL (34%).[20]

Studies from this and other laboratories have shown that LDL is heterogeneous in size and density.[21-25] LDL was found to consist of at least four subfractions, LDL-I, LDL-II, LDL-III, and LDL-IV, of which the latter two were determined to be Lp(a) positive.[26] These subfractions differ in some physical properties including hydrated density (1.027, 1.036, 1.050, and 1.061 g/mℓ, respectively) and molecular weight (3.32, 2.75, 3.47, and 3 to 5 $\times$ 10^6, respectively).

The amino acid composition of rhesus apo LDL (i.e., LDL-I or LDL-II) is very similar to its human counterpart.[21,27] Although apo LDL-III and apo LDL-IV have the same amino acid composition as apo LDL-I ad LDL-II, they differ in their high sialic acid content and their slower electrophoretic mobility on 4% gels (acrylamide or agarose) containing SDS.[25]

The HDL$_2$ and HDL$_3$ subfractions of normal rhesus serum isolated between the density intervals of 1.063 to 1.125 g/mℓ and 1.125 to 1.21 g/mℓ are also similar in size to those of humans and have molecular weights of 3.9 $\times$ 10^5 for HDL$_2$ and 1.97 $\times$ 10^5 for HDL$_3$.[20] However, unlike man whose major HDL species is HDL$_3$, rhesus monkeys contain mainly HDL$_2$.[20]

As in man, the two major apoproteins of rhesus HDL are apo A-I and apo A-II (28). Rhesus apo A-I is similar to that of human subjects, having a single chain of molecular weight 27,000 with aspartic acid as its NH$_2$-terminal and glutamic acid as its COOH-terminal residues. Rhesus apo A-II, unlike human apo A-II which is made of two identical chains of molecular weight 8500 linked together by a disulfide bridge in position 6 from the NH$_2$-terminus, consists of a single polypeptide chain of molecular weight 8500.[28] Sequence analyses of rhesus apo A-II revealed that a substitution of a serine residue for cysteine at position 6 was responsible for the monomeric form of this protein.[29] In addition to apo A-I and A-II, 10% of rhesus apo HDL is composed of C peptides which are separable, as in their human counterparts, into four chemically distinct fractions by DEAE-cellulose column chromatography.[28] However, their exact chemical and physical makeup have not been elucidated. A comparison of the migration distance of rhesus C proteins with those of humans on 8 *M* urea-PAGE indicated differences in mobility.[24] When apo HDL from monkeys fed a low cholesterol, high fat diet was applied to SDS-PAGE a minor apoprotein was found in the region of apo E, the arginine-rich apoprotein, but it failed to react against either human or patas anti apo E.[24] In control animals apo E is mainly found in VLDL.[30]

When rhesus monkeys are fed diets rich in saturated fats and cholesterol, their serum lipoproteins undergo profound changes in distribution, size, and chemical composition.[30-32] The size and concentration of LDL is increased, the particles become enriched in free cholesterol and cholesteryl ester, and buoyant density decreases. In contrast, the plasma

HDL concentration is reduced as a result of increased dietary cholesterol.[30] Furthermore, the concentration of HDL_2 decreases while that of HDL_3 increases. However, the apoprotein patterns of HDL from hypercholesterolemic animals appear similar to those of control animals. Increased dietary cholesterol also induces slight increases in arginine-rich protein that appear mainly in the IDL and LDL density range, probably as an integral part of HDL_c, a lipoprotein which has not been fully characterized in the rhesus monkey. However, in dogs and patas monkeys this cholesterol-rich particle has α2 mobility and is inducible by dietary cholesterol.[33] In addition to C peptides, HDL_3 contains apo E and apo A-I, but not apo B. Most of the induced arginine-rich protein is present in "slow-beta" VLDL particles which are rich in cholesteryl esters and contain the low-molecular-weight form of apo B, referred to as B-48. These hypercholesterolemic animals also have a second VLDL that has pre-β mobility and no apo B-48 or apo E, but has the high-molecular-weight apo B (B-100) and large amounts of apo C.[34]

Usefulness of GGE in the Analysis of Rhesus Lipoproteins and Apoproteins

GGE can be an important analytical tool for studying lipoprotein mixtures, especially when used in combination with other physical methods. Different lipoproteins often have overlapping ranges of buoyant density and thus cannot be separated by density gradient centrifugation. For example, the Lp(a) lipoprotein class spanning the density interval of 1.05 to 1.10 g/mℓ has particles with buoyant densities common to both LDL and HDL_2. Another case is HDL_c which exhibits great density heterogeneity and can be found co-floating with IDL (intermediate density lipoprotein), LDL, Lp(a), and HDL_2. With GGE these overlapping mixtures of isopycnic proteins can be easily separated, and additional information regarding their size can be obtained. GGE is also useful in characterizing fractions obtained by gel filtration or ion-exchange chromatography with respect to defining the degree of homogeneity of lipoproteins or apolipoproteins. Diet-induced changes in plasma lipoproteins may be easily studied with GGE by either electrophoresing prestained plasma[35-37] or by staining the gel with a fat-soluble dye.[14,37]

The commercially available apparatus can hold from one to four cassettes, thus allowing the analysis of up to 48 samples at a time under identical conditions. The gel slabs, whether homemade or purchased, can be stored at 4°C for weeks without deterioration. Furthermore, only one type of gel is necessary to perform GGE of lipoproteins and apolipoproteins; SDS-GGE is conducted by simply preelectrophoresing a conventional SDS-free gel with a SDS-containing electrophoresis buffer. With one kind of gradient gel, one can therefore both characterize lipoprotein heterogeneity and identify the apolipoprotein constituents.

We have provided several examples to illustrate the utility of GGE in the analysis of rhesus lipoproteins and their apoproteins. In Figure 1 we show a density gradient profile of the d 1.006 to 1.063 g/mℓ fraction of pooled serum from monkeys fed a normal Purina® primate chow diet which was applied to a 0 to 10% NaBr gradient and spun to isopycnic equilibrium in a SW-40 rotor at 39,000 rpm, 20° C, for 66 hr. The gradient was fractionated, and appropriate tubes were analyzed by GGE using a homemade, linear 3 to 25% acrylamide gel (Figure 1). LDL, which is the fast moving band in every lane, shows a direct relationship between its migration distance and its hydrated density indicating that as the density of LDL increases, particle size decreases. In lanes 5 and 6, Lp(a) appears as the slow moving band that is well separated from the smaller, faster moving LDL which has the same buoyant density. The other fraction in which heterogeneity was revealed when analyzed by GGE was run in lane one. The slow moving band is probably VLDL and the fast moving band IDL. The same fractions from the density gradient were analyzed by SDS-GGE using a continuous buffer system and a 2 to 16% polyacrylamide gradient gel (Figure 2). This figure illustrates the capacity of SDS-GGE to resolve apo Lp(a), which is the slow moving band in lanes 5 and 6, from the faster moving apo LDL.

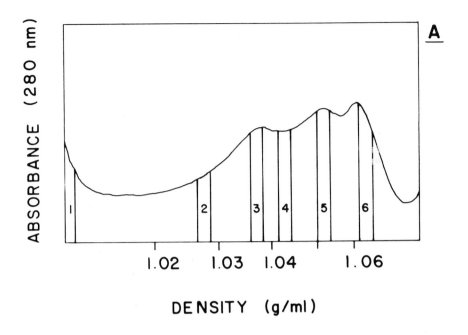

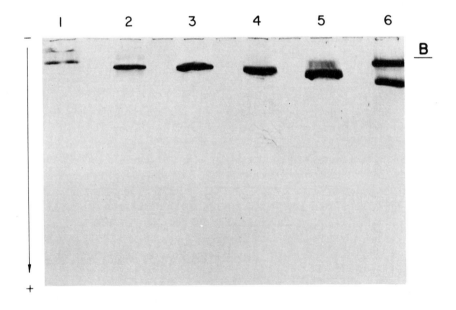

FIGURE 1.   Fractionation of normal rhesus LDL by density gradient ultracentrifugation and poly-acrylamide GGE. (A) Normal rhesus LDL was applied to a 0 to 10% NaBr gradient and spun at 39,000 r/min for 66 hr in the SW-40 rotor at 20°C. The gradient was fractionated with an ISCO density gradient fractionator and the absorbance profile at 280 nm was recorded with an ISCO UA-5 monitor. Densities of these fractions were obtained from refractometric measurements of a control gradient at 20°C. The fractions numbered 1 to 6 were taken for electrophoresis. (B) Gradient gel electrophoresis. A linear 3 to 25% homemade acrylamide gradient gel was subjected to preelectrophoresis for 20 min at 70 V. The samples were run into the gel at 70 V for 20 min before the voltage was increased to 150 V (constant voltage) and maintained there for 20 hr. Loading was approximately 20 μg protein per sample well.

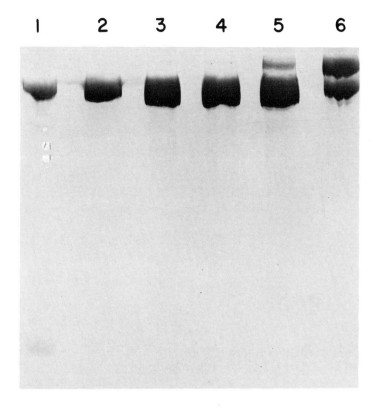

FIGURE 2. SDS polyacrylamide GGE of LDL and Lp(a) obtained by density gradient ultracentrifugation. Fractions 1 to 6, from the density gradient described in Figure 1A, were heated in 1% SDS for 1 hr at 50°C. The gel (2 to 16% concave acrylamide gradient, Pharmacia) was subjected to preelectrophoresis for 1 hr at 70 V before the samples were run into the gel for 15 min at 200 V. Electrophoresis was continued for 2 hr with 150 constant volts at 15°C. Loading was approximately 20 µg protein per sample well. A continuous electrophoresis buffer system was used.

Another dramatic example of the resolving power of GGE is given in Figure 3. Fractions obtained from a "single-spin" density gradient ultracentrifugation of serum from a monkey fed a normal monkey chow supplemented with 2% cholesterol and 25% coconut oil were analyzed by this technique. The fast moving band seen in lanes 2 through 6 is HDL$_c$ which migrated progressively further with increasing buoyant density, indicative of decreasing particle size. The broad bands appearing in lanes 7 and 8 are HDL$_2$ and HDL$_3$, respectively, the latter being considerably more heterogeneous than HDL$_2$. Lp(a) appears as the slowest bands in lanes 4 through 7. As in Figure 1, one sees an inverse relationship between size and buoyant density of LDL, which is represented by the slow moving bands in lanes 1, 2, and 3 and the intermediate bands in lanes 4 and 5. Thus, in lanes 4 and 5, GGE was able to resolve three lipoproteins of different sizes having similar buoyant densities.

GGE can also fractionate HDL according to particle weight. Pooled serum from normal rhesus monkeys was used to isolate the d 1.063 to 1.21 g/mℓ HDL fraction which was applied to a 0 to 20% NaBr gradient and spun to isopycnic equilibrium in the SW-40 rotor at 39,000 rpm, 66 hr, 20° C. The gradient was collected and the HDL peak was divided into four fractions which were analyzed by GGE (Figure 4). It is evident that with increasing buoyant density the migration distance of HDL increases which indicates an inverse correlation between density and particle size. In lanes 3 and 4, where the denser HDL fractions

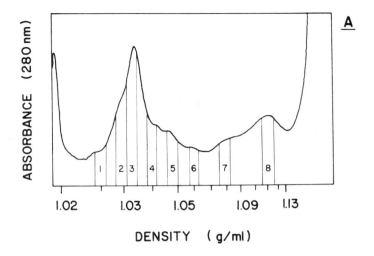

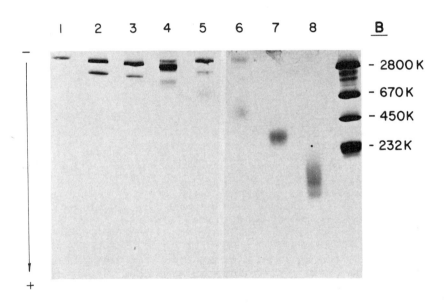

FIGURE 3.   Fractionation of lipoproteins from a hypercholesterolemic rhesus monkey by density gradient ultracentrifugation and polyacrylamide GGE. (A) "Single-spin" density gradient ultracentrifugation of hypercholesterolemic rhesus serum was carried out in the SW-40 rotor at 39,000 r/min for 66 hr at 20°C. The gradient was prepared by layering sequentially into a SW-40 tube: 0.5 g solid sucrose, 5.0 mℓ 4 M NaCl, 1 mℓ serum, and d 1.019 g/mℓ NaCl to the top of the tube. Densities of fractions from a control gradient were measured with a Mettler/Paar DMA-02 density meter at 20°C. The gradient was fractionated as described in Figure 1. (B) Polyacrylamide GGE (3 to 25% linear homemade acrylamide gradient) of fractions from above density gradient. Electrophoresis was carried out as described in Figure 1. Standards in the far right lane are: rhesus LDL II ($M = 2.8 \times 10^6$), thyroglobulin ($M = 6.7 \times 10^5$), apoferritin ($M = 4.5 \times 10^5$), and catalase ($M = 2.32 \times 10^5$).

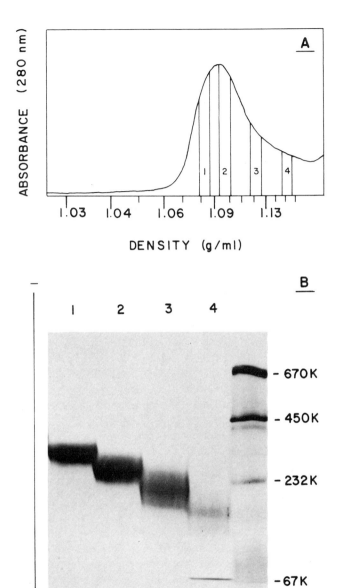

FIGURE 4. Characterization of rhesus HDL by density gradient ultracentrifugation and polyacrylamide GGE. (A) Normal rhesus HDL was applied to a 0 to 20% NaBr gradient containing 0.01% Na_2 EDTA and NaN_3, pH 7.0. Centrifugation conditions were 39,000 r/min for 66 hr in an SW-40 rotor at 20°C. The gradient was fractionated as described in Figure 1. (B) The fractions (1 to 4) from the density gradient were analyzed by GGE using a concave 4 to 30% acrylamide gradient gel (Pharmacia). The electrophoresis conditions were the same as those described in Figure 1. Standards in the right lane are: thyroglobulin ($M = 6.7 \times 10^5$), apoferritin ($M = 4.5 \times 10^5$), catalase ($M = 2.32 \times 10^5$), and bovine serum albumin ($M = 6.7 \times 10^4$).

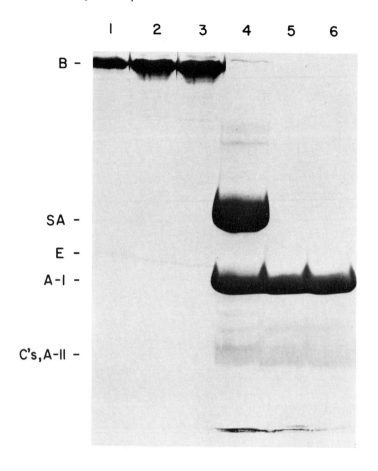

FIGURE 5.   SDS polyacrylamide GGE (concave 4 to 30% gel, Pharmacia) of various rhesus lipoprotein fractions using a continuous buffer system. Samples were denatured in 1% SDS for 5 min in a boiling water bath. Electrophoresis was carried out as described in Figure 2 except that the electrophoresis time was 3 hr. Loading was approximately 20 μg protein per well. Lane 1, normal rhesus LDL; lanes 2 and 3, rhesus LDL from a hypercholesterolemic monkey; lane 4, rhesus HDL from a hypercholesterolemic monkey (unpurified sample); lanes 5 and 6, normal rhesus HDL.

were applied, the bands are heterogeneous. The sharp, fast moving band in lane 4 is albumin. Unlike the standard proteins, HDL does not migrate in sharp and discrete bands but appears diffuse. This probably indicates that HDL consists of a population of particles with slightly differing molecular weights.

SDS-GGE in continuous buffer systems is an efficient method for characterizing the apoproteins of rhesus lipoproteins (Figure 5). In lane 1 we applied a sample of normal LDL from a pool of control monkeys and in lanes 2 and 3, LDL from hypercholesterolemic monkeys. In both cases, the major protein is apo B which is the heavily stained band near the top of the gel. The hypercholesterolemic LDL has two fast moving bands which are probably apo E and apo A-I from small amounts of HDL_c co-floating with LDL. The normal LDL fraction also has a fast moving band which can probably be attributed to apo E. Lanes 5 and 6 contain normal HDL and lane 4 contains a sample of impure hyperlipidemic HDL heavily contaminated with serum albumin. In all cases, apo A-I is the major protein component of HDL; apo A-II and the C peptides are present in minor amounts but are not

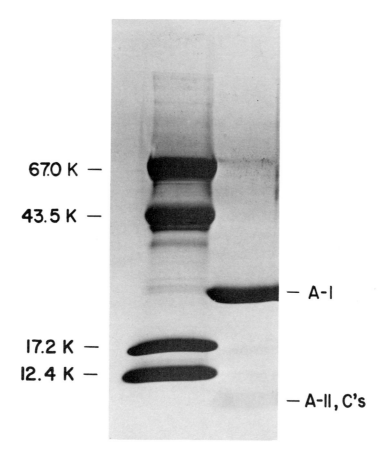

FIGURE 6.   SDS polyacrylamide GGE of HDL from a hypercholesterolemic rhesus monkey using the discontinuous buffer system of Laemmli. The running gel was constructed of a linear 3 to 25% acrylamide gradient; the stacking gel was made of 3% acrylamide. Samples and standards were denatured in 2% SDS and 5% 2-mercaptoethanol at 100° C for 5 min. Electrophoresis was carried out at 60 constant volts for 1 hr to allow the protein samples to migrate into the gel. The voltage was then increased to 100 V for 3 hr until the tracking dye reached the bottom of the gel. Standards are: bovine serum albumin ($M = 6.7 \times 10^4$), ovalbumin ($M = 4.35 \times 10^4$), myoglobin ($M = 1.72 \times 10^4$), and cytochrome C ($M = 1.24 \times 10^4$).

resolved from each other. When HDL was analyzed by SDS-GGE using the discontinuous buffer system of Laemmli with a linear 3 to 25% polyacrylamide running gel, the apoprotein bands were sharper (Figure 6). Three bands are visible in the region of the gel usually occupied by the C-peptides and apo A-II. Additionally, three other unknown bands of approximate molecular weight 17,000, 12,400, and 11,500 are detectable.

Method

The method described here is based partially on information presented in the *Pharmacia Handbook of Poly-acrylamide Gel Electrophoresis* and on our own experience.

I. Materials

Acrylamide, electrophoretic grade (Eastman Kodak® Co., Rochester, N.Y.)

Bisacrylamide, electrophoretic grade (Eastman Kodak®)

3-dimethylaminopropionitrile (DMAPN) (Eastman Kodak®)

Ammonium persulfate (Eastman Kodak®)

TRIS®-base (Sigma Chemical Co., St. Louis, Mo.)

Glycine (Fisher Scientific Co., Pittsburgh, Penna.)

Sodium acetate (Mallinckrodt® Inc., St. Louis, Mo.)

Na_2 EDTA (J. T. Baker Chemical Co., Phillipsburg, N.J.)

Na N_3 (Fisher)

Sodium dodecyl sulfate (SDS) (Matheson, Coleman and Bell, Norwood, Ohio)

Boric acid (J. T. Baker)

Hydrochloric acid (J. T. Baker)

Acetic acid (J. T. Baker)

Perchloric acid (Mallinckrodt®)

Sucrose (Mallinckrodt®)

2-mercaptoethanol (Eastman Kodak®)

Ethanol (U.S. Industrial Chemicals Co., New York, N.Y.)

Methanol (Mallinckrodt®)

Isopropanol (Mallinckrodt®)

Bromphenol blue (Fisher)

Coomassie blue G-250 (Bio-Rad® Laboratories, Richmond, Calif.)

Coomassie blue R-250 (Bio-Rad®)

Isoclean® (Isolab, Inc., Akron, Ohio)

Glass plates, 82 × 82 mm (Pharmacia, Inc., Uppsala, Sweden)

Spacers, 82 mm long × 3 mm wide (Pharmacia)

Sample applicators (Pharmacia)

Clamps

Waterproof tape (Pharmacia)

II. Equipment

Gel Electrophoresis Apparatus, GE-4 (Pharmacia)

Power Supply, model 3-1155 (Buchler Instruments Div., Fort Lee, N.J.)

Temperature regulating water bath, model FK (Haake, Inc., Saddle Brook, N.J.) or running cool water.

Gel slab casting apparatus, GSC-8 (Pharmacia).

Ultrograd® gradient mixer, model 11300 (LKB), for concave exponential gradients.

Gradient former, model 230 (Bio-Rad®), for linear gradients

Peristaltic pump, JP-4 (Brinkman Instruments, Inc., Westbury, N.Y.)

Magnetic stirrer, PC-353 (Corning Laboratory Products, Corning, New York)

Optional: Gel destainer, GD-4 and destainer power supply, DPS (Pharmacia)

III. Stock solutions

A. Plain electrophoresis buffer, pH 8.3

TRIS® Base (10.75 g)

Boric acid (5.04 g)

Na_2 EDTA (0.93 g)

Make up to 1 ℓ with distilled H_2O

B. 50% Acrylamide solution

Acrylamide (48.0 g)

Bisacrylamide (2.0 g)

Make up to 100 mℓ with buffer A.

(Warm to 30 to 40° to assist complete dissolution and filter)

C. 6% Acrylamide solution

Acrylamide (5.76 g)

Bisacrylamide (0.24 g)

Make up to 100 mℓ with buffer A.

D. Catalyst
  3-Dimethylaminopropionitrile (0.3 mℓ)
  Make up to 100 mℓ with buffer A.
E. Initiator
  Ammonium persulfate (0.3 g)
  Make up (fresh) to 100 mℓ with buffer A.
F. Density solution
  Sucrose (50.0 g)
  Bromphenol Blue (0.02 g)
  Sodium azide (0.02 g)
  Make up to 100 mℓ with H_2O.
G. Gel overlay
  Ethanol (20 mℓ)
  Make up to 100 mℓ with H_2O.
H. Protein stain
  Coomassie blue G-250 (0.4 g)
  70% Perchloric acid (50 mℓ)
  Methanol (200 mℓ)
  Make up to 1 ℓ with distilled H_2O and filter.
I. Destaining solution
  Methanol (250 mℓ)
  Acetic acid, glacial (100 mℓ)
  Make up to 1 ℓ with distilled H_2O.
J. Protein stain
  Coomassie blue R-250 (0.4 g)
  Acetic acid, glacial (100 mℓ)
  Isopropyl alcohol (250 mℓ)
  Make up to 1 ℓ with distilled H_2O and filter.
K. SDS Electrophoresis buffer (continuous), pH 7.4
  TRIS®-base (4.84 g)
  Sodium acetate (1.74 g)
  EDTA (0.59 g)
  SDS (2.00 g)
  Make up to 1 ℓ with distilled H_2O.
L. Lower gel buffer, pH 8.8
  TRIS®-base (18.15 g)
  6 N HCl (4.0 mℓ)
  SDS (0.4 g)
  Make up to 100 mℓ with distilled H_2O.
M. Upper gel buffer, pH 6.8
  TRIS®-base (6.0 g)
  6 N HCl (4 mℓ)
  SDS (0.4 g)
  Make up to 100 mℓ with distilled H_2O.
N. 33.3% Acrylamide solution
  Acrylamide (32.0 g)
  Bisacrylamide (1.33 g)
  Make up to 100 mℓ with distilled H_2O.
P. Initiator
  Ammonium persulfate (1 g)
  Make up fresh to 10 mℓ with distilled H_2O.
Q. Running buffer, pH 8.3
  TRIS®-base (3.03 g)
  Glycine (14.41 g)
  SDS (1 g)
  Make up to 1 ℓ with distilled H_2O.
R. Sample buffer, pH 6.8
  TRIS®-base (7.56 g)
  6 N -HCl (10 mℓ)
  Make up to 1 ℓ with distilled H_2O.

IV. Assembly of cassettes
1. Clean glass plates (82 mm × 82 mm) by immersing in Isoclean® or equivalent. Rinse with distilled H_2O and dry well.
2. Place two spacers (82 mm long × 3 mm wide) at opposite sides of one glass plate so that spacers are parallel to and flush with edges of plate.
3. Cover with second plate and clamp.
4. Tape plates together longitudinally along spacers, making sure that tape sticks to spacer as well as to glass.
5. In order to improve adhesion of tape, warm assembled cassettes in 60°C oven for 1 hr (especially when using buffers containing SDS).

V. A. Casting linear gradient gels in glass cassettes
1. Place assembled cassettes vertically into gel casting apparatus.
2. Connect in sequence gradient maker, peristaltic pump, and gel casting apparatus with silicone tubing (3 mm, i.d.) or equivalent.
3. Connect a 50-cc syringe containing 20 mℓ of solution F between the pump and the casting apparatus via a 3-way stopcock. Clear the stopcock of air.
4. At this point, determine the polymerization time of the two polyacrylamide solutions (20 to 30 min is a good time). If the solutions polymerize too slowly or quickly, adjust the persulfate and 3-dimethylamino-propionitrile concentrations appropriately.
5. While pumping approximately 25 mℓ of solution G into the gel casting apparatus, clear tubing of all air bubbles.
6. Mix stock solutions together in the following proportions and deaerate:

3% gel		**25% gel**	
C	2 parts	B	2 parts
D	1 part	D	1 part
E	1 part	E	1 part

When casting eight gels, approximately 65 mℓ of both the 3% and 25% polyacrylamide solutions are required.
7. Add the light solution to the mixing chamber of the gradient maker and the heavy solution to the other chamber. Stir.
8. Start the pump. To minimize convection in the casting apparatus, the flow rate should be less than 10 mℓ/min; we use a flow rate of 8 mℓ/min.
9. Stop the pump just before the last of the acrylamide has reached the three-way stopcock and clamp between pump and stopcock. Slowly introduce the 50% sucrose (solution F) into the line.
10. Clamp the tubing proximal to the casting apparatus when the sucrose has reached the bottom of the cassettes and the ethanol-acrylamide interface is 1 cm (2 cm for Laemmli gels) from the top of the cassette. Do not allow bubbles to enter the casting apparatus.
11. Flush tubing connecting the gradient maker, pump, and stockcock to remove traces of acrylamide.
12. Allow the gradient gel slabs to polymerize without disturbing the apparatus (at least 1 to 2 hr).
13. Wrap cassettes in electrophoresis wicks (Bio-Rad®) soaked in buffer A to which azide has been added (.02%). Store in a tight container at 4°C until used.

V. B. Making concave exponential gradients
1. Draw, cut out, and fit the gradient chart into the window of the LKB Ultrograd® gradient mixer according to the equation:

$$C = C_L + (C_H - C_L)\, e^{\dfrac{-(V_t - V)}{V_H}}$$

where C = concentration of acrylamide at any delivered volume, V
C_L = concentration of light acrylamide solution
C_H = concentration of heavy acrylamide solution
V_H = initial volume of heavy acrylamide solution
V_T = volume of heavy plus volume of light acrylamide solution

2. Follow steps 1 to 6 of Section V.A. In making a concave exponential gradient the volumes of light and heavy acrylamide solution are used approximately in the ratio of 3 to 1. To cast eight cassettes, about 100 mℓ of 3% and 33 mℓ of 25% acrylamide solution are needed.
3. Fill lines to the switching valve with both solutions. Fill the mixing chamber and pump-tubing with the 3% solution, making sure that no air bubbles are in the lines. Stop.

4. Start pump and Ultrograd® gradient mixer at the same time. Use a flow rate of less than 10 mℓ/min to minimize convection.
5. Continue with Steps 9 through 13 of Section V. A.

VI. A. Electrophoresis of lipoproteins on gradient gels
   1. Dialyze the lipoproteins against the electrophoresis buffer (solution A).
   2. Place plastic sample applicator on top of the gel, so that 1 to 2 mm of gel protrudes into the sample wells. As an alternative method for forming sample wells, insert pieces of 3 mm O.D. silicon or tygon tubing (1.5 cm long) perpendicular to and 1 to 2 mm below the gel surface at appropriate intervals.
   3. Fill the buffer vessels with 3.5 ℓ of buffer A. Insert gels and remove trapped air from the sample applicator by flushing buffer in and out of the sample wells with a Pasteur pipette.
   4. Pre-equilibrate the gradient gels for 20 min at 70 V (without sample), circulating the buffer between the upper and lower vessels.
   5. To 100-μℓ sample, add 5 μℓ solution F to increase sample density. *Note:* The amount of protein to be applied depends on the nature of the sample. With lipoproteins, 20 μg of protein per protein band give nice dense bands. However 1 to 5 μg protein can be easily visualized.
   6. Add samples to sample wells.
   7. Without circulating the buffer, electrophorese at 70 V for 15 to 20 min until proteins move into the gel.
   8. Circulate buffer to both upper and lower vessels, increase voltage to 150 V (constant voltage), and run for 16 to 24 hr.
   9. Maintain temperature between 15 to 25°C by running coolant through the apparatus.
   10. Stain and destain as in Section VIII.

VI. B. SDS electrophoresis of apoproteins and gradient gels
   1. Dialyze samples against buffer K (made up without SDS).
   2. Prepare gels as described in Section V.A.
   3. Place plastic sample applicator on top of gel or use 1.5 cm pieces of tubing (3 mm O.D.) as dividers in preparing sample wells.
   4. Fill electrophoresis apparatus with 3.5 ℓ of buffer K and pre-electrophorese the gradient gels for 1 hr at 70 V (constant voltage).
   5. Treat lipoprotein samples with SDS (2% final concentration) and 2-mercaptoethanol (5% final concentration) to dissociate lipids from apoproteins. VLDL and chylomicrons should be delipidated with diethyl ether or some other method before treatment with SDS. Incubate in 50°C water bath for 1 hr or immerse samples for 2 to 5 min in boiling water.
   6. Add sucrose solution F to samples (5 μℓ per 100 μℓ) and apply required volume to gels.
   7. Electrophorese samples into gels for 15 min at 200 V. Then reduce voltage to 150 V (constant voltage) and continue until the tracking dye has reached the bottom of the gel (2 to 2.5 hr). Depending on the sample and on the desired resolution, the electrophoresis may be continued after the tracking dye has migrated off the gel. However, care must be taken because peptides with molecular weights less than 25,000 will migrate off the gel (4 to 30%).
   8. Stain and destain as in Section VIII.

VII. SDS Electrophoresis of apolipoproteins on gradient gels using the discontinuous buffer system of Laemmli[10]
   1. Dialyze samples against buffer R.
   2. Follow Steps 1 through 5 of part V.A.
   3. To make gradient gels, mix the stock solutions together in the following proportions and deaerate:

	3% gel	**25% gel**
L	2.5 parts	2.5 parts
N	0.9 parts	7.5 parts
H₂O	6.6 parts	—

For eight gels, approximately 60 mℓ of both the 3% and 25% polyacrylamide solutions are needed.
   4. To initiate polymerization add 85 μℓ solution P and 8.5 μℓ DMAPN per 10 mℓ of the 3% acrylamide solution and 45 μℓ solution P and 4.5 μℓ DMAPN per 10 mℓ of the 25% acrylamide solution.
   5. Continue with Steps 7 through 12 of part V.A.
   6. Wrap and store gels in buffer Q until needed. Note that once the stacking gel is cast, the cassette must be used immediately.
   7. Before casting the stacking gel, rince lower gel with solution M diluted 1:3 with H₂O.
   8. For the stacking gel, mix the stock solutions together in the following proportions and deaerate:

M	2.5 parts
N	0.9 parts
H₂O	6.6 parts

About 1.5 mℓ is needed per cassette.

9. Initiate polymerization by adding 10 μℓ DMAPN and 100 μℓ of solution P per 10 mℓ of polymerization mixture.
10. Apply a layer of polymerization mixture 1 cm deep to the top of each cassette.
11. Immediately overlay the stacking gel with solution M diluted 1:3 with H$_2$O and containing 20% ethanol and allow to polymerize (about 1 hr).
12. To the lipoprotein samples add SDS (2% final concentration) and 2-mercaptoethanol (5%, final concentration) and boil 2 to 5 min.
13. Promptly proceed by filling the electrophoresis vessel with 3.5 ℓ of buffer Q. Insert sample applicators as described in Step 2 of Section VI.A, place gels into gaskets of upper chambers and flush air out of sample wells. Do not preelectrophorese.
14. Add 5 μℓ solution F per 100-μℓ sample and add samples to sample wells.
15. Electrophorese gels at 60 V until indicator dye has migrated completely into the gradient gel.
16. Increase voltage to 100 V and continue electrophoresis until the bromphenol blue marker has reached the bottom of the gel (about 4 hr).
17. Stain and destain as in Section VIII.

VIII. Staining and destaining

1. After electrophoresis, the cassette is opened with a razor blade and the gel is notched in the lower right hand corner for orientation purposes.
2. Place the gel into a small plastic box (13 × 5 × 9) containing 100 to 150 mℓ stain solution J for 4 to 16 hr. This stain can be used with both SDS-free and SDS-containing gels. A less sensitive but faster method which gives nice clear backgrounds is to stain with solution H for 1 to 2 hr. Do not use this stain with SDS-containing gels.
3. Destain by diffusion with solution I until the background is clear. An alternative method of destaining is to use electrophoresis. Place stained gel in gel holder screen and insert into apparatus containing destaining solution I and a dye-absorbent bag. Destain at 24 V for approximately 45 min. Check periodically for the progress of destaining.

REFERENCES

1. **Lewis, L. A. and Opplt, J. J., Eds.,** *CRC Handbook of Electrophoresis.,* Vol. 1, CRC Press, Boca Raton, Fla., 1980.
2. **Lewis, L. A. and Opplt, J. J., Eds.,** *CRC Handbook of Electrophoresis,* Vol. 2, CRC Press, Boca Raton, Fla., 1980.
3. **Margolis, J. and Kenrick, K. G.,** Electrophoresis in polyacrylamide concentration gradient, *Biochem. Biophys. Res. Commun.,* 27, 68, 1967.
4. **Slater, G. G.,** Stable pattern formation and determination of molecular size by pore-limit electrophoresis, *Anal. Chem.,* 41, 1039, 1969.
5. **Kopperschlager, G., Diezel, W., Bierwagen, B., and Hofmann, E.,** Molekulargewichtsbestimmungen durch Polyacrylamide Gel Elektrophorese unter Verwendung eines linearen Gel Gradienten, *FEBS Lett.,* 5, 221, 1969.
6. **Rodbard, D., Kapadia, G., and Chrambach, A.,** Pore gradient electrophoresis, *Anal. Biochem.,* 40, 135, 1971.
7. **Margolis, J.,** Practical system for polyacrylamide gradient gel electrophoresis, *Lab. Pract.,* 22, 107, 1973.
8. **Margolis, J. and Wrigley, C. W.,** Improvement of pore gradient electrophoresis by increasing the degree of cross-linking at high acrylamide concentrations, *J. Chromatogr.,* 106, 204, 1975.
9. **Ruechel, R., Mesecke, S., Wolfrum, D.-I., and Neuhoff, V.,** Microelectrophoresis in continuous-polyacrylamide-gradient gels. II. Fractionation and dissociation of sodium dodecylsulfate protein complexes, *Hoppe-Seyler's Z. Physiol. Chem.,* 355, 997, 1974.
10. **Laemmli, U. K.,** Cleavage of structural proteins during the assembly of the head of bacteriophage T4, *Nature (London),* 227, 680, 1970.
11. **Neville, D. M., Jr.,** Molecular weight determination of protein-dodecyl sulfate complexes by gel electrophoresis in a discontinuous buffer system, *J. Biol. Chem.,* 246, 6328, 1971.
12. **O'Farrell, P. H.,** High resolution two-dimensional electrophoresis of proteins, *J. Biol. Chem.,* 72, 4007, 1975.

13. **Melish, J. S. and Waterhouse, C.**, Concentration gradient electrophoresis of plasma from patients with hyperbetalipoproteinemia, *J. Lipid Res.*, 13, 193, 1972.

14. **Bautovich, G. J., Dash, M. J., Hensley, W. J., and Turtle, J. R.**, Gradient gel electrophoresis of human plasma lipoproteins, *Clin. Chem.*, 19, 415, 1973.

15. **Green, J.**, Gradient acrylamide gel electrophoresis of serum lipoproteins, *Scand. J. Clin. Lab. Invest.*, 29 (Suppl. 126), 1972.

16. **Anderson, D. W., Nichols, A. V., Forte, T. M., and Lindgren, F. T.**, Particle distribution of human serum high density lipoproteins, *Biochim. Biophys. Acta*, 493, 55, 1977.

17. **Blanche, P. J., Gong, E. L., Forte, T. M., and Nichols, A. V.**, Characterization of human high density lipoproteins by gradient gel electrophoresis, *Biochim. Biophys. Acta*, 665, 408, 1981.

18. **Krauss, R. M. and Burke, D. J.**, Identification of multiple subclasses of plasma low density lipoproteins in normal humans, *J. Lipid Res.*, 23, 97, 1982.

19. **Socorro, L. and Camejo, G.**, Preparation and properties of soluble immunoreactive apo LDL, *J. Lipid Res.*, 20, 631, 1979.

20. **Scanu, A. M., Edelstein, C., Vitello, L., Jones, R., and Wissler, R.**, The serum high density lipoproteins of Macacus rhesus. I. Isolation, composition and properties, *J. Biol. Chem.*, 248, 7648, 1973.

21. **Fless, G. M. and Scanu, A. M.**, Physicochemical characterization of rhesus low-density lipoproteins, *Biochemistry*, 14, 1783, 1975.

22. **Hill, P., Martin, W. G., and Douglas, J. F.**, Comparison of the lipoprotein profiles and the effect of *N*-phenylpropyl-*N*-benzyloxy acetamide in primates, *Proc. Soc. Exp. Biol. Med.*, 148, 41, 1975.

23. **Nelson, C. A. and Morris, M. D.**, A new serum lipoprotein found in many rhesus monkeys, *Biochem. Biophys. Res. Commun.*, 71, 438, 1976.

24. **Rudel, L. L., Greene, D. G., and Shah, R.**, Separation and characterization of plasma lipoproteins of rhesus monkeys *(Macaca mulatta)*, *J. Lipid Res.*, 18, 734, 1977.

25. **Fless, G. M. and Scanu, A. M.**, Isolation and characterization of the three major low density lipoproteins from normolipidemic rhesus monkeys *(Macaca mulatta)*, *J. Biol. Chem.*, 254, 8653, 1979.

26. **Fless, G. M., Kirchhausen, T., Fischer-Dzoga, K., Wissler, R. W., and Scanu, A. M.**, Relationship between the properties of the apo B containing low density lipoproteins (LDL) of normolipidemic rhesus monkeys and their mitogenic action on arterial smooth muscle cells grown *in vitro*, in *Atherosclerosis V*, Gotto, A. M., Jr., Smith, L. C., and Allan, B., Eds., Springer-Verlag, New York, 1980, 607.

27. **Chapman, M. J. and Goldstein, S.**, Comparison of the serum low density lipoprotein and of its apoprotein in the pig, rhesus monkey and baboon with that in man, *Atherosclerosis*, 25, 267, 1976.

28. **Edelstein, C., Lim, C. T., and Scanu, A. M.**, The serum high density lipoproteins of Macacus rhesus. II. Isolation, purification, and characterization of their two major polypeptides, *J. Biol. Chem.*, 248, 7653, 1973.

29. **Edelstein, C., Noyes, C., Keim, P., Heinrikson, R. L., Fellows, R. E., and Scanu, A. M.**, Covalent structure of apolipoprotein A-II from *Macaca mulatta* serum high-density lipoproteins, *Biochemistry*, 15, 1262, 1976.

30. **Rudel, L. L., Shah, R., and Greene, D. G.**, Study of the atherogenic dyslipoproteinemia induced by dietary cholesterol in rhesus monkeys *(Macaca mulatta)*, *J. Lipid Res.*, 20, 55, 1979.

31. **Fless, G. M., Wissler, R. W., and Scanu, A. M.**, Study of abnormal plasma low-density lipoprotein in rhesus monkeys with diet-induced hyperlipidemia, *Biochemistry*, 15, 5799, 1976.

32. **Lee, J. A. and Morris, M. D.**, The effect of cholesterol feeding on primate serum lipoproteins. I. Low density lipoprotein characterization from rhesus monkeys with high serum cholesterol, *Biochem. Med.*, 16, 116, 1976.

33. **Mahley, R. W.**, Alterations in plasma lipoproteins induced by cholesterol feeding in animals including man, in *Disturbances in Lipid and Lipoprotein Metabolism*, Dietschy, J. M., Gotto, A. M., Jr., and Ontko, J. A., Eds., American Physiological Society, Bethesda, Md., 1978, 181.

34. **Lusk, L., Chung, J., and Scanu, A. M.**, Properties and metabolic fate of two very low density lipoprotein subfractions from rhesus monkey serum, *Biochim. Biophys. Acta*, 710, 134, 1982.

35. **Allen, R. C.**, Electrophoretic separation of pre-stained lipoproteins on polyacrylamide gel slabs and their relationship to other plasma proteins, in *Electrophoresis and Isoelectric Focusing in Polyacrylamide Gel*, Allen, R. C. and Maurer, H. R., Eds., Walter de Gruyter, New York, 1974, 287.

36. **Naito, H. K. and Wada, M.**, The use of polyacrylamide gel electrophoresis for the detection of dyslipoproteinemia, in *CRC Handbook of Electrophoresis*, Vol. 1, Lewis, L. A. and Opplt, J. J., Eds., CRC Press, Boca Raton, Fla., 1980, 183.

37. **Narayan, K. A.**, Electrophoretic methods for the separation of serum lipoproteins, in *Analysis of Lipids and Lipoproteins*, Perkins, E. G., Ed., American Oil Chemists' Society, Champaign, Ill., 1975, 225.

LIPOPROTEINS OF CEBUS MONKEYS

Robert J. Nicolosi and K. C. Hayes

INTRODUCTION

The concentrations of circulating lipoproteins, especially low-density lipoproteins (LDL) and high-density lipoproteins (HDL), are thought to be among the factors which influence the development of atherosclerosis in humans and experimental animals. The metabolism of these lipoproteins has been extensively studied in the more common laboratory animals such as the rat, guinea pig, and rabbit. However, since these experimental models have lipoprotein patterns characteristically different than man, and therefore may not always be appropriate models, there has been increased interest in the utilization of Old and New World nonhuman primates for lipoprotein characterization and metabolic studies. Of the New World monkey species which include the squirrel monkey (*Saimiri sciureus*) and the cebus monkey *(Cebus albifrons* and *apella)*, both of the family Cebidae, the lipoproteins of the squirrel monkey have been more extensively studied.[1,2] However, in recent years, the responsiveness of the cebus monkey to diet-induced hypercholesterolemia[3,4] and its relative resistance to atherosclerosis compared to the squirrel monkey[4-6] have prompted us to initiate studies to investigate the character and metabolism of cebus lipoproteins. Since most of our studies have included both species, whenever significant species differences occur, the lipoproteins of the cebus and squirrel monkey will be compared.

LIPOPROTEIN DISTRIBUTION AND CHOLESTEROL LEVELS

As in most nonhuman primates, the lipoprotein distribution of chow-fed cebus monkeys reported by Rudel and Lofland[7] and Hill et al.[8] indicates a predominant HDL circulating class. Our own earlier studies[4] and more recent unpublished data (Table 1) in normocholesterolemic and hypercholesterolemic cebus monkeys fed physiological levels (31% kcal) of saturated fat (coconut oil) and polyunsaturated fat (corn oil) further establish the predominant HDL profile in this species. HDL represents 70 to 75% of the lipoproteins independent of plasma cholesterol levels. As expected, the saturated fat-induced hypercholesterolemia is associated with increases in the concentration of all lipoprotein classes. The concomitant increase in HDL cholesterol levels and total HDL concentration is in agreement with our earlier studies in cebus[4] and rhesus monkeys[9] but is in contrast to earlier reductions in HDL reported by Rudel et al.[10] for hypercholesterolemic rhesus monkeys. This discrepancy may be explained by the greater dietary cholesterol levels and resultant plasma cholesterol values reported in the studies of Rudel et al.[10]

HDL SUBCLASSES

Investigations of HDL subclasses in nonhuman primates are just being initiated. Nevertheless, it is apparent that most Old World monkey species have more HDL_3 than HDL_2.[11,12] In addition, earlier reports by Rudel and Lofland[7] and Illingworth[13] and our own recent investigations[11] (Figure 1, Table 2) reveal that the cebus monkey, unlike other New World species such as the squirrel monkey and marmoset, has relatively more HDL_3. This poses an interesting discrepancy since high HDL_2 levels in humans are inversely correlated with coronary heart disease, whereas the cebus monkey has relatively more HDL_3 and is reported to be more atherosclerosis resistant[14] than the susceptible squirrel monkey which has an HDL_2 profile. Obviously, more investigation of this relationship is needed.

Table 1
EFFECT OF DIETARY FAT SATURATION
ON THE PLASMA LIPOPROTEINS OF
THE CEBUS MONKEY

Dietary fat	Corn oil	Coconut oil
Plasma cholesterol (mg/dℓ)	175 ± 4[a]	241 ± 21
Lipoprotein cholesterol[b]		
VLDL and LDL (mg/dℓ)	77 ± 1	116 ± 32
HDL (mg/dℓ)	98 ± 5	125 ± 12
Lipoprotein concentration[c]		
VLDL	31 ± 3	99 ± 8
mg/dℓ		
%	6 ± 2	10 ± 3
LDL	104 ± 13	197 ± 33
mg/dℓ		
%	19 ± 4	20 ± 4
HDL	408 ± 38	690 ± 17
mg/dℓ		
%	75 ± 8	70 ± 10

[a] Values represent mean ± SD.
[b] Determined after heparin-Mn^{++} precipitation.
[c] Determined after ultracentrifugation as sum of lipid and apoprotein components.

From Nicolosi, R. J., Hojnacki, J. L., and Hayes, K. C., unpublished data.

The possible dietary manipulation of HDL subclasses (HDL_2 in particular) has been demonstrated in humans[15] and in the squirrel monkey.[7] Our own recent unpublished investigations indicate that the cebus monkey, in association with saturated fat-induced hypercholesterolemia, also expands its HDL_2 pool relative to HDL_3.

LIPOPROTEIN FLOTATION AND PARTICLE SIZE

Hill et al.[8] have reported that the flotation rates in Svedberg units measured in NaBr of d 1.20 at 20°C is 19.3 ± 0.6 for cebus LDL and 2.3 ± 0.1 for cebus HDL, both being comparable to values reported for human serum. Our earlier studies of the effect of dietary fat saturation on cebus monkey VLDL particle size determined by electron microscopy[16] and more recent electron microscopic investigations of cebus lipoproteins (Figures 2 to 4) indicate a range of 24 to 40 nm for VLDL, 18 to 20 nm for LDL, and 4 to 8 nm for HDL_3. All attempts at separating an HDL_2 particle by flotation ultracentrifugation revealed a contaminating LDL particle and therefore an electron micrograph is not included.

LIPOPROTEIN COMPOSITION

The composition of circulating lipoproteins in the cebus monkey which is similar to squirrel monkeys is also comparable to humans (Table 3). The saturated fat-induced hypercholesterolemia is not accompanied by significant alterations in lipoprotein composition, indicating that the moderate hypercholesterolemia in cebus or squirrel monkeys fed physiological levels of saturated fat is the result of an increase in the number of lipoprotein particles and not altered particle size as seen in studies that utilize diets inducing extreme elevations in plasma cholesterol.[17]

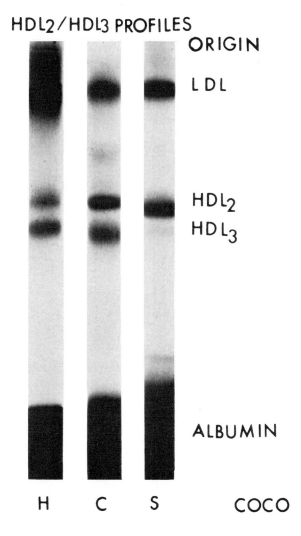

FIGURE 1. Electrophoretogram of HDL subclass profile in primates showing the greater HDL_2/HDL_3 ratios of the squirrel monkey (S) compared to the cebus (C) and human (H).

Table 2
ELECTROPHORETIC PROFILES OF HDL SUBCLASSES OF PRIMATES[a]

	%		HDL_2/HDL_3
	HDL_2	HDL_3	
Cebus apella (capuchin)	31.6 ± 2.9	63.1 ± 2.9	0.5 ± 0.04
Saimiri sciureus (squirrel-Brazil)	64.8 ± 2.2	32.4 ± 2.2	2.0 ± 0.06
Human primates			
Male	24.6 ± 3.3	75.4 ± 3.3	0.3 ± 0.02
Female	48.8 ± 1.0	51.2 ± 1.0	1.0 ± 0.03

[a] Values represent mean ± SE.

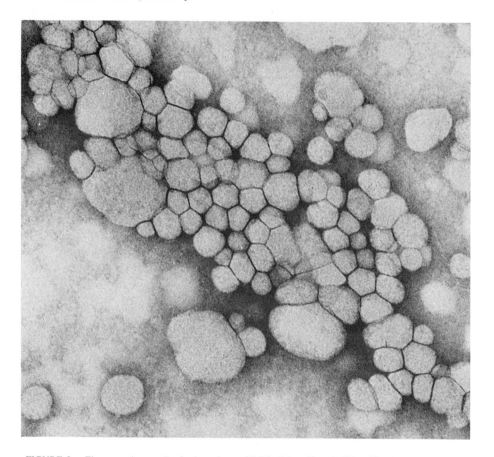

FIGURE 2.   Electron micrograph of cebus plasma VLDL (24 to 40 nm). (Magnification × 250,000.)

The phospholipid profile of cebus lipoproteins is not remarkably different from humans. Similar to the pattern for neutral lipids of the lipoproteins, data in Table 4 indicate that dietary fat had no effect on the relative distribution of the various phospholipids of either LDL or HDL. Phosphatidylcholine and sphingomyelin are the predominant phospholipids in both LDL and HDL.

LIPOPROTEIN FATTY ACIDS

Predictably, the fatty acid composition of lipoprotein lipid from cebus monkeys fed corn oil (Table 5) contains considerably more linoleate (18:2) and arachidonate (20:4) than coconut oil-fed monkeys whose fatty acid pattern is more saturated in nature.

The percentages of saturated fatty acids palmitate (16:0) and stearate (18:0) of LDL and HDL were unaltered by diet, with the most variable fatty acids being oleate (18:1) and linoleate (18:2). The cholesteryl ester, triglyceride and phospholipid 18:1/18:2 ratios (not shown) are consistently higher in the lipoproteins of coconut oil-fed monkeys. An estimation of actual differences in the circulating mass of fatty acids in monkeys fed dietary coconut oil indicates that 18:1 is considerably increased (140%) both in HDL phospholipids and cholesteryl esters of LDL and HDL while 18:2 is only moderately depressed (9 to 35%).

LIVER AND ADIPOSE LIPID FATTY ACIDS

The fatty acid composition of liver and adipose tissue lipids for both dietary groups is

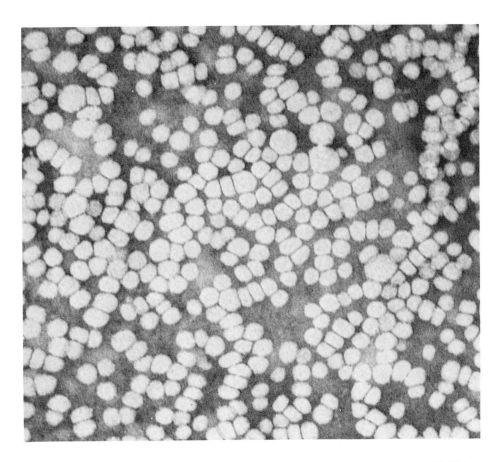

FIGURE 3. Electron micrograph of cebus plasma LDL (18 to 20 nm). (Magnification × 250,000.)

presented in Tables 6 and 7. Saturated fatty acids comprise more than 50% of the total acyl groups in cholesteryl esters, triglycerides, and free fatty acids in livers from coconut oil-fed monkeys while only 35% of the fatty acids in corresponding lipids from corn oil-fed animals are saturated. Similar to the lipoprotein phospholipid fatty acids (Table 5), the liver phospholipid fatty acids are highly saturated with the total percentage (50 to 55%) of these saturated fatty acids being relatively unchanged by dietary fat. As anticipated, saturated fatty acids constitute more than 80% of the total fatty acids esterified to triglycerides in the liver (Table 6) and adipose tissue (Table 7) of the monkeys fed coconut oil and are strikingly similar to patterns noted for diet (Table 8).

A remarkable difference between liver (Table 6) and adipose tissue (Table 7) triglyceride fatty acids and those of lipoprotein triglyceride (Table 5) was the high percentage of 14:0 in liver and 12:0 and 14:0 in adipose tissue not observed in the lipoproteins. This suggests that elongation of dietary saturated fatty acids and desaturation to form oleic acid (18:1) were actively contributing to the high 18:1/18:2 ratio in the liver and lipoproteins of the coconut oil-fed cebus monkeys.

LIPOPROTEIN APOPROTEINS

The apoproteins of cebus lipoproteins were isolated by density gradient ultracentrifugation, ten fractions collected, and initially characterized by sodium dodecyl sulfate (SDS)-polyacrylamide gel electrophoresis (PAGE) (Figure 5). Both cebus monkey and squirrel monkey major lipoproteins banded isopycnically on the gradient and had the following hydrated

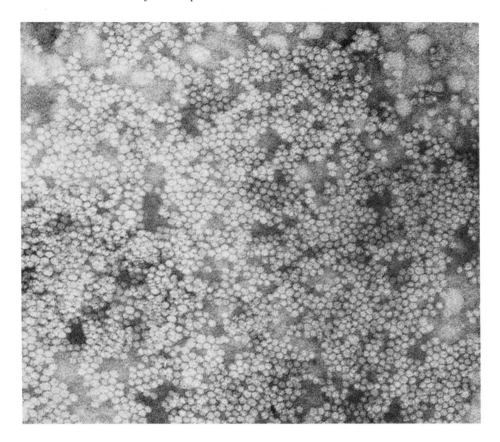

FIGURE 4.   Electron micrograph of cebus plasma HDL_3 (4 to 8 nm). (Magnification $\times$ 250,000.)

Table 3
PLASMA LIPOPROTEIN COMPOSITION OF CEBUS MONKEYS FED CORN OIL OR COCONUT OIL

	Cholesteryl ester	Free cholesterol	Triacylglycerol	Phospholipid	Protein
			% Distribution		
VLDL					
Corn oil	3.8 ± 1.7[a]	3.8 ± 0.1	65.9 ± 3.7	15.9 ± 1.1	10.7 ± 1.1
Coconut oil	2.5 ± 0.9	4.9 ± 1.2	58.9 ± 6.0	17.9 ± 2.7	15.8 ± 3.2
LDL					
Corn oil	31.1 ± 1.7	13.1 ± 1.4	10.7 ± 0.8	22.0 ± 2.1	23.1 ± 3.2
Coconut oil	32.2 ± 1.6	11.5 ± 1.8	8.9 ± 2.6	20.8 ± 2.3	26.5 ± 2.3
HDL					
Corn oil	10.3 ± 2.0	5.6 ± 0.6	10.1 ± 5.1	22.7 ± 1.3	51.2 ± 3.3
Coconut oil	12.8 ± 1.8	5.5 ± 0.7	6.5 ± 3.0	25.4 ± 2.9	49.8 ± 2.8

[a]   Values represent mean $\pm$ SD from at least four monkeys per dietary group.

From Nicolosi, R. J., Hojnacki, J. L., and Hayes, K. C., unpublished data.

Table 4
PHOSPHOLIPID COMPOSITION OF LOW-DENSITY AND HIGH-DENSITY LIPOPROTEINS FROM CEBUS MONKEYS FED CORN OIL OR COCONUT OIL

	Lipoprotein fraction and diet (%)			
	LDL		HDL	
Phospholipid	Corn oil	Coconut oil	Corn oil	Coconut oil
Lysophosphatidylcholine	—	—	1.7 ± 0.2	5.6 ± 3.8
Sphingomyelin	31.7 ± 3.7[a]	26.8 ± 0.9	11.5 ± 1.5	8.0 ± 0.2
Phosphatidylcholine	67.3 ± 3.3	70.2 ± 1.4	81.4 ± 2.1	80.8 ± 3.7
Phosphatidylinositol	0.5 ± 0.3	0.9 ± 0.5	2.8 ± 0.3	2.5 ± 0.3
Phosphatidylethanolamine	0.6 ± 0.6	2.1 ± 0.3	2.6 ± 1.8	3.1 ± 0.3

[a]   Values represent mean ± SE for three monkeys per diet.

From Nicolosi, R. J., Hojnacki, J. L., and Hayes, K. C., unpublished data.

Table 5
FATTY ACID ANALYSES OF PLASMA LIPOPROTEIN LIPIDS FROM CEBUS MONKEYS FED CORN OIL OR COCONUT OIL

	% Fatty acids					
Diet	14:0	16:0	18:0	18:1	18:2	20:4
Corn Oil						
CE-VLDL	1.9 ± 1.3[a]	30.3 ± 3.8	6.1 ± 0.4	22.0 ± 3.0	37.0 ± 2.5	2.6 ± 2.9
LDL	—[b]	10.8 ± 1.2	1.0 ± 0.2	23.9 ± 1.6	62.8 ± 0.8	1.5 ± 0.2
HDL	0.5 ± 0.2	14.4 ± 3.9	2.0 ± 0.7	22.4 ± 1.6	57.4 ± 9.0	3.3 ± 1.4
TG-VLDL	3.6 ± 5.2	42.1 ± 2.3	11.1 ± 2.5	19.7 ± 4.7	19.2 ± 5.8	3.8 ± 1.7
LDL	0.9 ± 0.2	28.4 ± 0.6	25.1 ± 2.6	28.8 ± 0.8	16.8 ± 1.2	1.7 ± 0.2
HDL	2.1 ± 0.6	20.0 ± 1.7	17.1 ± 1.6	27.0 ± 1.8	33.0 ± 3.0	0.8 ± 0.4
PL-VLDL	2.6 ± 0.2	36.8 ± 8.7	17.8 ± 1.8	20.3 ± 3.9	18.2 ± 5.5	3.4 ± 1.2
LDL	0.3 ± 0.2	37.9 ± 1.6	33.8 ± 3.6	10.9 ± 0.2	11.8 ± 1.8	5.5 ± 2.8
HDL	0.4 ± 0.1	29.4 ± 2.0	22.3 ± 3.0	12.3 ± 0.6	25.6 ± 2.6	10.0 ± 1.7
Coconut Oil						
CE-VLDL	7.7 ± 1.3	45.8 ± 3.8	4.4 ± 0.4	31.9 ± 3.0	10.0 ± 2.5	0.4 ± 2.9
LDL	4.2 ± 1.2	21.6 ± 1.2	2.9 ± 0.4	41.2 ± 1.2	29.0 ± 1.5	1.2 ± 0.5
HDL	7.7 ± 0.4	18.8 ± 1.0	3.1 ± 0.2	39.6 ± 1.0	28.6 ± 2.6	2.2 ± 0.1
TG-VLDL	11.8 ± 5.2	56.0 ± 2.3	11.6 ± 2.5	17.2 ± 4.2	1.6 ± 5.8	0.1 ± 1.7
LDL	11.6 ± 0.9	44.2 ± 0.5	15.3 ± 2.6	27.4 ± 0.2	0.5 ± 0.5	0.5 ± 0.5
HDL	15.5 ± 1.0	40.1 ± 2.2	5.5 ± 1.6	35.0 ± 2.3	4.0 ± 0.6	—[b]
PL-VLDL	5.2 ± 0.2	40.7 ± 8.7	20.2 ± 1.8	25.5 ± 3.9	6.0 ± 5.5	2.4 ± 1.2
LDL	2.0 ± 0.2	43.3 ± 0.9	26.1 ± 1.3	20.5 ± 0.8	6.6 ± 1.2	1.6 ± 0.4
HDL	1.8 ± 0.4	32.5 ± 0.7	20.8 ± 0.6	22.6 ± 1.3	12.5 ± 2.1	9.8 ± 2.1

[a]   Values are mean ± SE.
[b]   Not detected.

From Nicolosi, R. J., Hojnacki, J. L., and Hayes, K. C., unpublished data.

Table 6
COMPOSITION OF LIVER FATTY ACIDS OF CEBUS MONKEYS FED CORN OIL OR COCONUT OIL

Lipid class and dietary fat (mass %)

Fatty acid	Cholesteryl esters		Triglycerides		Phospholipids		Free fatty acids	
	Corn oil	Coconut oil	Corn oil	Coconut oil	Corn oil	Coconut oil	Corn oil	Coconut oil
12:0	—	0.1 ± 0.1[a]	—	1.6 ± 0.9	—	—	—	2.5 ± 1.8
14:0	0.2 ± 0.2	2.3 ± 1.2	0.4 ± 0.0	19.5 ± 5.3	—	1.4 ± 0.4	0.6 ± 0.2	13.4 ± 3.4
16:0	24.0 ± 3.9	38.8 ± 4.4	21.9 ± 0.8	56.9 ± 4.3	20.1 ± 0.9	29.8 ± 0.2	25.0 ± 0.6	38.4 ± 2.3
18:0	10.9 ± 1.9	11.1 ± 1.8	14.1 ± 2.0	6.6 ± 1.1	32.7 ± 0.3	25.8 ± 1.0	9.3 ± 2.5	9.2 ± 1.8
18:1	26.0 ± 0.1	38.6 ± 4.3	20.6 ± 0.4	14.9 ± 2.8	10.8 ± 0.3	20.9 ± 0.7	18.5 ± 3.4	28.7 ± 0.5
18:2	34.0 ± 2.2	8.7 ± 1.9	33.6 ± 2.5	0.5 ± 0.3	23.3 ± 1.1	10.2 ± 0.6	41.2 ± 3.0	6.2 ± 0.9
20:4	4.8 ± 3.1	0.4 ± 0.3	9.4 ± 0.6	0.1 ± 0.03	13.3 ± 1.6	12.0 ± 0.4	5.4 ± 2.6	1.6 ± 0.5

[a]   Values represent mean $\pm$ SE for three monkeys per diet.

From Nicolosi, R. J., Hojnacki, J. L., and Hayes, K. C., unpublished data.

Table 7
COMPOSITION OF ADIPOSE TISSUE FATTY ACIDS FROM CEBUS MONKEYS FED CORN OIL OR COCONUT OIL

Fatty acid	Diets[a] (mass %)	
	Corn oil	Coconut oil
12:0	0.1 ± 0.0[b]	29.8 ± 2.2
14:0	1.4 ± 0.1	25.2 ± 0.1
16:0	16.2 ± 1.3	23.4 ± 1.9
18:0	9.8 ± 0.4	4.6 ± 0.3
18:1	31.3 ± 0.6	14.5 ± 0.2
18:2	40.9 ± 1.3	2.4 ± 0.2
20:4	0.1 ± 0.0	0.2 ± 0.1

[a]   Values are mean $\pm$ SE.

From Nicolosi, R. J., Hojnacki, J. L., and Hayes, K. C., unpublished data.

densities in grams per milliliter: Fraction 1 = VLDL, d < 1.008; Fractions 2 to 4 = LDL, d = 1.017 to 1.065; Fractions 5 to 9 = HDL, d = 1.068 to 1.170. The distribution of apoproteins within each lipoprotein class is shown in Table 9.

The major VLDL apoproteins of both the cebus and squirrel monkey (not shown) were apo B, with an approximate molecular weight of 305 K; apo E, molecular weight = 39 K; and apo C complex, molecular weight = 8 to 12 K. Although the major apoprotein of LDL is apo B in both squirrel and cebus monkeys, it was possible to demonstrate on 3.5% SDS-PAGE (Figure 6) in preliminary experiments that cebus LDL apo B consisted of three molecular weight species, whereas squirrel LDL did not have the latter apo B subspecies. The significance of this difference and the origin of these apo B species remain to be

Table 8
FATTY ACID DISTRIBUTION IN
CORN OIL AND COCONUT OIL DIETS

Fatty acid	Diets (mass %)	
	Corn oil	Coconut oil
Caprylic (8:0)	—	0.8 ± 0.7[a]
Capric (10:0)	—	3.2 ± 1.4
Lauric (12:0)	—	49.3 ± 1.2
Myristic (14:0)	—	22.4 ± 1.9
Palmitic (16:0)	11.7 ± 0.0	11.2 ± 0.8
Stearic (18:0)	2.0 ± 0.0	3.0 ± 0.1
Oleic (18:1)	25.4 ± 0.0	8.3 ± 0.4
Linoleic (18:2 ω 6)	60.7 ± 0.0	1.9 ± 0.1
Linolenic (18:3)	0.3 ± 0.0	—
Oleic:linoleic	0.4 ± 0.0	4.4 ± 0.0
P:S ratio	4.5 ± 0.0	0.02 ± 0.0

[a]   Values represent mean ± SEM for duplicate analyses.

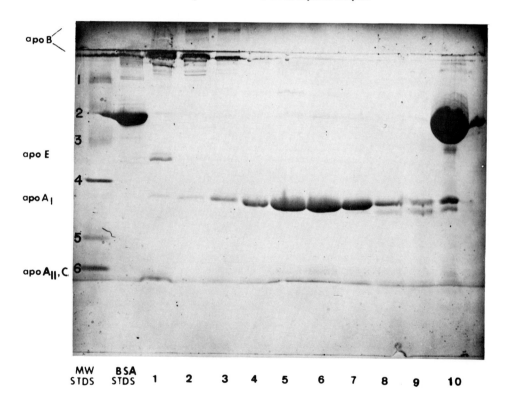

FIGURE 5.   10% SDS-PAGE of the ten density cuts taken of cebus plasma lipoproteins after ultracentrifugation showing the major apoproteins. Mol wt Standards are 1, phosphorylase B, 92,500; 2, bovine serum albumin, 66,200; 3, ovalbumin, 45,000; 4, carbonic anhydrase, 31,000; 5, soybean trypsin inhibitor, 21,500; 6, lysozyme, 14,400. Lanes 1, VLDL; 2 to 4, LDL; 5 to 9, HDL.

determined. The small amount of apo A-I, sometimes seen in LDL, is apparently due to the lack of a recentrifugation step (not performed to prevent apoprotein disassociation) that often accompanies multiple ultracentrifugations. The major apoprotein of HDL was apo A-

Table 9
LIPOPROTEIN APOPROTEIN DISTRIBUTION (%)
OF CEBUS MONKEYS

	apo B	apo E	apo A-I	apo A-II, C
VLDL	76.7 ± 3.7[a]	16.7 ± 2.5	ND[b]	6.7 ± 1.2
LDL	92.7 ± 0.8	6.9 ± 1.1	ND	0.2 ± 0.1
HDL	3.0 ± 0.4	3.1 ± 0.3	64.1 ± 5.3	29.8 ± 5.9

[a]   Values represent mean ± SD for six monkeys.
[b]   Not detected.

From Nicolosi, R. J., Hojnacki, J. L., and Hayes, K. C., unpublished data.

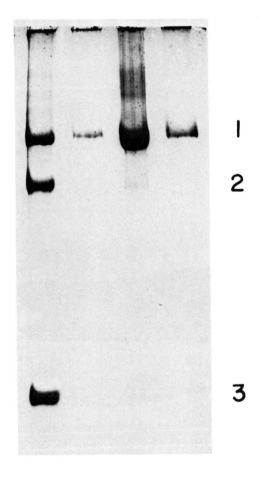

FIGURE 6.   3.5% SDS-PAGE of cebus and squirrel VLDL and LDL showing the species differences in apo B mol wt. Lanes A, cebus LDL; B, cebus VLDL; C, squirrel LDL; D, squirrel VLDL. 1 apo B molecular wt of 305,000; 2, apo B mol wt of 268,000; 3, apo B mol wt of 132,000.

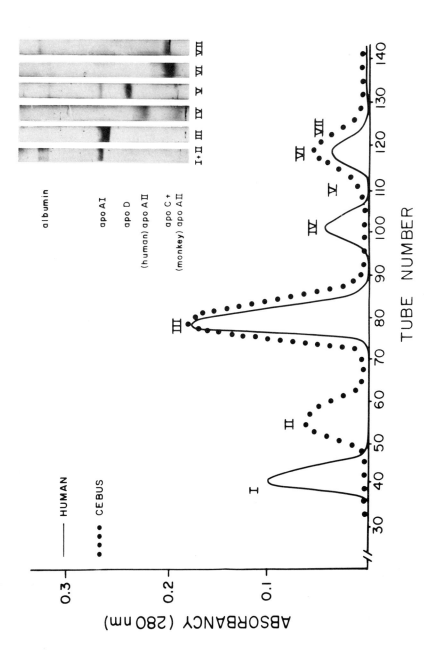

FIGURE 7. Elution profile of apo HDL of cebus and human separated on G-200 Sephadex column showing species difference in subfraction IV or apo A-II.

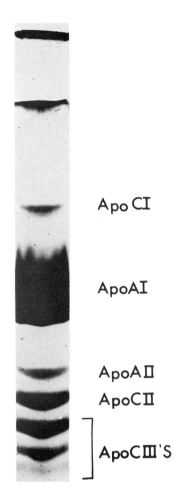

Apo CI

ApoAI

ApoA II
ApoC II

ApoC III'S

FIGURE 8.   Alkaline urea PAGE of cebus apo HDL further separating apo
A-II from the apo C complex.

I with an apparent molecular weight of 28 K. Unlike humans but like other nonhuman primates, SDS-PAGE with and without β-mercaptoethanol, indicated that apo A-II was a monomer of molecular weight 8500 and not a dimer. This was further verified by the comparative elution profiles of cebus and human apo HDL from a Sephadex G-200 column (Figure 7). Unlike humans, where three distinct HDL polypeptides can be isolated, i.e., Fraction III = apo A-I, Fraction IV = apo A-II, Fraction V = apo C, only two major polypeptides, apo A-I and apo A-II, C complex, are observed in the cebus. Alkaline urea PAGE of cebus apo HDL (Figure 8) permitted further separation of apo A-I and apo A-II from the apo Cs which were tentatively identified as apo C-I, apo C-II, and three forms of apo C-III.

Amino acid analyses of cebus apo A-I (Table 10) revealed a similar profile to human apo A-I except for possible differences in the content of methionine, histidine, and arginine. Thus, a combination of SDS, alkaline urea PAGE and gel chromatography indicates that the circulating lipoproteins, and in particular HDL, of the cebus monkey and the squirrel monkey (not shown) have a paucity of the arginine-rich peptide, apo E, thought to be a necessary prerequisite for hepatic recognition and uptake of lipoproteins from the circulation.

Table 10
AMINO ACID COMPOSITION
OF CEBUS MONKEY APO A-I

Amino acid[a]	I[b]	II[c]	III[d]
Aspartic acid	18.16	16.70	20.53
Threonine	7.38	7.20	9.64
Serine	12.89	13.82	14.72
Glutamic acid	45.20	40.10	45.71
Proline	9.87	8.11	10.43
Glycine	9.77	8.60	10.23
Alanine	18.08	16.40	18.99
Valine	15.00	13.60	13.32
Methionine	0.96	1.32	3.00
Leucine	38.64	35.20	37.42
Tyrosine	6.45	4.20	7.03
Phenylalanine	5.64	4.77	5.85
Lysine	17.55	16.52	20.95
Histidine	2.23	1.93	4.50
Arginine	8.95	7.72	15.67

[a]   All values expressed as mole/mole protein
     assuming a molecular weight of 28,000 for
     cebus apo A-I.
[b]   Values in column I were obtained after a
     24-hr hydrolysis in $6N$ HCl at $110° \pm 2°C$.
[c]   Values in column II were obtained after a
     48-hr hydrolysis in $6N$ HCl at $110° \pm 2°C$.
[d]   Values in column III are for human apo A-
     I.[33]

NASCENT LIPOPROTEIN SECRETION

In order to determine whether the monkey, like the rat,[18] is able to secrete a nascent
lipoprotein that is rich in apo E, both squirrel and cebus monkey livers were perfused in a
recirculating system containing ^{35}S-methionine, lipoproteins ultracentrifugally isolated from
the perfusate, and then subjected to SDS-PAGE and autoradiography. In both the cebus
monkey (Figure 9) and the squirrel monkey (Figure 10), lipoproteins secreted by the liver
contained considerable amounts of apo E. Utilizing criteria of density, elution profiles of
agarose column chromatography, EM particle sizing and compositional analyses, we have
tentatively identified each fraction as follows: Fraction 1 = VLDL, Fraction 2 = IDL,
Fractions 3 to 5 = LDL and Fractions 6 to 8 as HDL. In both species apo B and apo E are
major components of VLDL. However, there are species differences in the relative amounts
of the low-molecular-weight apoproteins of VLDL. There are also obvious species differences
in the amounts of apo E and apo C, especially for the fractions identified as HDL. As in
the rat,[18] nascent HDL secreted by the liver is also rich in apo B for which the existence of
both a low-molecular-weight and high-molecular-weight species remains to be determined.
Thus, these liver perfusion studies suggest that the livers of these two species of monkeys
secrete nascent lipoproteins rich in apo E, indicating that the paucity of lipoprotein apo E
in circulation is possibly due to rapid clearance of these particles.

While data on nascent lipoprotein lipid secretion from perfused livers of cebus monkeys
are preliminary, we have recently characterized the lipid composition of nascent VLDL
secreted by perfused livers of squirrel monkeys with low or high plasma cholesterol values.[19]
The two major findings were (1) livers of hypercholesterolemic monkeys tended to secrete
large VLDL particles that contained more lipid, and (2) the VLDL particles were enriched

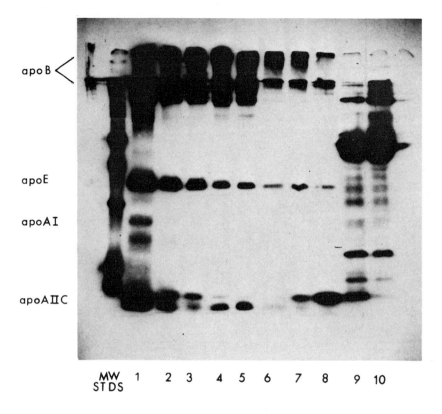

FIGURE 9.   Autoradiogram of 10% SDS-PAGE of cebus nascent hepatic lipoproteins showing their apoprotein composition. Note the paucity of apo A-I and the relative abundance of apo C and apo B in HDL. Lanes 1, VLDL; 2, IDL; 3 to 5, LDL; 6 to 8, HDL.

with cholesterol and had β-mobility indicating that squirrel monkey livers can secrete apo E-rich β-VLDL in proportion to their original plasma cholesterol concentration.

ENZYMES OF LIPOPROTEIN METABOLISM

Two major enzymes involved in lipoprotein metabolism are the (1) lecithinolesterol acyltransferase (LCAT) enzyme involved in (a) cholesterol esterification and formation of the plasma HDL particle and (b) conversion of the HDL_3 subclass to HDL_2 and (2) lipoprotein lipase enzyme involved in metabolism of triglyceride-rich lipoproteins with subsequent transfer of certain components to HDL. We have recently reported the level of LCAT activity in normocholesterolemic and hypercholesterolemic squirrel and cebus monkeys.[20] We have also completed preliminary measurements of LPL activity in these monkeys. Table 11 summarizes the data on both enzymes. The major points of these findings were (1) LCAT activity expressed as fractional rate (%/hr) which was twofold higher in monkeys compared to humans was significantly reduced in hypercholesterolemic monkeys; (2) LCAT activity expressed as molar rate was greater for coconut oil-fed cebus than squirrel monkeys; and (3) monkey LPL activity which was twofold greater than humans was significantly higher in squirrel monkeys compared to cebus. The greater LPL activity in monkeys compared to humans may explain the paucity of circulating VLDL in monkeys.[21] The greater LPL activity in squirrel monkeys compared to cebus may help explain the greater HDL_2 levels of the latter species since this enzyme is thought to be involved in HDL subclass conversion.[22]

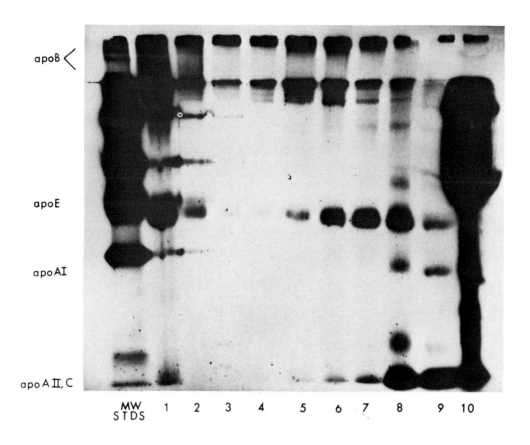

apoB

apoE

apoAI

apoAII,C

MW STDS   1   2   3   4   5   6   7   8   9   10

FIGURE 10. Autoradiogram of 10% SDS-PAGE of squirrel monkey nascent hepatic lipoproteins showing their apoprotein composition. Again, note that HDL is enriched in apo E and apo B with little apo A-I. Lanes 1, VLDL; 2, IDL; 3 to 4, LDL; 5 to 8, HDL.

Table 11
EFFECT OF DIET AND SPECIES ON LECITHIN CHOLESTERYL ACYL TRANSFERASE (LCAT) AND LIPOPROTEIN LIPASE (LPL) ACTIVITY

	LCAT activity		
	Fractional rate (% esterified/hr)	Molar rate (nmol esterified/mℓ/hr)	LPL activity (μmol FFA/hr/mℓ)
Corn oil			
Cebus	15.0 ± 1.2[a]	152 ± 17	14.2 ± 1.0[a]
Squirrel	14.4 ± 3.6	154 ± 39	31.1 ± 0.9
Coconut oil			
Cebus	10.2 ± 0.6	203 ± 18	ND[b]
Squirrel	7.8 ± 0.6	121 ± 28	ND
Human	6.0 ± 1.2	ND	7.8 ± 0.9

[a] Values represent mean ± SE.
[b] Not determined.

From Nicolosi, R. J., Hojnacki, J. L., and Hayes, K. C., unpublished data.

Table 12
EFFECT OF DIET AND SPECIES ON CHOLESTEROL METABOLISM IN CEBUS AND SQUIRREL MONKEYS[a]

	Cebus		Squirrel	
	SAF	COCO	SAF	COCO
t 1/2 A (days)	2.3 ± 0.4	2.3 ± 0.6	2.1 ± 0.3	2.0 ± 0.3
t 1/2 B (days)	17.3 ± 1.2	11.8 ± 1.0	11.5 ± 0.8	11.6 ± 0.8
M_A (mg/kg)	531 ± 146	645 ± 67	627 ± 158	533 ± 80
M_{B1} (mg/kg)	701 ± 106	499 ± 94	539 ± 80	542 ± 77
PR_A (mg/kg/day)	83 ± 22	96 ± 9	128 ± 14	105 ± 12
M_{AP} (mg/kg)	70 ± 5	109 ± 21	57 ± 6	107 ± 3
MCF (%)	120 ± 39	90 ± 13	229 ± 49	98 ± 9

Note: M_A, mass of pool A; M_B, mass of pool B; PR_A, production rate; M_{AP}, plasma cholesterol component of pool A; MCF, metabolic clearance factor.

[a] Values represent mean ± SD.

LIPOPROTEIN METABOLISM

Our present knowledge of lipoprotein metabolism in monkeys is very limited compared to other species. Nevertheless, recent studies in nonhuman primates[23,24] and in the squirrel monkey,[1] in particular, suggest that diet-induced hypercholesterolemia is often associated with alterations in the fractional catabolic rate (FCR) of lipoproteins. Studies in our laboratory in both normocholesterolemic and hypercholesterolemic cebus and squirrel monkeys which demonstrated a reduced metabolic clearance of circulating cholesterol[25] (Table 12), would tend to support this notion. Along these lines, we have also demonstrated that VLDL triglyceride secretion, which in steady-state conditions can also be interpreted as a measurement of clearance, was reduced in hypercholesterolemic monkeys,[26] a finding in agreement with the studies of Illingworth et al.[27] However, our liver perfusion studies[19] as well as those of others,[28,29] indicate that hypercholesterolemic animals also increase lipoprotein secretion indicating that both synthesis and/or catabolism of lipoproteins can be altered.

In studies of HDL metabolism recently completed in both squirrel and cebus monkeys[30] (Table 13, Figure 11), the reduction in HDL cholesterol, which accompanied the polyunsaturated fat-induced hypocholesterolemia, was associated with an increased FCR in contrast to human studies where a reduction in HDL synthesis was demonstrated. The reason for the discrepancy in the mechanisms demonstrated in the two studies is not known.

Although the mechanism(s) that determine how diet influences lipoprotein character and metabolism remain to be elucidated, previous studies have shown that dietary fat-induced changes in fatty acid composition such as those observed in Table 5, can influence lipoprotein catabolism. Our own studies of dietary influences on LCAT activity[20] and LPL activity, in agreement with other investigators[31,32] support this notion and indicate an important role for both of these enzymes in lipoprotein catabolism.

SUMMARY

The extensive effort being expended to characterize the lipoproteins of nonhuman primates has resulted in the observation that substantial species differences exist in the lipoprotein response to dietary perturbation. Careful utilization of these monkey models should permit us to investigate the mechanism(s) responsible for the wide diversity of responses to diet-induced hypercholesterolemia in humans. Although comparisons in the metabolism of normal

Table 13
EFFECT OF DIET AND FAT SATURATION ON KINETIC PARAMETERS OF
^{125}I-HDL METABOLISM

Diet and species	HDL Protein[a] (mg/dℓ)	Fractional catabolic rate (pools/day)	Absolute catabolic rate (mg/day)	Absolute catabolic rate (mg/kg/day)
Cebus monkey				
Corn oil (3)	294 ± 16[b]	0.54 ± 0.05[b]	198.4 ± 32.8[b]	63.4 ± 6.6[b]
Coconut oil (4)	376 ± 36[c]	0.43 ± 0.02[c]	187.5 ± 19.5[b]	64.3 ± 9.1[b]
Squirrel monkey				
Corn oil (3)	273 ± 54[b]	0.51 ± 0.02[b]	51.3 ± 7.8[b]	55.6 ± 8.6[b]
Coconut oil (4)	400 ± 70[c]	0.42 ± 0.02[c]	64.7 ± 14.8[b]	66.7 ± 11.0[b]
Grand mean[d]				
Corn oil (6)	283 ± 37[b]	0.53 ± 0.04[b]	—[e]	59.5 ± 8.1[b]
Coconut oil (8)	388 ± 53[c]	0.42 ± 0.02[c]	—[e]	65.5 ± 9.4[b]

[a]   Values represent mean ± SD for four monkeys of each species per dietary group.

[b,c]   Values with different superscripts are significantly different from each other, at least $p < 0.05$.

[d]   Calculated to determine the effect of diet, independent of species.

[e]   Not determined since there were species differences due to body weight.

From Nicolosi, R. J., Rodger, R., Ausman, L., Yuan, R., and Georas, S., Herbert, P. N., submitted.

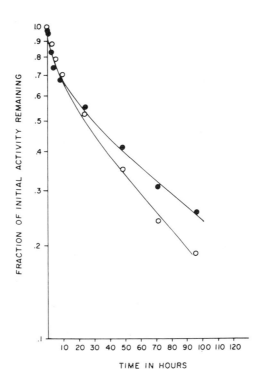

FIGURE 11. ^{125}I-HDL decay curve of cebus monkey showing faster turnover in corn oil-fed monkeys compared to coconut oil-fed animals.

and abnormal lipoproteins in monkeys are only in their infancy, it is already evident that much useful information is obtainable that should be more applicable to humans than other animal models.

ACKNOWLEDGMENTS

We acknowledge the expert technical assistance of Dr. Kui Song Chong, Ruth Yuan, and Dorothy Arrigo and secretarial assistance of Maureen Maguire.

This work was supported in part by NIH Grant No. HL-23792, USDA Grant No. 8000479, NIH Grant No. RR00168 from the Division of Research Resources, and a Grant from the R. J. Reynolds Foundation.

Animals used in this study were maintained in accordance with the guidelines of the Committee on Animals of the Harvard Medical School and those prepared by the Committee on Care and Use of Laboratory Animals of the Institute of Laboratory Animal Resources, National Research Council (DHEW publication no. [NIH] 78-23, revised 1978).

REFERENCES

1. **Portman, O. W., Alexander, M., Tanaka, N., and Soltys, P.,** The effects of dietary fat and cholesterol on the metabolism of plasma low density lipoprotein apoproteins in squirrel monkeys, *Biochim. Biophys. Acta,* 450, 185, 1976.
2. **Illingworth, D. R., Portman, O. W., and Whipple, L. E.,** Metabolic interrelationships between the lipids of very low, low and high density lipoproteins in the squirrel monkey, *Biochim. Biophys. Acta,* 369, 304, 1974.
3. **Corey, J. E., Hayes, K. C., Dorr, B., and Hegsted, D. M.,** Comparative lipid response of four primate species to dietary changes in fat and carbohydrate, *Atherosclerosis,* 19, 119, 1974.
4. **Nicolosi, R. J., Hojnacki, J. L., Llansa, N., and Hayes, K. C.,** Diet and lipoprotein influence on primate atherosclerosis, *Proc. Soc. Exp. Biol. Med.,* 156, 1, 1977.
5. **Portman, O. W. and Andrus, S. B.,** Comparative evaluation of three species of New World monkeys for studies of dietary factors, tissue lipids and atherogenesis, *J. Nutr.,* 87, 429, 1965.
6. **Clarkson, T. B., Lehner, N. D. M., Bullock, B. C., Lofland, H. B., and Wagner, W. D.,** Atherosclerosis in New World monkeys, in *Primates in Medicine,* Vol. 9, Strong, J. P., Ed., S. Karger, Basel, 1976, 90.
7. **Rudel, L. L. and Lofland, H. B.,** Circulating lipoproteins in nonhuman primates, in *Primates in Medicine,* Vol. 9, Strong, J. P., Ed., S. Karger, Basel, 1976, 224.
8. **Hill, P., Martin, W. G., and Douglas, J. F.,** Comparison of the lipoprotein profiles and the effect of *N*-phenylpropyl-*N*-benzyloxy acetamide in primates, *Proc. Soc. Exp. Biol. Med.,* 148, 41, 1975.
9. **Ershow, A. G., Nicolosi, R. J., and Hayes, K. C.,** Separation of the dietary fat and cholesterol influences on plasma lipoproteins of rhesus monkeys, *Am. J. Clin. Nutr.,* 34, 830, 1981.
10. **Rudel, L. L., Shah, R., and Greene, D. G.,** Study of the atherogenic dyslipoproteinemia induced by dietary cholesterol in rhesus monkeys *(Macaca mulatta), J. Lipid Res.,* 20, 55, 1979.
11. **Morello, A. M. and Nicolosi, R. J.,** Electrophoretic profiles of high density lipoprotein subclasses in primates, *Comp. Biochem. Physiol.,* 69B, 291, 1981.
12. **Nelson, C. A., Morris, M. D., and Geer, W. E.,** The distribution of high density lipoprotein subfractions in nonhuman primates, *Fed. Proc.,* 39, 1718a, 1980.
13. **Illingworth, D. R.,** Metabolism of lipoproteins in nonhuman primates. Studies on the origin of low density lipoprotein apoprotein in the plasma of the squirrel monkey, *Biochim. Biophys. Acta,* 388, 38, 1975.
14. **Kritchevsky, D.,** Experimental atherosclerosis in primates and other species, *Ann. N.Y. Acad. Sci.,* 162, 80, 1969.
15. **Shepherd, J., Packard, C. J., Gotto, A. M., Jr., and Taunton, O. D.,** Effects of dietary polyunsaturated and saturated fat on the properties of high density lipoproteins and the metabolism of apo lipoprotein AI, *J. Clin. Invest.,* 61, 1582, 1978.
16. **Hojnacki, J. L., Nicolosi, R. J., Hoover, G., Llansa, N., el Lozy, M., and Hayes, K. C.,** Effects of dietary fat on the composition and size of primate very low density lipoproteins, *Artery,* 3, 409, 1977.

17. **Rudel, L. L., Pitts, L. L., II, and Nelson, C. A.,** Characterization of plasma low density lipoproteins of nonhuman primates fed dietary cholesterol, *J. Lipid Res.,* 18, 211, 1977.

18. **Felker, T. E., Fainaru, M., Hamilton, R. L., and Havel, R. J.,** Secretion of the arginine-rich and A-I apolipoproteins by the isolated perfused rat liver, *J. Lipid Res.,* 18, 465, 1977.

19. **Nicolosi, R. J. and Hayes, K. C.,** Composition of plasma and nascent very low density lipoprotein from perfused livers of hypercholesterolemic squirrel monkeys, *Lipids,* 15, 549, 1980.

20. **Lichtenstein, A. H., Nicolosi, R. J., and Hayes, K. C.,** Dietary fat and cholesterol effects on plasma lecithinolesterol acyltransferase activity in cebus and squirrel monkeys, *Atherosclerosis,* 37, 603, 1980.

21. **Portman, O. W., Alexander, M., Tanaka, N., and Illingworth, D. R.,** Triacylglycerol and very low density lipoprotein secretion into plasma of squirrel monkeys, *Biochim. Biophys. Acta,* 486, 470, 1977.

22. **Patsch, J. R., Gotto, A. M., Jr., Olivecrona, T., and Eisenberg, S.,** Formation of high density lipoprotein-like particles during lipolysis of very low density lipoproteins in vitro, *Proc. Natl. Acad. Sci.,* 75, 4519, 1978.

23. **Kushwaha, R. S., Hazzard, W. R., Harker, L. A., and Engblom, J.,** Lipoprotein metabolism in baboons: effect of feeding cholesterol-rich diet, *Atherosclerosis,* 31, 65, 1978.

24. **Parks, J. S. and Rudel, L. L.,** Different kinetic fates of apolipoproteins A-I and A-II from lymph chylomicra of nonhuman primates. Effect of saturated versus polyunsaturated dietary fat, *J. Lipid Res.,* 23, 410, 1982.

25. **Corey, J. E., Nicolosi, R. J., and Hayes, K. C.,** Effect of dietary fat on cholesterol turnover in Old and New World monkeys, *Exp. Mol. Pathol.,* 25, 311, 1976.

26. **Nicolosi, R. J., Hayes, K. C., el Lozy, M., and Herrera, M. G.,** Hypercholesterolemia and triglyceride secretion rates in monkeys fed different dietary fats, *Lipids,* 12, 936, 1977.

27. **Illingworth, D. R., Whipple, L. E., and Portman, O. W.,** Metabolism of lipoproteins in nonhuman primates. Reduced secretion of very low density lipoproteins in squirrel monkeys with diet-induced hypercholesterolemia, *Atherosclerosis,* 22, 325, 1975.

28. **Noel, S. P., Wong, L., Dolphin, P. J., Dory, L., and Rubinstein, D.,** Secretion of cholesterol-rich lipoproteins by perfused livers of hypercholesterolemic rats, *J. Clin. Invest.,* 64, 674, 1979.

29. **Guo, L. S. S., Hamilton, R. L., Ostwald, R., and Havel, R. J.,** Secretion of nascent lipoproteins and apolipoproteins by perfused livers of normal and cholesterol-fed guinea pigs, *J. Lipid Res.,* 23, 543, 1982.

30. **Nicolosi, R. J., Yuan, R., Rodger, R., and Herbert, P.,** Effect of long-term feeding of dietary fats containing polyunsaturated fatty acids and saturated fatty acids on high density lipoprotein metabolism, *Fed. Proc.,* 41, 708a, 1982.

31. **Nestel, P. J., Carroll, K. F., and Havenstein, N.,** Plasma triglyceride response to carbohydrates, fats and caloric intake, *Metabolism,* 19, 1, 1970.

32. **Pawar, S. S. and Tidwell, H. D.,** Effect of ingestion of unsaturated fat on lipolytic activity of rat tissues, *J. Lipid Res.,* 9, 334, 1968.

33. **Brewer, H. B., Fairwell, T., LaRue, A., Ronan, R., Houser, A., and Bronzert, T. J.,** The amino acid sequence of human apo AI, an apolipoprotein isolated from high density lipoproteins, *Biochem. Biophys. Res. Commun.,* 80, 623—630, 1978.

34. **Nicolosi, R. J., Hojnacki, J. L., and Hayes, K. C.,** Lipoprotein and tissue lipids in cebus monkeys. I. Effect of dietary fat, *J. Lipid Res.,* submitted.

35. **Nicolosi, R. J., Rodger, R., Ausman, L., Yuan, R., Georas, S., and Herbert, P. N.,** Effects of long term feeding of polyunsaturated and saturated fat on the composition and metabolism of high density in nonhuman primates, submitted.

LIPOPROTEINS OF NONHUMAN PRIMATES: A REVIEW AND A STUDY OF THEIR RELATIONSHIP TO OXIDIZED CHOLESTEROL*

Shi-Kaung Peng, C. Bruce Taylor, and Lena A. Lewis

INTRODUCTION

There is a growing dissatisfaction with lower laboratory animals, such as chickens, rats, rabbits, and dogs as subjects for research designed to elucidate the pathogenesis of human atherosclerosis, because the lipoprotein spectra in these animals are different from those found in humans.[1,2] They have large amounts of α-lipoprotein (α-Lp) and much less β-lipoprotein (β-Lp). Lipid transport functions also seem different. On the other hand, non-human primates, having close similarities to human organisms in regard to anatomy of the body and vascular system, blood type, lipoprotein spectrum, lipid metabolism, and responses to an atherogenic diet,[3-9] provide an important animal model for experimental atherosclerosis. In a series of studies in rhesus monkeys, Taylor et al. produced the first and only known case of fatal myocardial infarction in an experimental animal, resulting from diet-induced hypercholesteremia.[8] It was also shown that the minimal critical level for atherogenesis was 200 mg/dℓ and that significant atherosclerosis developed 3 months after serum cholesterol levels exceeded 225 mg/dℓ.

Although the exact pathogenesis of atherosclerosis is still not known, many predisposing risk factors have been proposed. Among them, hypercholesteremia and hyperlipoprotememia are thought to be the major and obligate risk factors.[10] Cholesterol and other lipids are transported in the bloodstream by serum lipoproteins. The circulating lipoproteins are classified according to their densities and composition as chylomicrons, very low-density lipoproteins (VLDL), low-density lipoproteins (LDL), and high-density lipoproteins (HDL). The chylomicrons and VLDL primarily transport triglycerides. The chylomicrons consist mostly of triglycerides which originate in the intestinal epithelial cells. The liver clears chylomicrons from the blood by transferring the triglycerides to the VLDL which are then again released into the bloodstream. The remnants of VLDL after removing the bulk of triglycerides are converted to LDL. The main function of LDL has been shown to be the delivery of cholesterol to various peripheral tissues for cell growth and biogenesis of membranes. Glomset and Norum[11] suggested that the transport of cholesterol, from peripheral tissues to liver for subsequent catabolism and excretion, may be a function of HDL; unesterified cholesterol from peripheral tissue diffuses into the bloodstream and is esterified in contact with lecithin cholesterol acyl transferase (LCAT) and HDL.

Epidemiologic studies[12] have suggested that certain types of lipoproteins have been considered to be more atherogenic than others. Havel[2] observed that the lipoproteins containing β-apoprotein (namely LDL and VLDL) are linked to atherogenesis. Beyond this, however, properties of serum lipoproteins that are atherogenic are not well understood. In our previous studies,[13,14] we found that U.S.P. grade cholesterol and foods containing cholesterol (such as powdered egg yolk) are usually contaminated by spontaneously occurring oxidation products of cholesterol. In both in vivo and in vitro studies, these oxidized cholesterol derivatives, which are very toxic to aortic smooth muscle cells and endothelium, may be principally responsible for initiation of atherosclerosis rather than cholesterol per se, in fact, purified cholesterol is innocuous. Deposition of cholesterol and other lipids in atheromatous lesions may be merely a secondary phenomenon. It is also conceivable that these oxidized cholesterol

* Studies by authors cited were supported by the Veterans Administration and the American Heart Association of North Eastern New York, Grant #210030.

Table 1
PAPER LIPOPROTEIN ELECTROPHORESIS ON RHESUS MONKEYS

Monkey #	Sample #	Chylomicron	β-Lp[c]	α-Lp[c]
1	1[a]	0	2+	3+
	2[b]	0	4+	3+
	3[b]	±	4+	2+
2	1[a]	0	2+	3+
	2[b]	0	4+	4+

[a]   On control diet.

[b]   On atherogenic diet.

[c]   b-Lp 2+ represents 253 (160—440) mg/dℓ, 3+:358 (260—700) mg/dℓ, and 4+:742 (360—1000) mg/dℓ and α-Lp 2+ represents 220 (120—320) mg/dℓ, 3+:291 (180—460) mg/dℓ and 4+:392 (320—480) mg/dℓ. Data from Lewis, L. A., *Lipids*, 4, 60, 1969.

derivatives may be transported in a fashion different from that of cholesterol, presumably by lipoproteins carrying exogenous cholesterol (such as LDL and VLDL). On the other hand, HDL containing α-apoprotein which usually carries cholesterol of endogenous origin may carry no or miniscule amounts of oxidized cholesterol derivatives and therefore is not atherogenic. Nonhuman primates provided the most suitable animal due to the similarity of their lipoprotein profiles to those of humans. In addition, we have had experience with both rhesus monkeys *(Macaca mulatta)*[3,15] and squirrel monkeys *(Saimiri sciureus)*[16,17] in the production of experimental atherosclerosis and lipoprotein analysis. There are also many studies on lipoproteins in nonhuman primates done by other investigators for comparison.

REVIEW OF PERTINENT LITERATURE

Pioneer studies on the lipoprotein pattern on nonhuman primates were done by Cox et al.[3] using paper electrophoresis. By this technique using barbital buffer (pH 8.6) Sudan stains, no resolution of VLDL and LDL was observed in any of the rhesus monkey sera studied. HDL of rhesus monkeys on control diets were equal to or higher than LDL (Table 1). The value of LDL increased markedly on the atherogenic diets. LDL did not vary more than when they were on the control diet.

Hill et al.[18] studied three species of monkeys including squirrel, rhesus, and cebus monkeys *(Cebus appella)* using techniques of agarose gel electrophoresis as well as ultracentrifugation. Chylomicrons were not observed either by agarose gel electrophoresis or in ultracentrifugation in all three species of monkeys. Only one squirrel monkey displayed VLDL and LDL having only one component, while cebus and rhesus monkeys displayed two distinct LDL components. No clear separation of VLDL was found on agarose gel electrophoresis. HDL levels in these three species of monkeys were higher in concentration than those of LDL (194 mg/dℓ, 198 mg/dℓ and 169 mg/dℓ vs. 104 mg/dℓ, 158 mg/dℓ and 137 mg/dℓ for squirrel, rhesus, and cebus monkeys, respectively). Malinow et al.,[19] employing a poly-acrylamide gel electrophoretic (PAGE) technique, was able to separate VLDL, LDL, and HDL in cynomolgus monkeys *(Macaca fascicularis)*. When monkeys were fed control diets, lipoprotein profiles were 3% VLDL, 40% LDL, and 56% HDL. When they were fed a cholesterol-rich diet, there was a marked increase of VLDL and LDL of 15% and 65%, respectively, and a decrease of HDL to 17%. Hollander et al.[20] showed a similar result in cynomolgus monkeys using an ultracentrigation method, 28 mg/dℓ of VLDL, 127 mg/dℓ of LDL, and 222 mg/dℓ of HDL when they were on normal diets. The increases of VLDL and LDL were even more remarkable when they were fed atherogenic diets.

Table 2
CHOLESTEROL CONCENTRATION IN LIPOPROTEINS OF FOUR SPECIES OF MONKEYS

Monkey — species	Whole plasma	VLDL	ILDL	LDL	HDL
Carcopithecus aethiops (African Green)	144 ± 6[a]	0.4 ± 0.1	7 ± 1	48 ± 2	88 ± 4
Macaca fascicularis (cynomolgus)	140 ± 12	0.9 ± 0.3	11 ± 2	73 ± 12	55 ± 8
Macaca mulatta (rhesus)	191 ± 9	0.9 ± 0.2	31 ± 7	54 ± 2	98 ± 7
Saimiri sciureus (squirrel)	143 ± 15	1.5 ± 0.3	6 ± 1	61 ± 9	76 ± 9

[a] All data are expressed as mean ± SE in mg/dℓ.

Cholesterol contents in lipoproteins are different from the total lipid content which is usually reflected indiscriminately by electrophoretic techniques. Rudel[21] studied lipoproteins of four species of monkeys: African green monkey *(Carcopithecus aethiops),* cynomolgus, rhesus, and squirrel monkeys which were analyzed by an agarose column chromatographic technique for separation and purification. In addition to VLDL, LDL, and HDL, the intermediate sized low-density lipoprotein (ILDL) was also separated (Table 2). Most of these species of monkeys when fed a control or a low-cholesterol diet showed that cholesterol in HDL is slightly higher than cholesterol in LDL except in the case of the cynomolgus monkey. Low concentration of cholesterol in VLDL may be due to further separation of VLDL from ILDL and LDL by this technique. Monkeys respond variably to a high-cholesterol diet by an increase of VLDL or LDL or both. Cholesterol in HDL was relatively unchanged.

MATERIALS AND METHODS

Nonhuman primates were employed as models for investigation of the distribution of pure cholesterol as well as oxidized cholesterol in various lipoprotein fractions. In these studies we used ten squirrel monkeys; five of them were fed a single dose of 30 µCi of [4-^{14}C] cholesterol (specific activity 0.14 mCi/mg from New England Nuclear) per monkey. The other five monkeys were fed a single dose of 150 µCi of [^{3}H] 25-hydroxycholesterol (specific activity 1.54 mCi/mg from New England Nuclear) per monkey. From each monkey, 5 mℓ of blood were drawn via femoral vein 24 hr after administration of the radioactive material. Electrophoresis and ultracentrifugation were employed for lipoprotein analyses. Cellulose acetate electrophoresis was used for a qualitative assessment of the lipoprotein pattern. Blood samples collected in 0.1% EDTA were centrifuged. Cellulose acetate strips (Gelman Instrument Co.) were immersed in barbital buffer, blotted lightly between filter paper and placed in the electrophoresis chamber. A small aliquot of serum sample was applied at the cathodal end of the strip. Electrophoresis was performed at 200 V for 15 min at room temperature. The strip was blotted and stained overnight in an oil red O solution. The wet strip was scanned in the Gilman ADC-15 densitometer. Quantitative measurement of serum lipoproteins was done by ultracentrifugation, performed in a Spinco Model L2-65B using a Beckman type 50 rotor at a speed of 39,000 r/min (63,000 × *g*) at 4°C. The VLDL fraction was isolated at a serum density of 1.006 by ultracentrifugation for 24 hr. The remaining supernatant was adjusted to a density of 1.063 by the addition of solid NaCl or NaBr. The LDL fraction was then isolated by collecting the infranatant fraction after ultracentrifugation for 24 hr. For the HDL fraction, ultracentrifugation was carried out in the density range of 1.063 to 1.21 for 48 hr and then the infranatant fraction was collected. Serum cholesterol was measured by the method of Abell et al.[22] and protein by the method of Lowry et al.[23] The radioactivity of each lipoprotein was counted in a Packard Tri-Carb Scintillation counter with correction for background count and quenching.

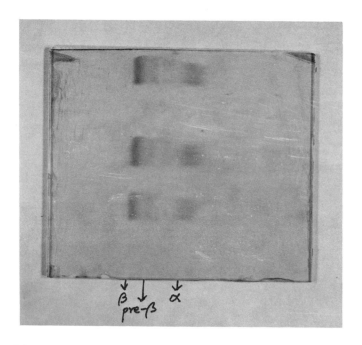

FIGURE 1.   Cellulose acetate lipoprotein electrophoresis on squirrel monkeys.

DISTRIBUTION OF OXIDIZED CHOLESTEROL IN SERUM LIPOPROTEINS[17]

Cellulose acetate electrophoresis of the serum lipoproteins of these squirrel monkeys, shown in Figure 1, is very similar to studies by other investigators.[18,19] Densitometric scanning showed 6% VLDL, 40% LDL, and 56% HDL. However, cellulose acetate does not yield the clear pattern seen with polyacrylamide gel, but it appears to be better than paper electrophoresis. The distribution of cholesterol content in various lipoproteins was determined primarily by ultracentrifugation. Serum cholesterol levels had an average of 154 mg/dℓ in which 6 mg/dℓ or 3.8% in VLDL, 69 mg/dℓ or 44.9% in LDL, and 79 mg/dℓ or 51.3% in HDL. Our results are very comparable to those of Rudel.[21] When monkeys were fed with radioactive cholesterol orally, the distribution of labeled cholesterol in the various lipoprotein fractions was such that 3.1% of the radioactivity was in the VLDL, 47.6% and 49.3% in the LDL and HDL, respectively. This distribution is very similar to the distribution of unlabeled cholesterol (Table 3). In other words, almost all cholesterol ingested exchanges with endogenous cholesterol in the body. The distribution of orally administered oxidized cholesterol (25-hydroxycholesterol), when compared with that of cholesterol, is totally different. Almost 90% of the total radioactivity of 25-hydroxycholesterol was distributed in LDL and VLDL (55.7% and 34.1%, respectively), and only 10.2% was present in HDL (Table 4). When the distribution of 25-hydroxycholesterol in various lipoproteins is compared with the cholesterol content of each lipoprotein and calculated on the basis of milligrams of cholesterol in the various lipoprotein micelles, the specific activity of 25-hydroxycholesterol in VLDL is shown to be the highest, approximately 90 times greater than that of HDL and 9 times greater than that of LDL. Also, the amount of 25-hydroxycholesterol in LDL is 10 times greater than that of HDL. In other words, VLDL and LDL contain significantly higher concentrations of 25-hydroxycholesterol than HDL, when compared with the distribution of cholesterol. The radioactivity of labeled cholesterol per unit of apoprotein was compared with that of labeled 25-hydroxycholesterol (Table 5). On this basis, the difference between the distribution of cholesterol and 25-hydroxycholesterol in various lipoproteins is even more

Table 3

**DISTRIBUTION OF UNLABELED AND LABELED CHOLESTEROL IN
VARIOUS LIPOPROTEINS**

Lipoprotein fractions	VLDL	LDL	HDL
Cholesterol conc (mg/dℓ)	7.0 ± 0.5	73.2 ± 7.7	82.8 ± 7.5
Distribution of cholesterol	4.3 ± 0.3%	44.9 ± 4.7%	50.8 ± 4.6%
Specific activity (dpm/mg chol.)	20.1 ± 2.6 × 10^3	30.1 ± 3.2 × 10^3	24.6 ± 2.1 × 10^3
Distribution of labeled chol.	3.1 ± 0.5%	47.6 ± 4.4%	49.3 ± 4.1%

Note: All data are expressed as means ± SE.

Table 4

**COMPARISON OF THE DISTRIBUTION OF LABELED
25-HYDROXYCHOLESTEROL WITH THAT OF UNLABELED CHOLESTEROL
IN VARIOUS LIPOPROTEINS**

Lipoprotein fractions	VLDL	LDL	HDL
Cholesterol conc. (mg/dℓ)	5.0 ± 0.6	65.6 ± 5.4	75.5 ± 5.5
Distribution of cholesterol	3.4 ± 0.4%	44.9 ± 3.7%	51.7 ± 3.7%
Specific activity (dpm/mg chol.)	202 ± 60 × 10^3	22.5 ± 5.5 × 10^3	2.2 ± 0.2 × 10^3
Distribution of labeled 25-OH chol.	34.1 ± 2.9%[a]	55.7 ± 2.9%[b]	10.2 ± 0.5%[a]

Note: All data are expressed as means ± SE.

[a] Significant difference between distribution of cholesterol and that of 25-hydroxycholesterol ($p < 0.01$).
[b] Relatively significant ($p < 0.05$).

Table 5

**COMPARISON OF THE AMOUNT OF CHOLESTEROL AND
25-HYDROXYCHOLESTEROL IN VARIOUS LIPOPROTEINS**

Lipoprotein fraction	VLDL	LDL	HDL
Apoprotein conc. (mg/mℓ)	1.07 ± 0.17	3.79 ± 0.61	28.9 ± 2.1
Specific activity of labeled chol. (dpm/mg prot.)	1942 ± 276	8930 ± 1056	1208 ± 124
Ratio of labeled chol. per mg prot.	1.6	.7.4	.1
Specific activity of labeled 25-OH chol. (dpm/mg prot)	10400 ± 990	4816 ± 423	116 ± 7
Ratio of labeled 25-OH chol. per mg protein	90	42	1

Note: All data are expressed as means ± SE.

striking. Normally, cholesterol has the highest concentration in LDL and is slightly higher in VLDL when compared with HDL (LDL 7.4 times and VLDL 1.6 times greater than HDL); however, the concentration of 25-hydroxycholesterol in LDL and VLDL is much higher than that in HDL (LDL 42 times and VLDL 90 times greater than HDL) (Table 5). It is obvious that the capacity of LDL and VLDL to carry 25-hydroxycholesterol is greater and more significant than that of HDL which carries only a trace amount of 25-hydroxycholesterol.

DISCUSSION AND CONCLUSION

From our experiments in the squirrel monkeys, 25-hydroxycholesterol, one of the most common and toxic oxidized cholesterol derivatives, is transported primarily by VLDL and LDL.[17] This is in agreement with our hypothesis since oxidized cholesterol derivatives are usually of dietary origin. It has been shown that squirrel monkeys have the greatest increase in VLDL in response to cholesterol feeding and a slight increase in LDL.[21] Cynomolgus and rhesus monkeys show quite a significant increase in both LDL and VLDL in response to dietary cholesterol. African green monkeys respond somewhat differently when challenged with dietary cholesterol. The lipoprotein that increased most in concentration was the LDL.[21] An earlier and very interesting study by Howard et al.[24] showed that lipoproteins of chimpanzees and baboons fed on atherogenic diet resemble those of Type II hyperlipidemia, spontaneously occurring in man. Results obtained by electrochromatography and by ultracentrifugation were similar and in given species and given diets were consistent. In chimpanzees, increase in cholesterol while on an atherogenic diet occurred only in β-Lp. In baboons and man α-Lp may be affected to some extent also. Increased phospholipid concentration, especially in sphingomyelin, occurred in hyperlipidemic animals and man. Even in most mammals, normally, the bulk of the serum cholesterol is carried in the HDL and relatively little is present in VLDL and LDL. When animals are subjected to cholesterol feeding there is a substantial increase in the concentration of lipoproteins bearing the β-apoproteins (VLDL and LDL) and they routinely develop atheromatous plaques.

Oxidized cholesterol derivatives such as 25-hydroxycholesterol have been found in a number of commonly consumed cholesterol-rich foods such as powdered eggs, a powdered custard mix, a pancake mix, cheeses kept at room temperature, and lard used for several weeks to french-fry potatoes. It is conceivable that in man oxidized cholesterol derivatives in these foods are absorbed in the body and transported by VLDL and LDL to peripheral tissue; they injure arterial endothelium and smooth muscle cells. Subsequently, arterial repair (principally by intimal proliferation) can accumulate lipid and calcium in the young myointimal cells if circulating LDL and VLDL are elevated.[25,26] Therefore, risk factors currently implicated in atherogenesis such as dietary cholesterol, VLDL and LDL may indeed have a single common cause, i.e., the presence of the spontaneously oxidized cholesterol derivatives.

REFERENCES

1. **Ho, K. J., Lawrence, W. D., Lewis, L. A., Liu, L. B., and Taylor, C. B.,** Hereditary hyperlipidemia in nonlaying chickens, *Arch. Pathol.,* 98, 161, 1974.
2. **Havel, R. J.,** Lipid and atherosclerosis, *Cardiovasc. Res. Center Bull.,* 15, 93, 1977.
3. **Cox, G. E., Taylor, C. B., Cox, L. G., and Counts, M. A.,** Atherosclerosis in rhesus monkeys. I. Hypercholesteremia induced by dietary fat and cholesterol, *Arch. Pathol.,* 66, 32, 1958.
4. **Taylor, C. B., Cox, G. E., Manalo-Estrella, P., and Southworth, J.,** Atherosclerosis in rhesus monkeys. II. Arterial lesions associated with hypercholesteremia induced by dietary fat and cholesterol, *Arch. Pathol.,* 74, 16, 1962.
5. **Taylor, C. B., Trueheart, R. E., and Cox, G. E.,** Atherosclerosis in rhesus monkeys. III. The role of increased thickness of arterial walls in atherogenesis, *Arch. Pathol.,* 76, 14, 1963.
6. **Cox, G. E., Trueheart, R. E., Kaplan, J., and Taylor, C. B.,** Atherosclerosis in rhesus monkeys. IV. Repair of arterial injury — an important secondary atherogenic factor, *Arch. Pathol.,* 76, 166, 1963.
7. **Taylor, C. B., Manalo-Estrella, P., and Cox, G. E.,** Atherosclerosis in rhesus monkeys. V. Marked diet-induced hypercholesteremia with xanthomatosis and severe atherosclerosis, *Arch. Pathol.,* 76, 239, 1963.
8. **Taylor, C. B., Patton, D. C., and Cox, G. E.,** Atherosclerosis in rhesus monkeys. VI. Fatal myocardial infarction in a monkey fed fat and cholesterol, *Arch. Pathol.,* 76, 404, 1963.

9. **Manalo-Estrella, P., Cox, G. E., and Taylor, C. B.,** Atherosclerosis in rhesus monkeys. VII. Mechanism of hypercholesteremia: hepatic cholesterologenesis and the hypercholesteremia threshold of dietary cholesterol, *Arch. Pathol.,* 76, 413, 1963.

10. **Taylor, C. B., Hass, G. M., Ho, K. J., and Liu, L. B.,** Risk factors in the pathogenesis of atherosclerotic heart disease and general atherosclerosis, *Ann. Clin. Lab. Sci.,* 2, 239, 1972.

11. **Glomset, J. A. and Norum, K. R.,** The metabolic role of lecithin cholesterol acyltransferase. Perspectives from pathology, *Adv. Lipid Res.,* 11, 1, 1973.

12. **Gordon, T., Castelli, W. P., Hjortland, M. C., Kannel, W. B., and Dawber, T. R.,** High density lipoprotein as a protective factor against coronary heart disease. The Framingham study, *Am. J. Med.,* 62, 707, 1977.

13. **Peng, S. K., Taylor, C. B., Tham, P., Werthessen, N. T., and Mikkelson, B.,** Effects of auto-oxidation products from U.S.P. grade cholesterol on aortic smooth muscle cells — in vitro study, *Arch. Pathol. Lab. Med.,* 102, 57, 1978.

14. **Taylor, C. B., Peng, S. K., Werthessen, N. T., Tham, P., and Lee, K. T.,** Spontaneously occurring angiotoxic derivatives of cholesterol, *Am. J. Clin. Nutr.,* 32, 40, 1979.

15. **Liu, L. B., Taylor, C. B., Peng, S. K., and Mikkelson, B.,** Experimental arteriosclerosis in rhesus monkeys induced by multiple risk factors: cholesterol, vitamin D and nicotine, *Art. Wall (Paroi Arterielle),* 5, 25, 1979.

16. **Peng, S. K., Taylor, C. B., Tham, P., and Mikkelson, B.,** Role of mild excesses of vitamin D_3 in arteriosclerosis, a study in squirrel monkey, *Art. Wall (Paroi Arterielle),* 4, 229, 1978.

17. **Peng, S. K., Taylor, C. B., Musbach, E. H., Huang, W. Y., Hill, J., and Mikkelson, B.,** Distribution of 25-hydroxycholesterol in plasma lipoproteins and its role in atherosclerosis — a study in squirrel monkeys, *Atherosclerosis,* 41, 395, 1982.

18. **Hill, P., Martin, W. G., and Douglas, J. F.,** Comparison of the lipoprotein profiles and the effect of *N*-phenylpropyl-*N*-benzyloxy acetamide in primates, *Proc. Soc. Exp. Biol. Med.,* 148, 41, 1975.

19. **Malinow, M. R., McLaughlin, P., McNulty, W. P., Naito, H. K., and Lewis, L. A.,** Treatment of established atherosclerosis during cholesterol feedings in monkeys, *Atherosclerosis,* 31, 185, 1978.

20. **Hollander, P., Prusty, S., Nagraj, S., Kirkpatric, B., Paddock, J., and Columbo, M.,** Comparative effects of cetaben (PHB) and dichlormethylene disphosphonate (Cl_2MDP) on the development of atherosclerosis in the cynomolgus monkey, *Atherosclerosis,* 31, 307, 1978.

21. **Rudel, L. L.,** Plasma lipoproteins in atherogenesis in nonhuman primates. The use of nonhuman primates in cardiovascular diseases. Kalter, S. S., Ed., University of Texas Press, Austin, 1980, 37.

22. **Abell, L. L., Levy, B. B., Brodie, B. B., and Kendall, F. E.,** A simplified method for the estimation of total cholesterol in serum and demonstration of its specificity, *J. Biol. Chem.,* 195, 357, 1952.

23. **Lowry, O. H., Rosebrough, N. J., Farr, A. L., and Randall, R. J.,** Protein measurement with the Folin phenol reagent, *J. Biol. Chem.,* 193, 265, 1951.

24. **Howard, A. N., Blaton, V., Vandamme, D., Landschoot, N. V., and Peeters, H.,** Lipid changes in the plasma lipoproteins of baboons given an atherogenic diet, *Atherosclerosis,* 16, 257, 1972.

25. **Peng, S. K. and Taylor, C. B.,** Atherogenic effect of oxidized cholesterol, in *Dietary Fats and Health,* Proc. Conf., Perkins, E. G., and Visek, W. J., Eds., American Oil Chemists' Society, Chicago, Ill. December 6 to 11, 1981.

26. **Taylor, C. B.,** The Reaction of Arteries to Injury by Physical Agents — with Discussion of Arterial Repair and its Relationship to Atherosclerosis, National Research Council Publication, No. 338, National Academy of Sciences, New York, 1954, 74.

THE SERUM LIPOPROTEINS AND APOLIPOPROTEINS IN THE BABOON

Arthur W. Kruski

INTRODUCTION

This review will be limited to a presentation of the chemical and physical characteristics of baboon plasma (serum) lipoproteins (LP) and their component apolipoproteins (apo LP). The reader is referred to several other review articles for information about either other nonhuman primates[1-3] or for studies of atherosclerosis in the baboon.[4-8]

Common baboons (genus *Papio*) are found throughout most of Africa south of the Sahara. Common baboons are a single polytypic species with at least five different interbreeding subspecies: the guinea baboon (*Papio cynocephalus papio* = *Papio papio*), olive baboons (*Papio cynocephalus anubis* = *Papio anubis*), hamadryas baboons (*Papio cynocephalus hamadryas*), yellow baboons (*Papio cynocephalus cynocephalus* = *Papio cynocephalus*), and chacma baboons (*Papio cynocephalus ursinus* = *Papio ursinus*).

CLASSIFICATION AND SEPARATION

The serum lipoproteins (in this review, no distinction will be made between either serum or plasma derived lipoproteins) found in normal, healthy baboons fed a regular chow, with a low fat and cholesterol content, are similar to those found in humans. The nomenclature for baboon lipoproteins, like that of human lipoproteins, is derived from the type of separation technique used for their isolation (Tables 1 and 2). The ultracentrifugally derived nomenclature will be used in this review, i.e., very low-density lipoproteins (VLDL), low-density lipoproteins (LDL), and high-density lipoproteins (HDL).

The sequential isolation of each lipoprotein class by preparative ultracentrifugation, using the same density cutoffs as used for human serum, is the most common separation method.[9-22] The electrophoretic[14,18,21,22,30,31] and precipitation[7,23-30] techniques are faster, less expensive, and are less technically demanding. A technique which yields a continuous, rather than a discontinuous (sequential preparative ultracentrifugal or precipitation) lipoprotein profile is more informative with sera from unknown or dyslipoproteinemic sources. These sources may either have different density cutoffs compared to normal subjects or may give a nonquantitative precipitation. For these reasons the electrophoretic, gel filtration,[12] high performance liquid chromatographic (HPLC)[35] or single-spin density gradient ultracentrifugation methods are recommended. The gradient gel electrophoretic (GGE) technique has been used to characterize human HDL.[36] HPLC is a rapid (less than 1 hr per sample) and quantitative method[37,38] requiring only small quantities of sample. In our laboratory, the single-spin density gradient ultracentrifugal method is routinely used to quantitate all the lipoproteins in normal and dyslipoproteinemic baboons (Figures 1 and 2). The method is reproducible, relatively rapid (24 hr), quantitative, and sparing of sample (0.2 mℓ serum). The separated lipoproteins can be collected for further characterization by scaling up the amount of initial serum.

PHYSICAL PROPERTIES

The apolipoproteins, present in each of the serum lipoprotein classes found in baboons, are similar in many chemical and physical properties (Table 3) to their counterparts in the human serum lipoproteins. Apo B, present in both VLDL and LDL, is a difficult protein to characterize because of its uncertain molecular weight and/or subunit composition.[12,18]

Table 1
CLASSIFCATION OF BABOON SERUM LIPOPROTEINS BY SEPARATION TECHNIQUES

Separation technique	Lipoprotein nomenclature	Medium density (g/mℓ)	Flotational characteristic		Comment	Ref.
			$S_f(1.063)$	$S_f(1.21)$		
Ultracentrifugal	Chylomicrons	<1.006				9—15
	VLDL	<1.006	>400			16
	IDL	1.006—1.019	20—400			9,11,12,14-17
						16
	LDL	1.006—1.063	12-20			10,12,13,14
		1.019—1.063				9,11,14,15
		1.024—1.045				18
			0—12			16,18
			Peak range 6.8—7.9			18
	$F^0$9—28	1.037—1.064		28—56		16
	HDL-total	1.063—1.21		9—28		18
		1.080—1.21				9—11,13,19
						20
				0—9		16
	HDL₂	1.063—1.125				10,12,14,15
						16
	HDL₃	1.125—1.21		3.5—9		10,12,14,15
				0—3.5		16
Electrophoretic	Pre-beta, beta, alpha				Agarose	14,21,22
	LDL				Agarose	18
Precipitation	VLDL + LDL, HDL				Heparin-MgCl₂	24
					Heparin-MnCl₂	29
					Dextran sulfate-CaCl₂	7,23—28,30

Table 2
OTHER SEPARATION METHODS USED FOR BABOON SERUM LIPOPROTEINS

Method	Ref.
Electrochromatography on filter paper	14, 30, 31
Paper electrophoresis	14, 31
Gradient gel electrophoresis (GGE)	16, 32
Single-spin density gradient ultracentrifugation	16, 33, 34
Gel filtration on either Bio-Gel A-5m or Sepharose 4-B columns	12
High performance liquid chromatography (HPLC)	35

Density Gradient Ultracentrifugation of Typical Baboon and Human Serum

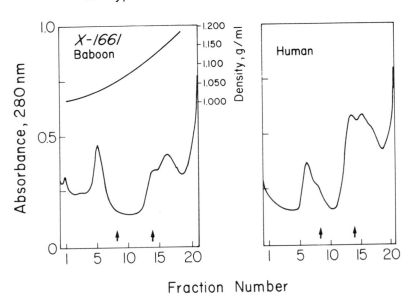

FIGURE 1.   Density gradient profile of 0.2 mℓ of a male baboon serum (left) and 0.2 mℓ of a female human serum (right). The abscissa is the fraction number, the lower ordinate the optical density at 280 nm, and the upper ordinate the density in grams per milliliter. The density gradient was prepared according to the method of Kruski and Kelley.[33] Centrifugation was at 38,000 r/min for 24 hr at 14°C in the six-place (volume = 13.2 mℓ/tube) Beckman SW-40 swinging bucket rotor. The human sample had a serum total cholesterol of 201 mg/100 mℓ and a HDL cholesterol of 111 mg/100 mℓ. The broad peak in fractions 2 to 3 corresponds to IDL, while the peaks found in fractions 4 to 6, 14 to 15, 16 to 18 correspond to LDL, HDL_2, and HDL_3, respectively, in the baboon. In the human profile, the peak found in fractions 5 to 6, and fractions 7 to 8, with average densities of 1.026 and 1.038 g/mℓ, respectively, are two populations of LDL. The shoulder found in fractions 13 to 15 (average density 1.12 g/mℓ) and the peak in fractions 16 to 20 (average density 1.16 g/mℓ) correspond to HDL_2 and HDL_3, respectively. A small amount of VLDL is found at the top (fraction number 1) while the serum proteins start in fraction 21. The elutions were terminated when the optical density due to the bottom fraction (serum proteins) went off scale (sensitivity was at 0.1). The arrows near fractions 10 and 15 correspond to the traditional d = 1.063 and 1.125 g/mℓ peparative ultracentrifugal "cuts" for LDL and HDL_2, respectively.

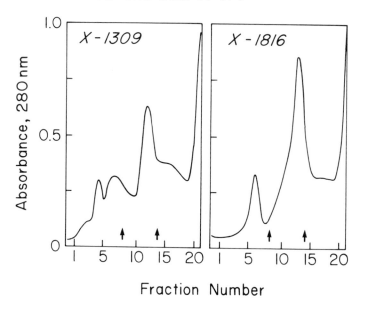

Density Gradient Ultracentrifugation
of Baboon Serum — Progeny of Sire
X-102 and Dam A-177

Fraction Number

FIGURE 2.   Density gradient profile of 0.2 mℓ of baboon X-1309 (left) and baboon X-1816 (right). These animals are progeny of a selected mating, many of whose progeny had the F°9-28 lipoprotein. In baboon X-1309, fractions 1 to 2, 3 to 4, 5 to 9, 10 to 14, and 15 to 18 correspond to IDL, LDL, F°9-28, HDL$_2$, and HDL$_3$, respectively. In baboon X-1816, fractions 5 to 7, 8 to 10, 11 to 15, and 16 to 19 correspond to LDL, F°9-28, HDL$_2$, and HDL$_3$, respectively. Note that the F°9-28 lipoprotein does not appear as a distinguishable peak but rather as broad, ill-defined shoulder on the less dense side of HDL$_2$. The presence of HDL$_3$ in all the baboons in Figures 1 and 2 is unusual because most baboons do not have any noticeable HDL$_3$. For further details, refer to legend to Figure 1.

When sodium dodecyl sulfate-polyacrylamide gel electrophoresis (SDS-PAGE) is performed on either SDS-treated baboon LDL or delipidated LDL, a major protein band of about 300,000 to 350,000 daltons is found along with several uncharacterized and minor proteins in the150,000 to 300,000 dalton range. The apo HDLs are fractionated and characterized by SDS-PAGE (or urea-PAGE) or a combination of gel filtration (usually Sephadex G-200) and ion-exchange (DEAE cellulose) chromatography in 6 M guanidine·HCl or 8 M urea.[12,14,19,20,40] The major HDL protein is designated apo A-I, analogous to that found in human HDL. Apo A-I has a molecular weight of about 28,000 daltons, is present as dimers and tetramers when delipidated, and has glutamine[14,19] as the carboxyl-terminal amino acid. With the exception of two changes, the amino acid sequence of baboon apo A-I is identical to human apo A-I through the first 30 residues. Baboon apo HDL contains a monomeric apo A-II with the mobility of monomeric human apo A-II (human apo A-II is present as a dimer) and a molecular weight of 8500. The structure of baboon apo A-I is highly ordered with a helical content of 67% compared to 55% for human apo A-I.[19] The amount of helicity present in baboon apo A-II is concentration dependent and reaches 37% for the higher-molecular-weight species found at higher concentrations. Recombination of baboon apo HDL with dimyristoylphosphatidyl choline indicates similar thermodynamic binding properties to those of human apo HDL.[19] Microcalorimetric studies show that the maximal-

Table 3
SOME PHYSICAL PROPERTIES OF BABOON SERUM LIPOPROTEINS AND APOLIPOPROTEINS

Property	Value	Lipoprotein or apolipoprotein	Ref.
Diameter	245 ± 32 Å	LDL	18
	240—284 Å	LDL	16
	119—235 Å	F°9—28	16
	104—120 Å	HDL_{2b}	16
Molecular weight	300,000—350,000 daltons main band and 150,000—300,000 smaller mol wt bands	Apo LDL	18
	28,000—29,000 daltons	Apo A-I	12,20
	8,500 daltons	Apo A-II	12,19
	15,400 daltons	Apo D	39
Self association	53,000 and 105,000 daltons (dimers and tetramers)	Apo A-I	20
	53,000 dimers bind best to lysolecithin	Apo A-I	20
Density	1.035 g/ml	LDL	12
	1.04—1.07 g/ml	F°9—28	12,16
	1.110 g/ml	HDL_{2b}	12
	1.125 g/ml	HDL_{2a}	12
	1.15 g/ml	HDL_3	12
Electrophoretic mobility	$R_f = 1.35$ relative to human	LDL	18
Enthalpy change with binding to dimyristoyl-phosphatidylcholine	210 kcal/mol	Apo HDL	19
	190 kcal/mol	Apo A-I	19
	40 kcal/mol	Apo A-II	19
Helical structure	67% alpha helix	Apo A-I	19
	37% alpha helix which is concentration dependent	Apo A-II	19
	54% alpha helix	Apo A-II complexed to dimyristoyl-phosphatidylcholine	19
Isoelectric points	4.67—4.88	Apo-I had 8 subfractions	40

binding enthalpies for the phospholipid to either apo A-I or apo A-II were lower for the baboon than for their human counterparts.

COMPOSITION AND CONCENTRATION

In general, the lipid and protein compositions of baboon serum lipoproteins resembles the corresponding lipoproteins in humans (Table 4). The VLDL found in baboons is generally present in small amounts (Table 5). Compared to humans, baboons have a lower concentration of LDL and a higher concentration of HDL, especially the HDL_2.[12,14,41] In our laboratory, either one asymmetric HDL band corresponding to HDL_2 or two overlapping bands similar to HDL_{2a} and HDL_{2b} are found by the single-spin density gradient ultracentrifugal method[33] (Figures 1 and 2). A HDL_3 is usually present to a small extent in baboon serum.

The VLDL in baboon has a larger percentage of cholesteryl esters and a lower percentage of glycerides compared to human VLDL.[10] Baboon LDL, however (generally reported as LDL-cholesterol concentration), has a greater percentage of glycerides and a slightly lesser amount of cholesteryl esters than found in human LDL.[10,18] This may account for the slightly larger diameter of baboon LDL compared to human LDL.[16,18]

Recently, an unusual lipoprotein class has been identified in some baboons. This lipoprotein is operationally defined as F°9-28 because it has flotational characteristics intermediate between LDL and HDL.[16] This lipoprotein contains apo A-I and in some instances

Table 4
COMPOSITION OF THE LIPOPROTEINS FROM NORMAL BABOON SERUM (MEAN WEIGHT %)

Lipoprotein	Protein	Phospholipid	Free cholesterol	Cholesteryl ester	Glycerides	Ref.
VLDL	9.2	6.5	10.2	27.7	46.2	10
LDL	23.5	21.3	14.2	30.2	10.8	18
	24.7	12.6	12.2	37.3	13.1	10
F°9-28						
1.030-1.040 g/mℓ	24	32	8	35	1	16
1.043-1.057 g/mℓ	31	31	7	31	Negligible	16
HDL	49.9	20.5	4.9	13.6	10.9	10
	48.8	28.8	3.6	15.4	3.4	19
LP-D	52	13.3	5.2	13.2	16.4	39

Table 5
CONCENTRATION OF BABOON LIPOPROTEINS AND LIPID COMPONENTS (mg/100 mℓ)

Serum cholesterol				Lipoproteins					
Total	Beta + pre-beta	Alpha	Triglyceride	Beta	Pre-beta	Alpha	Sex	No. of animals	Ref.
105	34	72		72	3	167	M/F	12	21
				118	3	345	M/F	23	17
108	36	59	47	131	29	334	M/F	5	13
65—120			140—220	60—106	2—10		M	2	18
104			60	128	11	268	F	14	10
129	57	72	29				F	15	9
111	39	72	28				M	15	9
117	47	53	95				M/F	6	22
116			36				M/F	18	14
108	37	69					M/F	24	27
110	56	54	77				M	3	29
116	57	59	78				F	3	29
116	30	86	12				M	18	12
117	57	60	38				M/F	18	30,41
113			57				M/F	60	7
98	57	41	30				M	4	11
125	62	63					M	2	42

apo E, occurs in a wide range of particle sizes (120 to 220 Å in diameter), is rich in phospholipids, may be genetically controlled, and is elicited by a saturated fat diet but not necessarily by dietary cholesterol. A minor lipoprotein, lipoprotein-D (LP-D), is reported present in baboon serum.[39] It is characterized by its large percentages of protein and glycerides with a suprisingly small percentage of phospholipid compared to baboon HDL (Table 4). Apo D is a low-molecular-weight (15,400 dalton) protein (Table 3).

Phosphatidylcholine is the major phospholipid class present in both man and the baboon and it is present at about the same percent in comparable lipoproteins from both species (Table 6). Sphingomyelin, on the other hand, is present to a much lesser extent in baboon lipoproteins than in human lipoproteins. Phosphatidylethanolamine, lysolecithin, phosphatidylinositol, and phosphatidylserine are generally present in greater percentages in baboon compared to human lipoproteins. Baboon LP-D has a much larger percentage of phospha-

Table 6
PHOSPHOLIPID COMPOSITION OF NORMAL BABOON
SERUM LIPOPROTEINS (%)

Sample	Lyso PC	Sph	PC	PI	PS	PE	Total[a] PL	Ref.
Plasma	6.2	7.3	73.9	3.7	1.2	5.7		14
HDL	0.7	5.9	81.2	3.6	0.4	8.3		14
Plasma	8.6	9.8	70.7			10.9	122	41
Alpha-LP	8.1	8.5	73.9			9.4	82	41
Beta-LP	9.9	11.4	67.0			11.6	40	41
LP-D	11.2	47.0	57.2			4.6		39
HDL₃	3.9	6.2	73.8			16.0		12

Note: Abbreviations: Lyso Pl, lysophosphatidylcholine; Sph, sphingomyelin; PC, phosphatidylcholine; PI, phosphatidylinositol; PS, phosphatidylserine; PE, phosphatidylethanolamine; PL, phospholipid.

[a] In mg/100 mℓ plasma.

Table 7
FATTY ACID COMPOSITION OF THE PHOSPHOLIPID CLASSES FOUND IN
NORMAL BABOON SERUM LIPOPROTEINS (%)

Lipoprotein	Phospholipid	Fatty acids							Total unsaturated	Ref.
		14:0	16:0	16:1	18:0	18:1	18:2	20:4		
Beta-LP	Lyso PC	4.1	40.5	2.2	23.7	25.2	4.3	—	31.7	41
	Sph	4.9	35.3	9.2	17.2	23.9	7.1	2.4	42.6	
	PC	0.7	25.6	1.6	25.9	12.6	20.3	13.3	47.8	
	PE	2.1	23.0	2.5	31.7	13.8	9.7	17.2	43.2	
Alpha-LP	Lyso PC	2.0	37.6	2.9	27.6	20.2	7.3	2.4	32.8	41
	Sph	3.9	39.5	4.7	21.6	23.8	6.5	—	35.0	
	PC	1.0	24.9	1.0	25.7	13.7	19.0	14.7	48.4	
	PE	1.2	21.7	1.7	35.2	13.3	9.9	17.0	41.5	
HDL₂	Lyso PC	0.8	25.2	—	31.6	29.3	10.2	3.5	43.0	12
	Sph	0.5	26.5	—	35.8	25.4	11.3	—	36.7	
	PC	0.3	26.0	—	40.0	12.6	14.7	6.4	33.7	
	PE	1.6	23.8	—	39.9	21.9	9.3	3.1	34.3	
HDL₃	Lyso PC	1.4	18.2	—	31.4	35.9	10.4	3.1	49.4	12
	Sph	0.9	15.8	—	33.1	38.1	9.3	3.2	50.6	
	PC	1.2	23.5	—	49.8	14.0	7.9	3.6	25.5	
	PE	0.4	18.6	—	43.7	27.0	9.4	1.1	37.5	

Note: Lyso PL, lysophosphatidylcholine; Sph, sphingomyelin; PC, phosphatidylcholine; PI, phosphatidylinositol; PS, phosphatidylserine; PE, phosphatidylethanolamine; PL, phospholipid.

tidylcholine and sphingomyelin than found in human LP-D. The structural and functional aspects of the different phospholipid class compositions in these lipoproteins is an unanswered question at present.

FATTY ACID COMPOSITION

In general, the fatty acid composition of alpha-PC (alpha = HDL and beta = LDL in the electrophoretically derived nomenclature) and beta-PC (LDL-PC), as well as alpha-PE and beta-PE were identical, suggesting that the phospholipids in each of the lipoproteins originated from the same body pool[41] (Table 7). The relatively large percentage of arachidonic

Table 8
FATTY ACID COMPOSITION OF THE LIPID CLASSES FOUND IN NORMAL BABOON SERUM LIPOPROTEINS (%)[a]

Lipoprotein	Lipid[b]	Fatty acids								Total unsaturated	Ref.
		14:0	16:0	16:1	18:0	18:1	18:2	18:3	20:4		
Beta-LP											30
	CE	1.3	15.3	3.1	4.5	23.5	46.8	—	5.3	78.7	
	PL	0.3	25.7	1.7	26.0	12.9	20.5	—	12.9	48.0	
	TG	1.2	30.8	4.4	5.3	35.4	21.6	—	1.2	62.6	
Alpha-LP											30
	CE	1.1	10.4	4.0	2.9	26.2	49.6	—	5.7	85.5	
	PL	0.3	26.8	1.3	27.4	10.7	19.8	—	13.7	45.5	
	TG	1.3	31.0	3.8	8.5	32.7	20.9	—	1.7	59.1	
HDL											19
	CE	1.5	11.5	5.2	2.6	28.0	41.7	1.0	6.1	82.0	
	PL	0.5	22.7	1.6	22.9	15.0	22.7	0.5	9.6	49.4	
	TG	1.1	23.3	5.2	7.3	38.8	20.3	2.4	0.9	67.6	
HDL$_2$											12
	CE	1.6	25.4	—	6.9	17.4	45.5	—	3.1	66.0	
	TG	1.8	43.5	—	20.2	20.3	14.1	—	—	34.4	
HDL$_3$											12
	CE	4.7	39.8	—	16.0	13.0	23.8	—	5.0	41.8	
	TG	5.9	30.0	—	24.3	21.9	9.8	—	8.2	39.9	

[a] Sum of percentages in one lipid class may not equal 100% due to rounding or due to presence of unidentified minor components.

[b] Abbreviations: CE, cholesteryl ester; PL, phospholipid; TG, triglyceride.

acid in the one study was probably due to the diet.[41] The fatty acid composition of the other lipid classes is given in Table 8. Cholesteryl esters in baboon lipoprotein, like in man, contain about 45% of its fatty acids as linoleic acid.[12,19,30] In addition, the cholesteryl ester fatty acid composition of LDL and HDL is nearly identical and contains 65 to 85% unsaturated fatty acids. The triglyceride fatty acid composition is also similar between LDL and HDL, but contains less linoleic and more oleic and palmitic acid than the cholesteryl esters.

APOLIPOPROTEINS

The baboon apo LPs are most easily identified by comparison of their polyacrylamide gel patterns with those of human apo LPs. The 8 *M* urea and the 0.1 to 1.0% SDS-PAGE systems are each best used to identify the different low-molecular-weight apo Cs or to obtain an estimate of the molecular weight of the constituent apo LPs, respectively (Table 9). Apo LPs resembling the apo Cs of man are found in small concentrations in both LDL and HDL.[12,15,18]

A protein of molecular weight 30,000 to 35,000 daltons is occasionally observed in VLDL, LDL, or F°9-28, and less frequently in HDL.[12,14,16,33] This protein is analogous to human apo E on the basis of molecular weight. To date, there have been no reports of either the amino acid composition or any isoelectric focusing gel electrophoretic data of individual baboon apo Cs or apo E.

IMMUNOCHEMISTRY

Immunological characterization of baboon serum lipoproteins with monospecific antisera to human apolipoproteins showed the presence of considerable heterogeneity with respect

Table 9
BABOON APOLIPOPROTEINS IDENTIFIED BY
POLYACRYLAMIDE GEL ELECTROPHORESIS

Lipoprotein	Apolipoproteins identified	Gel system[a]	Ref.
LDL	Apo B, apo E	10% A, 0.1% SDS	12
LDL	Apo B and other proteins of mol wt 150,000—300,000 daltons	10% A, 0.1% SDS	18
LDL	Apo Cs (less than 5% of total Apo B), proteins of similar mobilities as human apo C-I, apo C-II, apo C-III-1; apo C-III-2	7.5% A, 8 M Urea	18
F°9-28	Apo B (major protein), apo E in fractions < d = 1.034 g/mℓ; apo A-I (major protein) small amount apo E and small mol wt (12,300—13,700 daltons) in fractions > d 1.034 g/mℓ	Contains SDS	16
HDL$_2$	Apo A-I, apo A-II, apo Cs	7.5% A, 8 M Urea	15
HDL$_2$	Apo A-I, apo A-II, apo E	10.9% A, 0.1% SDS	14
HDL$_2$	Apo A-I, apo A-II, apo Cs	10% A, 0.1% SDS	12
HDL$_3$	Apo A-I, apo A-II, albumin, apo Cs	10% A, 0.1% SDS	12
HDL$_3$	Apo A-I, apo A-II, apo Cs	7.5% A, 8 M Urea	15
HDL$_3$	Apo A-I, apo A-II, apo E	10.9% A, 0.1% SDS	14

[a]   Abbreviation: A, acrylamide; SDS, sodium dodecyl sulfate.

to antigenic composition[10] (Table 10). Antisera to human apolipoproteins A-I, A-II, C-I, C-II, C-III, D, and E cross-reacted with baboon lipoproteins indicating a close analogy between the apolipoproteins of these two species. Baboon apo B reacted with antisera to human apo B at about 84% of the level shown between human apo B and its homologous antiserum.[18] Baboon LP-D was isolated by a combination of chromatography on an immunoabsorber containing antibodies to human apo D followed by chromatography on a hydroxylapatite column.[39] This purified baboon LP-D or delipidated LP-D, i.e., apo D gave a single precipitin line when tested with antiserum to human apo D. The apo B concentration of chow-fed baboons was 47 mg/100 mℓ serum determined by electroimmunoassay using antisera to baboon apo B.[27] Diet regimens containing saturated fat and/or supplemented dietary cholesterol resulted in elevated serum apo B concentrations. In one study, baboon sera cross-reacted with sheep and rabbit antihuman antiserum specific for apo A and apo B, however in more than half the sera apo C was absent and LP(a) was never detected.[14] In another study, 4 of 13 baboons gave precipitin rings by a standard radial immunodiffusion assay using rabbit antihuman Lp(a) sera.[45] The mean concentration of the positive baboon sera was 23 ± 12 mg/dℓ serum.

AMINO ACID COMPOSITION

The amino acid composition of the LDL$_2$ (d = 1.019 to 1.063 g/mℓ) is nearly identical for the baboon, chimpanzee, and man[14,18,43] (Table 11). Small differences in the amino acid composition of LDL$_1$ (d = 1.006 to 1.019 g/mℓ) of baboon compared to man were attributed to the presence of apo C.[14] Apo Cs were removed by using the void volume peak (apo B) from detergent solubilized apo LDL separated by gel filtration chromatography.[18] A typical amino acid profile of baboon or human apo B has a high percentage (8% or greater of total moles amino acids) of lysine, aspartic and glutamic acids, serine, and leucine and a low percentage (less than 5%) of histidine, arginine, proline, methionine, and tyrosine.[43] The amino acid composition of baboon apo A-I was nearly identical with that of human apo A-

Table 10

**IMMUNOCHEMICAL STUDIES WITH NORMAL BABOON
SERUM LIPOPROTEINS OR APOLIPOPROTEINS**

Sample	Technique[a]	Comment	Ref.
VLDL	ID/IE	Apo B, apo D present	10
LDL	ID/IE	Apo A-I, A-II, B, D, E	10
HDL	ID/IE	Apo A-I, A-II, B, C-III	10
Lp(a)	ID/IE	Apo D, E; (C-I, C-III)	10
LDL	ID/IE	80—85% cross-reaction with human LDL	18
LDL	Precipitation and inhibition tests	71.2% and 51.2% cross-reactivity to human LDL and apo B, respectively	43
Apo B	Precipitation and inhibition tests	About 63% cross-reactivity to either human LDL or human apo B	43
Apo D	ID/IE	Cross-reacts with human apo D	39
Serum	Electroimmunoassay	Normal = 47 mg apo B/100 mℓ serum lowered with diets of low cholesterol/polyunsaturated fats but elevated on diets of high cholesand/or saturated fats	27
Apo E	ID	Anti Rhesus apo E serum cross-reacts with baboon serum	44
Sera	(ID)	Cross-reaction with human apo A, apo B but apo C absent in about 50% of baboons tested, Lp(a) never detected	14
Sera	Radioimmunoassay	Cross-reacts with antihuman Lp(a)	45
Sera	Gel diffusion	Cross-reacts with antihuman Lp(a)	46

[a] Abbreviations: ID, immunodiffusion; IE, immunoelectrophoresis.

Table 11

**AMINO ACID ANALYSIS OF THE
APOLIPOPROTEINS FROM NORMAL BABOON
SERUM (MOL/1000 MOL OF AMINO ACIDS)**

Amino acids	LDL$_1$	LDL$_2$		Apo A-I	Apo A-II	Apo D
Lys	74	65	91	83	96	76
His	26	22	24	30	0	27
Arg	37	34	34	63	13	60
Trp	3	ND[a]	ND	13	0	17
Asp	105	104	109	79	55	99
Thr	65	68	61	45	76	63
Ser	84	128	80	65	81	50
Glu	134	121	122	189	221	147
Pro	36	49	40	42	51	42
Gly	49	89	53	45	27	52
Ala	62	75	64	74	80	81
Cys/2	7	ND	ND	0	0	ND
Val	54	45	52	61	84	66
Met	27	ND	16	12	18	15
Ile	46	42	52	0	0	42
Leu	118	91	116	152	104	94
Tyr	28	25	30	26	45	43
Phe	45	42	53	21	49	40
Ref.	14	14	18,43	14	14	39

[a] ND, not determined.

Table 12
CARBOHYDRATE COMPOSITION OF NORMAL BABOON SERUM LIPOPROTEINS (mg/100 mg APO LP)

Lipoprotein	Sialic acid	Fucose	Glucosamine	Mannose	Galactose	Glucose	Total	Ref.
LDL$_2$	0.17	0.19	0.42	0.35	0.43	0.13	1.69	15
HDL$_2$	0.22	0.33	0.50	0.11	0.31	0.26	1.73	15
HDL$_3$	0.54	0.53	0.95	0.26	0.46	0.18	2.92	15
LP-D	ND[a]	—	4.20	7.10	4.20	—	15.50	39

[a] ND, not determined.

I, each containing a characteristically large percentage of glutamic and aspartic acids, leucine, and lysine along with an absence of isoleucine.[14] The monomeric apo A-II found in baboon HDL has a moderately different amino acid composition compared to either human or chimpanzee dimeric apo A-II, the only two species known to have the dimeric form.[14] Baboon apo A-II, unlike human apo A-II, contains some arginine but has no cystine or isoleucine. The amino acid profile of baboon apo D is similar to human apo D except that it consists of more histidine, arginine, and glutamic acid with correspondingly less aspartic acid, proline, and isoleucine.[39] The minimum molecular weight of apo D calculated from the amino acid composition is 15,400 compared to 22,100 for human apo D.

CARBOHYDRATE COMPOSITION

Human apo LDL$_2$ has twice the total carbohydrate content as baboon apo LDL$_2$, 3.69% vs. 1.69%, respectively.[15] Sialic acid, fucose, and glucose are minor carbohydrate components of baboon apo LDL$_2$ while mannose and glucose are minor constituents of baboon apo HDL$_2$ and apo HDL$_3$ (Table 12). Presumably, in the HDL, the carbohydrate moieties reside on baboon apo Cs. The apo LDL$_2$ found in baboon consists of relatively lower proportions of hexose and glucosamine to fucose than either man or several other nonhuman primates. The difference in carbohydrate composition between baboon apo HDL$_2$ and apo HDL$_3$ suggests either a different composition or different amounts of the same glycoprotein. It was speculated that the sugar content of the baboon apo LPs may modulate the catabolism of the lipoproteins in vivo. Baboon apo D is a glycoprotein containing 15% carbohydrate by weight with twice as much mannose as found in human apo D.[39]

DIETARY STUDIES

Although the main focus of this review was the chemical and physical properties of serum lipoproteins from healthy baboons fed a chow diet, the effects of various diets and or treatments on the serum lipoproteins (Table 13) or on serum lipid metabolism (Table 14) has added substantially to our knowledge of baboon lipoproteins and because of this will be briefly presented in this review. The lowest serum cholesterol concentrations in healthy baboons are found in animals fed chow diets which contain little fat or cholesterol. The supplementation of fat (whether saturated or polyunsaturated), cholesterol[11,13,14,25,27,30,41,49] or simple carbohydrates (fructose, sucrose, glucose, starch, or lactose[22,29]) elevates the serum cholesterol above that found in baboons fed chow diets (Table 13). In our laboratory, baboons fed a diet containing 1.7 mg/kcal cholesterol and 40% of calories as saturated fat (this combination of high saturated fat and high cholesterol resulted in the largest elevations) result in a change of the serum total cholesterol concentration from about 100 mg/dℓ to an average 200 to 250 mg/dℓ. Rarely do baboons have serum cholesterol concentrations greater than 300 mg/dℓ and the fasted animals rarely have elevated serum triglyceride levels when

Table 13
EFFECT OF DIET ON BABOON LIPOPROTEINS AND SERUM
LIPID LEVELS (SELECTED)

| | Effect (change from control) | | | | |
| | Serum | | Total cholesterol | | |
Diet treatment or agent	TC	TG	LDL	HDL	Ref.
HF, 3.3% C	↑	NC			14
HF, 0.55% C	↑		↑	↑	30
HF, 2.69% C	↑	↓	↑	↑	11
HF, 1.5 mg C/kcal with cigarette smoking	NC[a]	NC[a]	NC[a]	NC[a]	25
HPUF, LC	SL ↑		SL ↑	SL ↑	2714
HPUF, HC	↑		↑	↑	27
HSF, LC, or HC	↑		↑	↑	27
HC	↑	↑	↑	↑	13
HF, high protein	↑	↓	↑	↓	13
Terbufibrol	↓		↓	↓	13
Clofibrate	↓		↑	↓	13
Terbufibrol, HC or high protein	NC		↓	↓	13
0.55% C	↑		↑	↑	41
Diabetes by pancreas removal	↑	↑	↑	↓	10
Ascorbic acid	NC	SL ↓			47
L/SU/F/S/G, 0.1% C	↑	NC	↑	↓	22
SU/F/S/G	↑	↑	SL ↑	↑	29
Ovariectomy or estrogens, 1.5 mg C/kcal	NC[a]	SL ↑	NC[a]	NC[a]	48
Infant/juvenile study, HF, 1.0 mg C/kcal	↑		↑	↑	49
HF, HC, BSA injections	↑				50

Note: Abbreviations: HF, high fat; C, cholesterol; NC, no change; HPUF, high polyunsaturated
fat; LC, low cholesterol; SL, slight; HC, high cholesterol; HSF, high saturated fat; L/
SU/F/S/G, lactose/sucrose/fructose/starch/glucose; SU/F/S/G, sucrose/fructose/starch/
glucose; BSA, bovine serum albumin.

[a] Contrasted with control animals on a HF, HC diet.

fed the supplemented diets. The serum cholesterol elevations occur in both the LDL and
HDL cholesterol concentrations. The response to diets which elevate serum cholesterol
concentrations is less than that of most other nonhuman primates used in experimental
atherosclerosis research, but it is within the same range of responses found in humans.[7] In
baboons, elevated VLDL + LDL cholesterol levels were highly correlated with atherosclerotic lesions, whereas HDL cholesterol levels were inversely correlated.[7]

Dietary cholesterol elevated VLDL + LDL cholesterol when combined with either a
previous saturated or unsaturated fat-supplemented diet, but caused a greater rise when
combined with an unsaturated fat diet than with a saturated fat diet.[27] Dietary saturated fat
elevated VLDL + LDL cholesterol when dietary cholesterol was low, but not when dietary
cholesterol was high. Saturated fat also elevated HDL cholesterol more than did unsaturated
fat or dietary cholesterol. The serum total cholesterol, LDL and HDL cholesterol, and serum
triglycerides were all elevated when a chow diet was supplemented with either sucrose,
fructose, starch, or glucose as 40% of total calories whether the carbohydrates were supplemented with or without 0.1% cholesterol.[22,29] Although all the carbohydrates resulted in
greater aortic lesion formation compared to controls, a lactose diet resulted in the largest
and most marked increase in lesion formation.[22] Breast feeding or feeding formulas containing
several levels of cholesterol for 3 months during infancy did not result in significant dif-

Table 14
SELECTED REFERENCES ON SERUM LIPID METABOLISM IN BABOONS

Study	Comments	Ref.
VLDL turnover	VLDL decays biphasically to IDL and LDL, hypercholesterolemic animals have lowest fractional catabolic rates	11
Genetic effects on serum cholesterol levels and cholesterol metabolism	Serum cholesterol levels, cholesterol turnover rate and cholesterol turnover rate are moderately to highly heritable	26,51—53
Biosynthesis and transport of cholesterol	Total cholesterol specificity activity is similar in LDL and HDL	28
Blood flow in brain gray matter	LDL and 5-hydroxytryptamine cause an increase	54
Cholesterol metabolism	Labeled acetate and mevalonic acid metabolism followed in serum, liver, and aorta	55
Effects of diet on serum lipid and cholesterol absorption in infants	Infant cholesterol intake does not affect subsequent serum cholesterol at 15 months. Genetic influences noted on lipid metabolism	56,57
Relationship between cholesterol absorption and cholesterol synthesis	Liver is subject to negative feedback; intestine has incomplete feedback	58
Cholesterol metabolism	On a cholesterol-rich diet, the baboon absorbs little cholesterol	59
Alcoholic liver injury	Ethanol produces fatty liver, cirrhosis, hyperlipidemia, and triglyceride accumulation in the liver	60
Cholesterol metabolism in old and young animals	There is little difference between elevations in serum triglyceride or cholesterol with cholesterol-rich diets between old vs. young	61

ferences in total serum cholesterol concentration from baboons 4 to 6 years of age.[49] There was, however, a significant interaction of type of infant diet and HDL cholesterol concentration at 4 to 6 years of age.

METABOLIC STUDIES

In lipoprotein metabolism studies, radioactivity from VLDL apo B in normolipidemic donors and recipients decayed biphasically with an immediate appearance of most of it in the IDL and LDL.[11] The first phase (t = 0.53 hr) was much more rapid than the second (t = 20.5 hr). Almost all of the VLDL apo B decayed during the rapid first phase. A high line and a low line of baboon progeny were produced from 6 male and 134 adult female baboons assigned to high or low lines depending on their response to a high-cholesterol, high-saturated fat diet.[26] The mean serum cholesterol of the progeny produced from each line were significantly different from each other at the end of their first year indicating a heritability for serum cholesterol levels. The cholesterol turnover rate, cholesterol production rate, and several cholesterol pool parameters derived from a two-pool model were also heritable.[51] Recent evidence indicates that besides the serum cholesterol, HDL cholesterol, apo A-I, VLDL + LDL cholesterol, but not apo B serum concentrations are also heritable.[52] A strong negative genetic correlation (r_g = −0.95) was shown between apo A-I and cholesterol turnover in that study. These results suggest a close physiological relationship between apo A-I concentrations and cholesterol turnover. There was a positive genetic correlation (r_g = 0.62) between VLDL + LDL cholesterol and HDL concentrations which may be related to serum cholesterol turnover.

A similar mechanism of cholesterol transport was found in man and baboons by the use of radioactively labeled acetate, mevalonic acid, or cholesterol.[28] Peak specific activities of free and esterified cholesterol occurred within 1 to 3 days and both activities were equal over a 10-day period. The total specific activity of the alpha (HDL) and beta (LDL) lipoproteins was about the same during the study. Experiments, showing that the baboon is similar to man in its modes of cholesterol synthesis and degradation, indicate that the baboon is a suitable animal model for atherosclerosis.[55]

Double isotope studies have shown, that in baboons fed low-cholesterol diets, hepatic cholesterol synthesis accounts for about three fourths of total endogenous cholesterol production.[58] Ingestion of cholesterol-supplemented diets caused complete inhibition of liver synthesis only when absorption approximates the 400 to 500 mg/day of cholesterol synthesized when no cholesterol is fed. The intestine, which has incomplete negative feedback control of cholesterol synthesis when cholesterol is fed, may be a significant cholesterol synthesis site in cholesterol-fed baboons. In baboons fed an atherogenic diet (basal diet supplemented with saturated fat and cholesterol) the percentage of dietary cholesterol absorbed and the fraction of total body cholesterol pool that was derived from the diet was less than the squirrel monkey or the rhesus monkey.[59] It was concluded that of the nonhuman primates studied, the baboon's cholesterol metabolism was most like that in man.

A baboon model, for the administration of ethanol (50% of total calories) as part of a nutritionally adequate liquid diet, has proved a useful tool for studies of the pathogenesis and treatment of alcoholic liver injury.[60] The diet resulted in the production of fatty livers, alcoholic hepatitis and cirrhosis, hyperlipemia, large accumulations of triglyceride in the liver, and enhanced activities of microsomal enzymes. Cholesterol metabolism and serum lipids were studied in young (3- to 4-year olds) and old (> 6-year old) baboons fed a basal, low-fat chow followed by adding cholesterol and saturated fat to the chow.[61] The serum cholesterol and triglyceride concentration changes with diet were the same for both age groups. The change in cholesteryl ester fatty acids, fractional contribution of dietary cholesterol to the body pool, and the suppression of endogenous cholesterol synthesis were greater in the younger animals after the change to the high-cholesterol diet. The percentage of absorbed dietary cholesterol and the increase in fractional and absolute rate of cholesterol turnover were similar for both groups.

In summary, both the articles cited,[62] as well as the many publications outside of the scope of this review, show that the baboon is an excellent nonhuman primate model for the study of serum lipoprotein structure, function, metabolism, and role in the genesis of atherosclerosis.

REFERENCES

1. **Rudel, L. L. and Lofland, H. B.,** Circulating lipoproteins in nonhuman primates, *Prim. Med.,* 9, 224, 1976.
2. **Chapman, M. J.,** Animal lipoproteins: chemistry, structure, and comparative aspects, *J. Lipid Res.,* 21, 789, 1980.
3. **Mills, G. L.,** Lipoproteins in animals, in *Handbuch der Inneren Medizin, Band VII/4 Fettstoffwechsel,* Schettler, G., Greten, H., Schlierf, G., and Seidel, D., Eds., Springer-Verlag, Berlin, 1976, 173.
4. **Vagtborg, H., Ed.,** The baboon in medical research, in *Proc. 1st Int. Symp. on the Baboon and Its Uses as an Experimental Animal,* University of Texas Press, Austin, 1965.
5. **Strong, J. P. and McGill, H. C., Jr.,** Diet and experimental atherosclerosis in baboons, *Am. J. Pathol.,* 50, 669, 1967.
6. **Geer, J. C., Catsulis, C., McGill, H. C., Jr., and Strong, J. P.,** Fine structure of the baboon aortic fatty streak, *Am. J. Pathol.,* 52, 265, 1968.

7. **McGill, H. C., Jr., McMahan, C. A., Kruski, A. W., and Mott, G. E.,** Relationship of lipoprotein cholesterol concentrations to experimental atherosclerosis in baboons, *Arteriosclerosis,* 1, 3, 1981.

8. **McGill, H. C., Jr., Mott, G. E., and Bramblett, C. A.,** Experimental atherosclerosis in the baboon, *Prim. Med.,* 9, 41, 1976.

9. **Pena, A. DeLa, Matthijssen, C., and Goldzieher, J. W.,** Normal values for blood constituents of the baboon. II, *Lab. Anim. Sci.,* 22, 249, 1972.

10. **Bojanovski, D., Alaupovic, P., Kelley, J. L., and Stout, C.,** Isolation and characterization of the major lipoprotein density classes of normal and diabetic baboon (*Papio Anubis*) plasma, *Atherosclerosis,* 31, 481, 1978.

11. **Kushwaha, R. S., Hazzard, W. R., Harker, L. A., and Engblom, J.,** Lipoprotein metabolism in baboons, *Atherosclerosis,* 31, 65, 1978.

12. **Kruski, A. W.,** manuscript in preparation.

13. **Howard, A. N., Zschocke, R., Loser, R., and Hofrichter, G.,** The hypocholesterolemic effect of terbufibrol and other drugs in normal and hypercholesterolemic baboons, *Atherosclerosis,* 32, 367, 1979.

14. **Blaton, V. and Peeters, H.,** The nonhuman primates as models for studying human atherosclerosis: studies on the chimpanzee, the baboon and the rhesus macacus, *Adv. Exp. Med. Biol.,* 67, 33, 1976.

15. **Pargaonkar, P. S., Radhakrishnamurthy, B., Srinivasan, S. R., and Berenson, G. S.,** Carbohydrate composition of serum low and high density lipoproteins of nonhuman primate species, *Comp. Biochem. Physiol.,* 56B, 293, 1977.

16. **Kuehl, T. J., Gong, E. L., Babiak, J., Blanche, P. J., Mott, G. E., Forte, T. M., McGill, H. C., Jr., and Nichols, A. V.,** submitted.

17. **Howard, A. N., Gresham, G. A., Boyer, D. E., and Lindgren, F. T.,** Aortic and coronary atherosclerosis in baboons, *Prog. Biochem. Pharmacol.,* 4, 438, 1968.

18. **Chapman, M. J. and Goldstein, S.,** Comparison of the serum low density lipoprotein and of its apoprotein in the pig, rhesus monkey, and the baboon with that in man, *Atherosclerosis,* 25, 267, 1977.

19. **Blaton, V., Vercaemst, R., Rosseneu, M., Mortelmans, J., Jackson, R. L., Gotto, A. M., Jr., and Peeters, H.,** Characterization of baboon plasma high-density lipoproteins and of their major apoproteins, *Biochemistry,* 16, 2157, 1977.

20. **Rosseneu, M., Blaton, V., Vercaemst, R., Soetewey, F., and Peeters, H.,** Phospholipid binding and self-association of the major apoprotein of human and baboon high-density lipoproteins, *Eur. J. Biochem.,* 74, 83, 1977.

21. **Srinivasan, S. R., McBride, J. R., Jr., Radhakrishnamurthy, B., and Berenson, G. S.,** Comparative studies on serum lipoprotein and lipid profiles in subhuman primates, *Comp. Biochem. Physiol.,* 47B, 711, 1974.

22. **Kritchevsky, D., Davidson, L. M., Kim, H. K., Krendel, D. A., Malhotra, S., Mendelsohn, D., van der Watt, J. J., du Plessis, J. P., and Winter, P. A. D.,** Influence of type of carbohydrate on atherosclerosis in baboons fed semipurified diets plus 0.1% cholesterol, *Am. J. Clin. Nutr.,* 33, 1869, 1980.

23. **Jennings, C., Weaver, D. M., and Kruski, A. W.,** Effects of freeze-thawing on determinations of cholesterol and high-density lipoproteins in baboon sera, *Clin. Chem.,* 26, 490, 1980.

24. **Kruski, A. W. and Weaver, D. M.,** Separating lipoproteins in 50 μL of serum: heparin-MnCl$_2$ and dextran sulfate-CaCl$_2$ procedures compared, *Clin. Chem.,* 26, 1103, 1980.

25. **Rogers, W. R., Bass, R. L., III, Johnson, D. E., Kruski, A. W., McMahan, C. A., Montiel, M. M., Mott, G. E., Wilbur, R. L., and McGill, H. C., Jr.,** Atherosclerosis-related responses to cigarette smoking in the baboon, *Circulation,* 61, 1188, 1980.

26. **Flow, B. L., Cartwright, T. C., Kuehl, T. J., Mott, G. E., Kraemer, D. C., Kruski, A. W., Williams, J. D., and McGill, H. C., Jr.,** Genetic effects on serum cholesterol concentrations in baboons, *J. Hered.,* 72, 97, 1981.

27. **McGill, H. C., Jr., McMahan, C. A., Kruski, A. W., Kelley, J. L., and Mott, G. E.,** Responses of serum lipoproteins to dietary cholesterol and type of fat in the baboon, *Arteriosclerosis,* 1, 337, 1981.

28. **Kritchevsky, D., Werthessen, N. T., and Shapiro, I. L.,** Studies on the biosynthesis of lipids in the baboon. Biosynthesis and transport of cholesterol, *Clin. Chim. Acta,* 11, 44, 1965.

29. **Kritchevsky, D., Davidson, L. M., Shapiro, I. L., Kim, H. K., Kitagawa, M., Malhotra, S., Nair, P. P., Clarkson, T. B., Bersohn, I., and Winter, P. A. D.,** Lipid metabolism and experimental atherosclerosis in baboons: influence of cholesterol-free, semi-synthetic diets, *Am. J. Clin. Nutr.,* 27, 29, 1974.

30. **Blaton, V., Howard, A. N., Gresham, G. A., Vandamme, D., and Peeters, H.,** Lipid changes in the plasma lipoproteins of baboons given an atherogenic diet. I. Changes in the lipids of total plasma and of alpha- and beta-lipoproteins, *Atherosclerosis,* 11, 497, 1970.

31. **Howard, A. N., Blaton, V., Vandamme, D., Van Landshoot, N., and Peeters, H.,** Lipid changes in the plasma lipoproteins of baboons given an atherogenic diet. III, *Atherosclerosis,* 16, 257, 1972.

32. **VandeBerg, J. L.,** personal communication, 1982.

33. **Kruski, A. W. and Kelley, J. L.,** manuscript in preparation.

34. **Kruski, A. W.,** Analysis of the single spin ultracentrifugal method for plasma lipoprotein separation, in Conference: New Frontiers on the Relationships of lipids, lipoproteins and the arterial wall in cardiovascular disease, San Francisco, 1978.

35. **McGill, H. C., Jr. and Williams, M.,** personal communication, 1982.

36. **Blanche, P. J., Gong, E. L., Forte, T. M., and Nichols, A. V.,** Characterization of human high density lipoproteins by gradient gel electrophoresis, *Biochim. Biophys. Acta,* 665, 408, 1981.

37. **Payne, M., Jasheway, D., and Busbee, D.,** Separation and detection of prestained human plasma lipoproteins using gel filtration high performance liquid chromatography, *Altex Chromatogram,* 4, No. 2, 3, 1981.

38. **Okazaki, M., Ohno, Y., and Hara, I.,** High performance aqueous gel permeation chromatography of human serum lipoproteins, *J. Chromatogr.,* 221, 257, 1980.

39. **Bojanovski, D., Alaupovic, P., McConathy, W. J., and Kelley, J. L.,** Isolation and partial characterization of apolipoprotein D and lipoprotein D from baboon plasma, *FEBS Lett.,* 112, 251, 1980.

40. **Goldsworthy, P. D., Lim, G., Glomset, J. A., and Volwiler, W.,** Polypeptides of baboon serum high density lipoprotein, *Fed. Proc. Fed. Am. Soc. Exp. Biol.,* 32, 547A, 1973.

41. **Peeters, H., Blaton, V., Declercq, B., Howard, A. N., and Gresham, G. A.,** Lipid changes in the plasma lipoproteins of baboons given an atherogenic diet. II. Changes in the phospholipid classes of total plasma and of alpha and beta lipoproteins, *Atherosclerosis,* 12, 283, 1970.

42. **Shapiro, I. L., Jastremsky, J. A., Eggen, D. A., and Kritchevsky, D.,** Cholesterol metabolism in the baboon, *Lipids,* 3, 136, 1968.

43. **Goldstein, S., Chapman, J. M., and Mills, G. L.,** Biochemical and immunological evidence for the presence of an apolipoprotein B-like component in the serum low-density lipoproteins of several animal species, *Atherosclerosis,* 28, 93, 1977.

44. **Getz, G.,** personal communication, 1982.

45. **Albers, J. J., Adolphson, J. L., and Hazzard, W. R.,** Radioimmunoassay of human plasma Lp (a) lipoprotein, *J. Lipid Res.,* 18, 331, 1977.

46. **Berg, K.,** The Lp system, *Ser. Haematol.,* 1, 111, 1968.

47. **Kotze, J. P., Menne, I. V., Spies, J. H., and De Klerk, W. A.,** Effect of ascorbic acid on serum lipid levels and depot cholesterol of the baboon *(Papio ursinus), S.A. Med. J.,* 49, 906, 1975.

48. **McGill, H. C., Jr., Axelrod, L. R., McMahan, C. A., Wigodsky, H. S., and Mott, G. E.,** Estrogens and experimental atherosclerosis in the baboon *(Papio cynocephalus), Circulation,* 56, 657, 1977.

49. **Mott, G. E., McMahan, C. A., Kelley, J. L., Farley, C. M., and McGill, H. C., Jr.,** Influence of infant and juvenile diets on serum cholesterol, lipoprotein and apolipoprotein concentrations in juvenile baboons *(Papio* sp.), *Atherosclerosis,* in press.

50. **Howard, A. N., Patelski, J., Bowyer, D. E., and Gresham, G. A.,** Atherosclerosis induced in hypercholesterolaemic baboons by immunological injury: and the effects of intravenous polyunsaturated phosphatidyl choline, *Atherosclerosis,* 14, 17, 1971.

51. **Flow, B. L. and Mott, G. E.,** Genetic mediation of cholesterol metabolism in the baboon *(Papio cynocephalus), Atherosclerosis,* 41, 403, 1982.

52. **Flow, B. L., Mott, G. E., and Kelley, J. L.,** Genetic mediation of lipoprotein cholesterol and apoprotein concentrations in the baboon *(Papio* sp.), *Atherosclerosis,* 43, 83, 1982.

53. **Mott, G. E., McMahan, C. A., and McGill, H. C., Jr.,** Diet and sire effects on serum cholesterol and cholesterol absorption in infant baboons, *Circ. Res.,* 43, 364, 1978.

54. **Eidelman, B. H., Mendelow, A. D., McCalden, T. A., and Bloom, D. S.,** Potentiation of the cerebrovascular response to intra-arterial 5-hydroxytryptamine, *S. A. J. Physiol.,* 234, PH300-4, 1978.

55. **Kritchevsky, D.,** Cholesterol metabolism in the baboon, *Trans. N.Y. Acad. Sci.,* 32, (No. 7), 821, 1970.

56. **McGill, H. C., Jr., Mott, G. E., and Kuehl, T. J.,** Effects of diet and heredity on serum cholesterol and triglyceride concentrations and on cholesterol absorption of infant and juvenile baboons *(Papio cynocephalus),* in *Childhood Prevention of Atherosclerosis and Hypertension,* R. M. Lauer and R. B. Shekelle, Eds., Raven Press, New York, 1980, 113.

57. **Mott, G. E., Jackson, E. M., and Morris, M. D.,** Cholesterol absorption in baboons, *J. Lipid Res.,* 21, 635, 1980.

58. **Wilson, J. D.,** The relation between cholesterol absorption and cholesterol synthesis in the baboon, *J. Clin. Invest.,* 51, 1450, 1972.

59. **Eggen, D. A.,** Cholesterol metabolism in rhesus monkey, squirrel monkey, and baboon, *J. Lipid Res.,* 15, 139, 1974.

60. **Lieber, C. S. and DeCarli, L. M.,** Alcoholic liver injury: experimental models in rats and baboons, *Adv. Exp. Med. Biol.,* 59, 379, 1975.

61. **Eggen, D. A. and Strong, J. P.,** Diet and cholesterol metabolism in young and old baboons, *Atherosclerosis,* 12, 359, 1970.

62. **Howard, A. N., Blaton, V., Gresham, G. A., Vandamme, D., and Peeters, H.,** The lipoproteins in hyperlipidaemic primates as a model for human atherosclerosis, *Protides Biol. Fluids,* 19, 341, 1971.

Lipoproteins of Aquatic Mammalian Species

MARINE MAMMALIAN LIPOPROTEINS

Donald L. Puppione

INTRODUCTION

There are many mammals whose neutral fat depots serve both as a thermal insulator and as a source of metabolic energy. For a variety of marine mammals with blubber, these fat stores provide these animals with buoyancy, in addition to serving as an insulator and an energy store. From a physiological standpoint, knowledge of lipid transport in the blood of these animals is fundamental to understanding how they develop and utilize their depot fat.

In preparing this chapter on marine mammalian lipoproteins, I have primarily drawn upon a limited number of studies carried out on small groups of mammals which have been kept in captivity in marine facilities. Data are presented on cetacean (dolphins, porpoises, and whales),[1] pinniped (seal, sea lions, and walruses),[1] polar bear,[2] and manatee lipoproteins, but none could be found for dugongs or sea otters. The cetacean data are limited to a few members of the suborder Odontoceti. No data have been obtained on cetaceans belonging to the suborder Mysticeti. In pinniped studies, lipoprotein analyses were carried out on a single walrus (family Odobenidae); however, animals belonging to several species of both Otarids and Phocids were studied. The total of different species studied only comprise one fourth of the various types of pinnipeds.

The data will be presented in the following four sections: Serum Lipid Concentrations; Analytical Ultracentrifugal Data; Composition of Ultracentrifugal Fractions; Other Lipoprotein Data. In the tables for the first three sections, I have numbered the entries whenever more than one member of a given species and sex has been studied. For ease of comparisons, the number designation for a given animal remains the same for all the tables in this chapter.

The majority of analytical ultracentrifugal data cited in the tables in the second section have been obtained using the published techniques of Lindgren and co-workers.[3] Because lipoproteins will be classified in the text in terms of their physicochemical properties, a brief explanation of the nomenclature will be given. Circulating triglyceride-rich lipoproteins will be referred to in terms of either their ultracentrifugal density fraction, i.e., d $<$ 1.006 g/mℓ, or their flotation rates in a salt solution having a density of 1.063 g/mℓ. The symbol S_f^o is used to define the flotation rate as being measured at 26°C in a medium of 1.745 molal NaCl (d $=$ 1.063 g/mℓ). The superscript "o" indicates that the flotation rates have been corrected for effects associated with concentration dependence. The large triglyceride carriers or chylomicra which have an S_f^o value greater than 400 cannot be measured using schlieren analysis.[3] Lipoproteins, isolated in the density interval 1.006 to 1.063 g/mℓ can be measured within two subclasses having flotation rate intervals S_f^o 0 to 12 and 12 to 20. Lipoproteins with densities between 1.063 to 1.21 g/mℓ are designated high-density lipoproteins, or HDL. To measure the concentrations and the flotation rates of these lipoproteins, analyses are done in a salt solution having a density of 1.216 g/mℓ. The measured flotation rates are corrected and then expressed as being determined in a solution of density 1.20 g/mℓ. With the correction, flotation rates of HDL are expressed in terms of $F_{1.20}$ values. The units for both S_f^o and $F_{1.20}$ values are Svedbergs of flotation. In the tables which appear in the second section, the concentrations of lipoproteins will be given for specified flotation rate intervals. However, in referring to these lipoproteins in the text, I will use VLDL (very low-density lipoproteins) for the S_f^o 20 to 400 interval, IDL (intermediate density lipoproteins) for the S_f^o 12 to 20 interval, LDL (low-density lipoproteins) for the S_f^o 0 to 12 interval, HDL_1 for the $F_{1.20}$ 9 to 20 interval, HDL_2 for the $F_{1.20}$ 3.5 to 9 interval, and HDL_3 for the $F_{1.20}$ 0 to 3.5 interval. A few representative schlieren patterns derived from computer printouts will also be presented.

In the fourth section, the distribution of lipoproteins separated electrophoretically will be discussed. Photographs of both paper and agarose electrophoretograms are included to illustrate the general features. Paper electrophoresis was done according to Hatch and Lees[4] and the agarose electrophoresis was done according to Nobel et al.[5] As is the case in the plasma of several terrestrial mammals, α-Lp of marine mammals are highly polydisperse in size and density, with the large α-Lp having densities less than 1.063 g/mℓ. Rather than considering this group of lipoproteins as belonging to HDL_1 subclass (as other authors have done) I, by combining their properties of density and electrophoretic mobility, have defined them to be alpha LDL. In addition to emphasizing the polydispersity of the alpha lipoproteins, this approach eliminates the confusion which arises by referring to these lipoproteins as HDL_1 when they are comparable in size and density to LDL with beta electrophoretic mobility, i.e., β LDL. The other data presented in this section were obtained using a modification[6] of the heparin manganese precipitation techniques of Burstein and Scholnick.[7]

In the final portion of this chapter there is a general discussion of the lipoprotein data which have been presented, together with comments concerning lipid metabolism in marine mammals.

SERUM LIPID CONCENTRATIONS

In developing Tables 1 and 2, an entry was made only if the serum concentration of all of the lipid classes had been determined. The Weddell and fur seal samples were obtained from animals which were not being kept in oceanariums or marine facilities. These animals were caught and bled on Antarctica (Weddell seals) and the Pribilof Islands (fur seals). Compositional analyses did not reveal any major differences between these animals and pinnipeds being kept in captivity. The data for the Northern fur seals, Steller sea lion pups, and the Weddell seals were pooled (the means and the standard deviations are shown in Table 2 for the serum concentrations as well as the percent content of each lipid class for each group of pinnipeds).

The serum lipid data in Tables 1 and 2 indicated that cholesteryl esters equaled or exceeded the levels of phospholipids in the sera of bottlenose dolphins and killer whales, whereas phospholipids were the predominant class among the serum lipids of other marine mammals. After correcting for the fatty acid content (assumed to be 40% by weight in the cholesteryl esters), the calculated values for total cholesterol concentrations among cetaceans ranged from 61 mg/dℓ (Dall's porpoise) to 382 mg/dℓ in male killer whale 2. Among pinnipeds, a high value was found for female Weddell seals, with a mean for total cholesterol of 420 mg/dℓ. Triglyceride levels were typically low in fasting serum. During alimentary lipemia, the levels of triglycerides did become elevated, and as indicated by the postprandial data in Table 2, the percentage content of triglycerides among serum lipids increased. In some marine mammals, the period of alimentary lipemia was longer than 13 hr. As a result, whenever a percentage triglyceride content having a value greater than 15% appears in Tables 1 or 2, one can assume that chylomicra were still being formed and metabolized. Finally it should be noted that the levels of free fatty acids (FFA) in marine mammals were quite high, attaining a value of 97 mg/dℓ (approximately 3.5 meq/ℓ) in male bottlenose dolphin 5. In general, the FFA levels in sera of cetaceans were higher than in pinniped sera.

ANALYTICAL ULTRACENTRIFUGE DATA

The concentrations of lipoproteins as determined in the analytical ultracentrifuge have been measured for six different subclasses as shown in Table 3 for cetaceans and in Table 4 for pinnipeds.

Table 1
CONCENTRATION AND COMPOSITION OF LIPIDS IN
CETACEAN SERUM

Mammal	mg/dℓ	Percent composition					Ref.
		CE	PL	TG	UC	FFA	
Bottlenose Dolphin							
(Tursiops truncatus)							
Female							
1	565	38.1	42.4	8.0	8.1	3.4	8
2	524	39.3	39.8	9.7	9.7	1.7	8
3	648	41.5	36.5	9.8	8.1	4.0	8
4	839	34.7	30.5	19.4	7.2	5.8	9
5	705	37.5	23.8	19.6	6.7	6.2	9
Male							
1	690	39.8	35.8	13.6	8.8	2.0	8
2	557	44.7	38.2	6.7	7.5	3.0	8
3	533	45.8	35.4	7.0	9.0	2.8	10
4	698	25.6	23.0	31.2	8.5	4.5	9
5	785	27.5	31.7	19.7	8.0	12.3	9
Common dolphin							
(Delphinus delphis)							
Male	1080	38.3	44.2	7.7	7.1	2.6	10
Dall's Porpoise							
(Phocoenoides dalli)							
Male	912	13.5	27.3	50.1	6.1	3.0	10
Sex not determined	274	18.7	46.5	19.8	11.2	3.7	10
Killer whale							
(Orcinus orca)							
Female							
1	589	32.2	32.1	17.3	14.8	3.6	11
2	1010	36.8	31.1	19.9	11.6	0.7	11
3	592	42.3	27.2	16.5	9.0	5.0	11
4	547	46.0	27.2	9.2	13.8	3.7	1
Male							
1	853	39.9	26.5	20.9	9.9	2.8	11
2	1050	43.2	29.1	14.5	10.5	2.8	11
3	838	35.4	31.7	22.8	9.3	0.9	11
4	1110	41.0	36.4	12.9	8.1	1.6	11
Pacific white-sided dolphin							
(Lagenorhynchus obliquidens)							
Female 1	328	26.2	44.2	13.3	11.0	5.4	10

Note: Abbreviations used in Table 1 and subsequent tables: CE, cholesteryl esters; PL, phospholipids; TG, triglycerides; UC, unesterified cholesterol; FFA, free fatty acid.

HDL Data

The data shown in Tables 3 and 4 revealed that HDL were the major lipid carriers in both cetaceans and pinnipeds. An exception was female killer whale 4 which was dying of pneumonia at the time the blood was drawn.[1] When comparing the data for pinnipeds and cetaceans distinctive features in the HDL distribution can be seen. Although the levels of HDL subclasses varied among the animals, cetaceans tended to have approximately the same percentage of HDL_2 and HDL_3. On the other hand, the major lipoproteins of pinnipeds for the most part were HDL_3. In almost all animals studied, a discernible schlieren pattern was measured in the HDL_1 subclass.

LDL Data

The LDL data revealed concentration differences between killer whales and the other

Table 2
CONCENTRATION AND COMPOSITION OF LIPIDS IN SERUM OF MARINE CARNIVORES

Carnivore	mg/dℓ	CE	PL	TG	UC	FFA	Ref.
		Percent composition					
		Pinnipedia					
Family Otariidae							
California sea lion							
(*Zalophus californianus*)							
Male	717	35.4	51.1	4.5	8.4	0.6	8
Northern Steller Sea Lion							
(*Eumetopias jubata*)							
Female	853	36.4	50.0	5.6	7.4	0.5	10
8 Female pups	752	32.5	49.0	3.1	9.7	2.9	1
	±120	±2.9	±1.5	±2.6	±1.2	±2.1	
4 Male pups	755	34.4	49.0	4.7	8.6	3.4	1
	±64	±0.8	±2.0	±2.0	±0.5	±0.8	
Northern fur seals							
(*Callorhinus ursinus*)							
5 Males	798	34.0	50.6	4.1	7.5	3.0	1
	±190	±2.9	±2.6	±0.7	±0.7	±2.0	
Family Odobenidae							
Walrus							
(*Odobenus rosmarus*)							
Female pup	1140	38.0	45.3	3.4	9.4	3.8	8
Family Phocidae							
Elephant seal							
(*Mirounga angustirostris*)							
Female 1							
1 (5th hr)	922	28.7	42.5	32.0	6.7	0.7	10
Male							
1	950	33.6	51.3	4.5	8.4	0.6	1
2 (10th hr)	1220	20.7	39.1	32.4	7.0	0.8	12
3 (10th hr)	898	19.0	37.9	34.9	7.5	0.7	12
Gray seal							
(*Halichoerus grypus*)							
Female	1180	38.3	47.9	5.3	5.8	2.8	9
Male	890	32.9	51.1	4.4	7.7	2.6	9
Harbor seal							
(*Phoca vitulina*)							
Female							
1	1050	32.1	55.0	4.2	7.2	1.5	10
2	897	32.8	55.9	3.2	6.9	1.1	8
3	971	32.9	54.5	4.8	6.7	1.0	10
4	963	35.4	52.1	4.8	6.8	0.6	10
1 (5th hr)	1290	27.7	45.2	18.3	7.6	1.2	12
4 (5th hr)	1270	26.4	48.4	18.3	5.2	1.7	1
Weddell seal							
(*Leptonychotes weddelli*)							
4 Females	1320	39.5	47.7	3.4	8.1	1.8	1
	±70	±1.0	±1.0	±1.0	±0.3	±0.6	
4 Fetuses	832	36.5	51.3	2.6	9.0	0.7	1
	±340	±4.0	±3.2	±1.5	±1.2	±0.2	

Table 2 (continued)
CONCENTRATION AND COMPOSITION OF LIPIDS IN SERUM OF MARINE CARNIVORES

Carnivore	mg/dℓ	Percent composition CE	PL	TG	UC	FFA	Ref.
		Carnivora					
Family Ursidae							
(*Ursinus maritimus*)							
Female	969	21.5	41.8	23.3	13.5		2
27.9	±64	±0.6	±0.3	±0.9	±0.0		
Male	794	19.7	46.1	21.4	12.8		2
	±62	±0.0	±0.2	±0.1	±0.4		

Note: In Table 2 and subsequent tables, times in parentheses designate the postprandial hours at which blood was drawn from elephant and harbor seals.

Table 3
CONCENTRATION (mg/dℓ) OF CETACEAN LIPOPROTEINS WITHIN MAJOR FLOTATION RATE INTERVALS

Species	S_f^o 20—400	12—20	0—12	$F_{1.20}$ 9—20	3.5—9	0—3.5	Ref.
Bottlenose dolphin							
Female							
1	48	19	25	82	303	270	8
2	17	32	62	13	239	180	8
3	82	10	16	83	387	343	8
6	19	100	150	8	218	236	10
7	3	15	63	34	184	216	10
Male							
1	44	46	113		195	283	8
2	62	33	53	57	328	229	8
6	22	6	189	223	304	256	10
Killer whale							
Female							
1	83	147	191	119	171	173	11
2	106	115	379	81	251	297	11
3	26	189	122	56	163	205	11
4	31	76	369	15	94	76	1
5	54	21	24		304	95	12
Male							
1	128	106	221	11	221	300	11
2	102	78	221	153	328	386	11
3	156	87	241	6	228	322	11
4	60	147	381	217	495	379	11
Pacific white-sided dolphin							
Female							
2	34	39	114	2	202	200	1
Male	3	2	84		139	183	1

Table 4

CONCENTRATION (mg/dℓ) OF PINNIPED LIPOPROTEINS
WITHIN MAJOR FLOTATION RATE INTERVALS

Species	S_f^o			$F_{1.20}$			Ref.
	20—400	12—20	0—12	9—20	3.5—9	0—3.5	
California sea lion							
Male	21	4	98	84	366	428	8
Walrus							
Female pup	26	100	361	172	391	318	8
Elephant seal							
Male							
1	47		186	201	594	308	1
4		1	22	7	450	438	10
2 (10th hr)	71				408	699	13
3 (10th hr)	43				187	591	13
Harbor seal							
Female							
1	9	15	185	68	531	757	1
2	3		138		402	740	8
3	17	7	133		226	953	10
Male	10	23	202	42	526	732	14
1 (5th hr)	33	10	218	153	590	728	13
2 (5th hr)	5		131	1	410	782	10
4 (5th hr)	43	9	157	9	403	799	13

marine mammals. In both male and female animals, the concentrations of LDL were higher than values observed in the other marine mammals and in some cases even exceeded normal human LDL levels. Interestingly, female killer whale 5 which did not have elevated LDL died shortly after the blood sample was obtained and post-mortem analyses revealed that the animal had atherosclerosis.[12]

In general, the major LDL peaks had a higher flotation rate in cetaceans than in pinnipeds. The LDL of two species of pinnipeds, harbor seal and the California sea lion, were found to have two distinct schlieren components. The major slower floating component had an S_f^o value of approximately 2. The minor component had an S_f^o value of 7.

LDL Patterns in Young Animals

Although belonging to distinct mammalian groups, three young animals (a dolphin, an elephant seal, and a walrus) had in common an LDL component with a slow flotation rate. In each case, the concentration of this component was higher than levels in adult animals.

The LDL of the walrus pup exhibited two peaks with flotation rates comparable to those found in harbor seals and the Calfiornia sea lion. However, unlike the other pinnipeds, the concentrations of these two LDL components were nearly the same. Male bottlenose dolphin 6 also had an LDL pattern with the lipoprotein mass equally distributed between components with a fast and slow flotation rate. The walrus pup was being fed on a milk formula diet and the male dolphin was still obtaining milk from its mother.

Elephant seal 1 had a single component with a peak S_f^o value of 1.4. Elephant seals typically had either no discernible schlieren pattern at all within the S_f^o 0 to 12 interval or a very small peak, as was detected in elephant seal 4. Elephant seal 1, although no longer being fed milk at this time, was a young animal less than a year in age. The possible significance of this slower floating LDL components in young animals will be discussed at the end of the chapter.

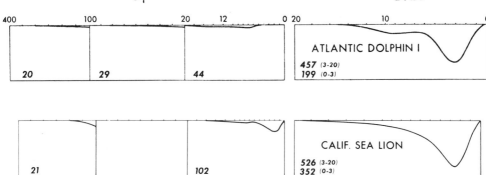

FIGURE 1. Computer-derived plot of the schlieren patterns obtained during analytical ultracentrifugal analyses of lipoproteins floating in salt solution of d = 1.063 g/mℓ (VLDL, IDL, and LDL distribution) and d = 1.21 g/mℓ (HDL distribution). Values for serum concentrations of lipoproteins in milligrams per deciliter are given in the lower left hand corner. Two values in the HDL distribution represent the concentrations of the $F_{1.20}$ intervals 0 to 3 and 3 to 20.

General Features of the Schlieren Data

Some general features which were characteristic of the ultracentrifugal distribution of cetacean and pinniped lipoproteins can be seen in Figure 1. As noted above, the flotation rate of the major LDL component is higher in cetaceans than in pinnipeds. This is illustrated in Figure 1 where the major peak can be seen to be higher (S_f^o value 7.5) in the dolphin than in the California sea lion (S_f^o value 1.9).

Examining the HDL distributions on the right in Figure 1, one finds that both animals have discernible material in the HDL_1 subclass. In cetaceans part of this was due to the presence of a minor peak, distributed between HDL_1 and HDL_2. An example of such a minor HDL component can be seen in the dolphin pattern. The major peak flotation rate for HDL data presented in Table 3 and 4 varied between 3.5 and 2.5 in both groups of mammals; however, in most cases, the bulk of pinniped HDL was distributed within the HDL_3 subclass.

Postprandial Data

The concentration of VLDL was low in both pinnipeds and cetaceans. Even in postprandial studies, only small changes in the VLDL distributions of harbor seals and elephant seals were observed. In fasting animals, the VLDL had S_f values primarily between 20 and 100. Following fat ingestion there was a shift in the distribution, with an increase in the concentration of lipoproteins with S_f values between 100 and 400. However, at no postprandial period could the increased concentration of lipoproteins with S_f values between 20 and 400 account for the mass of triglycerides associated with lipoproteins with densities less than 1.006 g/mℓ. The majority of triglycerides were transported by chylomicra with S_f values greater than 400. No major postprandial change was observed in the distribution of lipoproteins with densities greater than 1.006 g/mℓ.

LIPID COMPOSITION OF MAJOR LIPOPROTEIN FRACTIONS

The lipid compositions of the major lipoprotein fractions, given in Tables 5, 6, and 7,

Table 5
CONCENTRATION AND COMPOSITION OF LIPID
COMPONENTS OF TRIGLYCERIDE-RICH
LIPOPROTEINS (D < 1.006 g/mℓ) IN SERA OF
MARINE MAMMALS

Species	Serum·conc (mg/dℓ)	Percent composition				Ref.
		CE	PL	TG	UC	
Bottlenose dolphin						
Female						
1	59	15.0	27.9	45.1	11.3	8
3	83	9.1	21.7	56.0	12.1	8
Male 2	52	27.8	19.7	40.7	11.7	8
Killer whale						
Female						
1	79	22.5	21.2	45.5	10.4	11
2	110	5.7	18.9	62.9	12.7	11
3	52	10.4	19.5	55.7	13.7	11
Male						
1	142	10.4	18.1	63.8	8.0	11
2	128	14.3	21.2	55.9	8.7	11
3	155	6.5	20.2	65.8	7.2	11
4	92	9.7	23.8	58.3	7.8	11
California sea lion						
Male	28	8.6	19.9	57.5	12.4	8
Elephant seal						
Female						
1 (5th hr)	442	2.7	11.0	82.8	3.1	10
2	31	18.4	22.7	44.7	12.6	10
Male						
3	75	3.0	19.1	70.4	6.5	1
2 (10th hr)	497	1.3	6.3	84.4	8.0	13
3 (10th hr)	248	1.4	7.2	83.5	7.6	13
Harbor seal						
Female						
4	32	25.2	23.4	44.3	5.9	10
1 (5th hr)	188	4.3	7.5	78.4	9.1	1
4 (5th hr)	203	4.5	11.6	78.1	5.1	1
Polar bear	46		15.2	65.2	19.6	2

revealed in general a reciprocal relation between the percent content of neutral lipids and the particle density.

Triglyceride-Rich Lipoproteins

In both cetacean and pinniped sera, the bulk of the triglyceride in the circulation were associated with lipoproteins with densities less than 1.006 g/mℓ. This was not the case for the polar bear (see Table 6) and the manatee (see Table 8). Because polar bear lipoproteins in the d 1.006 to 1.063 g/mℓ fraction had a high triglyceride content (42%), it was suggested that the lipoproteins in this density fraction were primarily IDL[2]. However, no further ultracentrifugal fractionation was carried out.

Triglyceride concentrations in the sera of fasting pinnipeds and cetaceans were less than 60 mg/dℓ. In fasting sera, the triglyceride content of the VLDL varied between 40 and 60% (see Table 5). During alimentary lipemia, when chylomicra, which are larger and less dense than VLDL, enter the plasma, the triglyceride concentration in sera increased to values above 200 mg/dℓ and the triglyceride content of lipoproteins in the d < 1.006 g/mℓ fraction was between 75 and 85%.

Table 6

CONCENTRATION AND COMPOSITION OF LIPID
COMPONENTS OF LIPOPROTEINS IN THE DENSITY
FRACTION (1.006—1.063 g/mℓ)

Species	Serum conc (mg/dℓ)	Percent composition				Ref.
		CE	PL	TG	UC	
Bottlenose dolphin						
Female						
1	80	31.6	40.4	13.1	14.1	8
2	100	26.4	36.0	21.6	15.6	8
Male 2	91	45.4	30.2	10.1	14.0	8
Killer whale						
Female						
1	264	40.2	24.4	18.0	16.9	11
2	431	38.5	25.5	20.3	15.5	11
3	268	36.7	24.2	22.8	16.0	11
4	362	46.7	24.6	11.2	17.0	1
Male						
1	252	37.2	24.6	24.0	13.8	11
2	320	45.8	27.1	11.4	14.3	11
3	267	40.1	26.3	19.6	13.8	11
4	455	39.8	30.9	17.0	11.9	11
California sea lion						
Male	131	37.4	41.7	7.0	13.5	8
Northern fur seal						
Male 1[a]	153	23.0	40.5	20.6	15.0	10
Steller sea lion						
Female pup 1[a]	213	35.5	41.7	7.1	14.9	10
Male pup						
1[a]	170	38.1	34.0	12.8	13.2	10
2[a]	218	33.3	37.7	14.4	14.1	10
Walrus						
Female pup[a]	526	41.6	37.4	9.0	11.5	8
Elephant seal						
Male						
1	216	39.9	46.2	2.0	11.7	1
2 (10th hr)	38	25.8	44.2	14.6	13.2	1
3 (10th hr)	31	17.5	39.2	29.3	13.2	1
Harbor seal						
Female						
2	130	31.4	41.8	12.9	13.6	8
4	178	29.2	38.1	20.1	12.3	10
1 (5th hr)	219	35.9	44.2	6.6	12.9	1
4 (5th hr)	148	37.0	42.4	7.5	12.6	1
Polar bear	399	25.8	17.5	41.9	14.8	2

[a] Analyses were carried out on lipids isolated from lipoproteins with densities less than 1.063 g/mℓ. Also included were low amounts of triglyceride-rich lipoproteins.

Density Fraction 1.006 to 1.063 g/mℓ

Among the animals listed in Table 6, polar bears and killer whales had the highest concentrations of lipids in the 1.006 to 1.063 g/mℓ density class. In general, for lipoproteins in this density interval, the content of neutral lipids was slightly higher in cetaceans in comparison with pinnipeds. This is consistent with the bulk of cetacean LDL having a higher flotation rate, and presumably a larger size, than pinniped LDL.

Table 7
CONCENTRATION AND COMPOSITION OF LIPID
COMPONENTS OF HDL

Species	Serum conc (mg/dℓ)	Percent composition				Ref.
		CE	PL	TG	UC	
Bottlenose dolphin						
Female						
1	356	50.2	41.9	1.2	6.0	8
2	249	49.6	42.0	1.4	6.8	8
3	466	54.8	36.5	0.8	6.4	8
6	230	51.5	39.7	1.2	6.7	10
Male						
1	304	52.6	39.4	1.2	6.6	8
2	345	52.2	36.5	1.1	6.9	8
Common dolphin						
Male	763	44.8	45.1	2.4	7.1	1
Dall's porpoise						
Male	278	38.7	50.0	3.1	6.5	1
Killer whale						
Female						
1	222	61.0	29.2	1.7	6.9	11
2	302	55.0	36.3	2.7	5.6	11
3	204	63.9	27.2	1.7	5.7	11
4	117	60.4	25.2	1.6	11.4	1
Male						
1	286	57.7	33.3	2.1	5.5	11
2	451	57.5	33.1	1.3	6.4	11
3	287	53.8	37.8	2.6	5.2	11
4	371	58.6	33.1	1.7	5.7	11
California sea lion						
Male	536	36.6	54.7	0.9	7.5	8
Northern Steller sea lion						
Female						
1	481	32.7	55.4	1.1	7.7	10
2	534	34.4	55.4	0.9	7.9	10
3	199	31.6	56.2	1.6	9.0	10
Male						
1	455	33.7	56.9	0.5	8.2	10
2	420	36.5	53.9	0.6	6.7	10
Northern fur seals						
6 Males	591	36.1	55.1	1.0	6.6	10
	± 146	± 2.7	± 1.6	± 0.4	± 0.9	
Walrus						
Male pup	495	34.2	55.7	0.6	7.5	8
Elephant seal						
Female						
1 (5th hr)	738	41.7	50.5	1.1	6.2	10
2	448	40.3	52.4	1.3	5.1	10
Male						
1	630	41.2	50.4	0.4	7.1	1
3	279	37.8	52.9	1.5	5.5	10
2 (10th hr)	664	34.0	56.8	2.3	6.3	1
3 (10th hr)	465	33.1	58.8	2.3	5.4	1
Harbor seal						
Female						
2	730	31.5	61.2	0.8	6.2	8
1 (5th hr)	709	33.6	58.3	0.9	6.3	1
4 (5th hr)	805	34.2	57.9	1.5	5.2	1
Polar bear	704	38.7	53.9	0.9	6.5	2

Table 8
DISTRIBUTION OF LIPOPROTEIN
CHOLESTEROL DETERMINED FOLLOWING
HEPARIN MANGANESE PRECIPITATION[16]

Species	Conc (mg/dℓ)				
	TG	TC	d > 1.006	α	β
Elephant seal					
Fasting					
4	12	97		88	7[a]
5	61	227	224	209	15
Fed					
5	567	211	193	181	12
6	174	166	158	159	
7	69	197	191	180	11
8	630	160	157	141	16
9	114	129		121	
Bottlenose dolphins					
Fasting					
Female					
6	87	196		131	48[a]
7	112	235		173	40[a]
Male					
6	68	154		114	26[a]
7	75	126		103	8[a]
8	77	139		120	4[a]
Beluga whale					
(*Delphinapterus leucas*)					
Female	116	154		96	54[a]
Male					
1	152	203		111	62[a]
2	159	200		120	4
Manatee					
(*Trichechus manatus*)					
Male	50	174			
Female					
1	100	178			
2	62	151			
3	46	122			
Pooled sample	50	149	147	65	82

Note: d > 1.006: cholesterol concentration of lipoproteins with densities greater than 1.006 g/mℓ. α : cholesterol concentration in heparin manganese supernatant. β : calculated concentration of cholesterol associated with beta lipoproteins.

[a] β-Lp lipoprotein cholesterol calculated using the relation: TC (total plasma cholesterol) $- (\alpha + TG/5) = \beta$

HDL Class

A similar relationship also can be seen when comparing the composition and the ultracentrifugal distribution of cetacean and pinniped HDL. As the data in Table 7 show, the cetaceans, in particular bottlenose dolphins and killer whales, have a higher HDL cholesteryl ester content (50 to 64%) than pinnipeds (30 to 42%). The triglyceride content of marine mammalian HDL was 3% or less, even for samples obtained during alimentary lipemia.

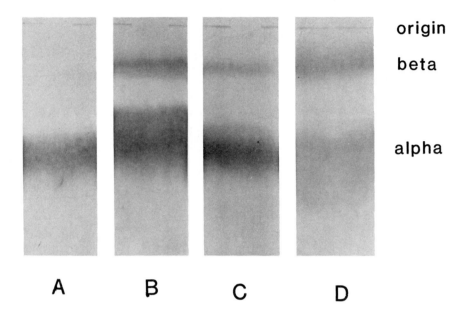

FIGURE 2.   Paper electrophoretograms of pinniped lipoproteins. (A) Elephant seal, (B) Harp seal, (C) Harbor seal, and (D) Walrus pup.

OTHER LIPOPROTEIN DATA

Electrophoretic Data

Electrophoretic studies[8,11,13] done on marine mammalian sera have shown that the four major electrophoretic bands, chylomicron, pre-beta, beta, and alpha lipoproteins can be detected using both agarose gels and paper strips. Paper electrophoretograms of serum lipoproteins of four different species of pinnipeds are shown in Figure 2. A broad alpha lipoprotein (α-Lp) band was found on each of the strips. Beta lipoproteins (β-Lp) were clearly discernible on three of the four strips. The absence of a demonstrable β-Lp band in elephant seal sera agreed with the analytical ultracentrifugal data reported in the second section. A similar result shown in Figure 3 was obtained when electrophoretic studies were done on the serum of female bottlenose dolphin 3 which also had low levels of LDL. When dolphins with higher levels of LDL were studied, β-Lp were detected.[8] At the other end of the spectrum, killer whale sera which have comparatively high levels of LDL, were all found to have an intensely staining β-Lp band as shown in Figure 3.

Because paper electrophoresis failed to demonstrate β-Lp in the sera of seven different elephant seals, a more sensitive technique of agarose gel electrophoresis has been used to study the electrophoretic distribution of lipoproteins in these animals. Unequivocal results have yet to be obtained. In a few, but not all cases, a faintly staining band with pre-beta mobility has been observed in whole plasma and the d $>$ 1.006 g/mℓ ultracentrifugal fractions, as shown in Figure 4. Whether this corresponds to β-Lp with greater mobility than normal or to a form of Lp(a)[15] remains to be determined.

In postprandial studies, chylomicron bands at the origin have been detected in electrophoretograms of both cetacean[16] and pinniped[13,16] plasma samples. These bands were also present when electrophoretic analyses were done on the d $<$ 1.006 density fractions isolated from the same samples.

In contrast to other marine mammals, electrophoretograms of manatee plasma revealed β-Lp to be the major band, as shown in Figure 5. Dangerfield et al.[17] reported similar results for sirenian serum, but the authors did not specify whether the sample had been obtained

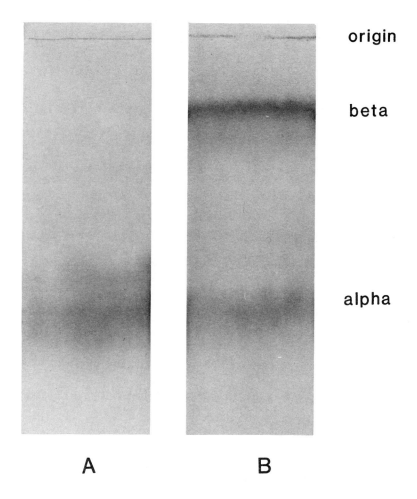

origin

beta

alpha

A   B

FIGURE 3.   Paper electrophoretograms of cetacean lipoproteins. (A) Bottlenose dolphin and (B) Killer whale.

from a dugong or a manatee. Interestingly, these same authors[17] also reported β-Lp to be the major lipid carrier in sera of elephants that are thought to be linked evolutionarily to sirenians. Concentrations of cholesterol and triglycerides in manatee plasma are given in Table 8. In all four animals the triglyceride concentrations were higher than levels found in other herbivorous mammals.[18]

Heparin Manganese Precipitation

Heparin manganese precipitation also was used to examine the levels of beta lipoproteins in elephant seal serum. As seen from the results given in Table 8, values of β-Lp cholesterol varied between 0 and 16 mg/mℓ. The values were obtained in most cases by subtracting the cholesterol concentration in the supernatant from the cholesterol concentration in the d > 1.006 g/mℓ ultracentrifugal fraction. Otherwise, the cholesterol associated with the triglyceride-rich lipoproteins was estimated by dividing the plasma triglyceride concentration by five and subtracting the resulting value together with the supernatant cholesterol concentration from the total plasma cholesterol concentration. Only α-Lp were present in the supernatant. As can be seen from the data for elephant seal 9, the β-Lp cholesterol can be underestimated if chylomicra are present in the plasma.

Data on cholesterol distribution also are presented in Table 8 for bottlenose dolphins,

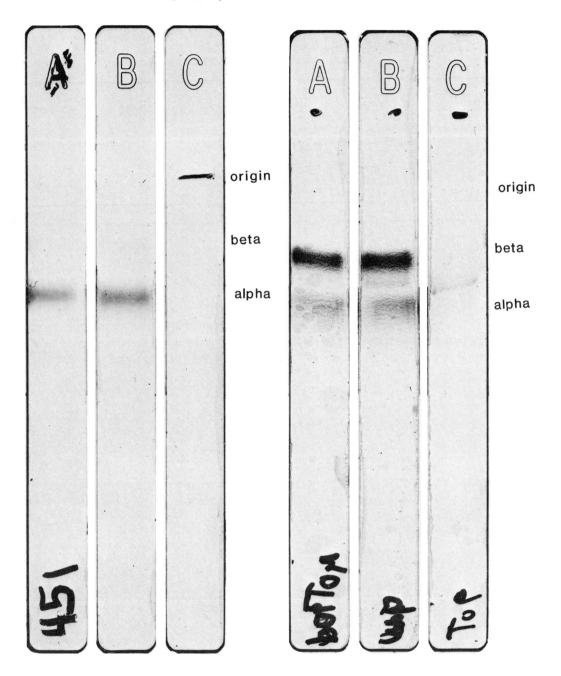

FIGURE 4.   Agarose electrophoretograms of elephant seal lipoproteins. (A) d > 1.006 g/mℓ, (B) whole plasma, and (C) d < 1.006 g/mℓ. Note the presence of a very faint band with pre-beta mobility in A and B.

FIGURE 5.   Agarose electrophoretograms of manatee lipoproteins. (A) d > 1.006 g/mℓ, (B) whole plasma, and (C) d < 1.006 g/mℓ.

beluga whales, and manatees. Consistent with α-Lp being the principal lipid cariers, the major portion of the serum cholesterol was found in the supernatant of all animals except manatees. Manatee α-Lp also were uniquely different from other marine mammals with a relatively high triglyceride to cholesterol ratio of 0.4. Furthermore, the majority (57%) of triglycerides in manatee plasma were associated with lipoproteins that sedimented rather than floated in a solution of density of 1.006 g/mℓ.

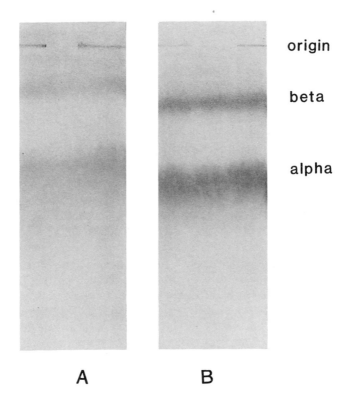

FIGURE 6. Paper electrophoretograms of lipoproteins with densities less than 1.063 g/mℓ. (A) Bottlenose dolphin pup and (B) Harbor seal. Note the presence of both α- and β-Lp.

Evidence for α-LDL

In analytical ultracentrifugal studies, two components with a high and a low flotation rate were observed in the schlieren patterns for the LDL of certain cetaceans and pinnipeds. Electrophoretic analyses have provided additional information concerning the nature of these two LDL components. When ultracentrifugal fractions with density less than 1.063 g/mℓ were examined, both β- and α-Lp bands were detected. An example of this is shown in Figure 6 for both the young male dolphin 6 and female harbor seal 4. Following separation by preparative ultracentrifugation the α-Lp were isolated almost exclusively in the density interval 1.04 to 1.06 g/mℓ.[14] In other words, the component with the high S_f^o values (8 to 9) consists primarily of β-Lp and the component with low S_f^o values (0 to 3) consists of α-Lp. In describing these lipoproteins in physicochemical terms, a distinction therefore must be made between alpha LDL and beta LDL.

DISCUSSION

Importance of Fat Stores

To survive in the ocean, most marine mammals build up vast stores of fat which provide them with buoyancy, thermal insulation, and energy.[19] Metabolized fat also is the major source of water for these animals. Being able to ingest large amounts of fat and to transport triglyceride-rich lipoproteins in the circulation enables the animal to maintain and utilize these all important fat deposits.

Because of the extreme cold which the animals encounter in the ocean, the subcutaneous layers of fat reduce loss of body heat. The thermal conductivity of water is higher than that of air and the need for such insulation is apparent. Among fur seals, this reliance on blubber is less than in other pinnipeds. Also the sea otter has very little in the way of fat stores.[20] Still, fat, in the form of oils secreted by sebaceous glands, provides a coat to the fur, enabling it to remain dry. Without this oily coating, fur would quickly become a very poor insulator in the ocean.[21]

The utilization of adipose tissue for essential endergonic processes is best illustrated by those marine mammals which fast for prolonged periods of time. In contrast to hibernating animals that make physiological adjustments in body temperature and pulse rate and remain in a dormant state during the winter months,[22] pinnipeds, on which most behavioral observations have been made, can be quite active during their fast. Harem bulls, having large stores of fat, are able to go without food for as long as 2 months, during which time they mate frequently and engage in combat to maintain their harems.[23-25] While lactating, female pinnipeds will fast for long periods, in some cases for as much as 5 weeks. A study done on a Laysan monk seal (*Monachus schauinslandi*) estimated that the mother lost 200 lb while the pup was gaining 100 lb.[24] The rapid gain in weight observed in neonates is due to the ingestion of milk which has a high fat content (20 to 50%) but little or no carbohydrate.[26-30] Cetaceans, which are born in the water without blubber, must rapidly develop adipose stores following birth. Harp seals, are born with a protective coat of fur which they shed at the age of 2 weeks by which time they have developed additional blubber.[31] Although not as active as the adults, certain species of pinniped pups remain on land following weaning and fast from 2 weeks to nearly 3 months.[32] After undergoing a perhaps critical developmental change, they enter the water and begin to hunt for food. Data on periods of fasting among cetaceans are limited. Several species of large Mysticeti are believed to fast while migrating over distances of several thousand miles.[33] The female gray whale (*Eschrichtius robustus*) even lactates while returning from Baja California to the Arctic Circle.[34]

Role and Nature of Dietary Fats

Marine mammals maintain their fat stores on a diet consisting mostly of protein and fat, with little or no carbohydrate. Dietary protein is undoubtedly utilized for gluconeogenesis and synthesis of proteins in the body, rather than being converted to fat. The walrus and sea otters, which have a diet principally of shell fish, do ingest carbohydrates. The glycogen content of some mollusks can exceed 10%.[35] The sea otter, lacking extensive fat stores, is not confronted with the problem of depositing large amounts of triglycerides. This is obviously not the case for the blubber-rich walrus. A diet containing large amounts of shell fish might be thought to be unable to provide sufficient quantities of fat for the development of such large adipose stores. However, fatty acid analyses of blubber fat have indicated that the walrus does have a considerable amount of fat in its diet.[36] The ω3 fatty acids are found in tissue lipids at every level of the marine food chain.[36-39] Being unable to synthesize these fatty acids, the presence of polyenoic acids, such as 20:5, 22:5, and 22:6, in mammalian tissue is a reflection of the animal's diet. The polyenoic ω3 fatty acids comprised 30% or more of the total in walrus blubber.[37] Fatty acids of the ω9 series were also present and these could have originated either from *de novo* synthesis or modification of dietary lipids. If *de novo* synthesis were the major source of fatty acids, one also would expect to find high levels of polyenoic acids of the ω9 series in order for the blubber to have a low enough melting temperature to be compatible with the environment. Because only monoenoic ω9 acids, mostly 18:1 were detected, steps involving elongation and desaturation to form polyenoic acids for adipose stores are probably not major pathways in the walrus. It would then appear that the walrus, like other marine mammals, relies on dietary fats to build up adipose stores with the required fluidity needed in a cold environment. The fate of the dietary carbohydrate in the walrus is unknown. Because other pinnipeds have been shown to lack intestinal disaccharidase activity,[40-42] it is very probable that the walrus must rely on the assistance of intestinal microorganisms to reduce the carbohydrate in its diet to monosaccharides or other absorbable metabolites. Existing on such a high fat diet, little if any *de novo* fatty acid synthesis would be expected to occur in marine mammals. Although measured in a single species, studies of tissue enzymes in Weddell seal support, but do not prove, this point. Fried et al.[43] reported that the activity of glucose-6-phosphate dehydrogenase in seal tissue was less than found in terrestrial mammals. This enzyme is involved in the first step of the shunt pathway for the direct oxidation of glucose to yield nicotinamide adenine dinucleotide phosphate, a co-factor required for fatty acid synthesis.

Sirenians, the Exception to the Rule

Studies done on dugongs and manatees indicate that a very low percentage of dietary polyenoic acids are absorbed intact by these animals. In an early anatomical study,[44] Murie reported that manatees were distinctly different and had fat, comparable in consistency to tallow or lard rather than the oily fat of seals and whales. Consistent with this observation, fatty acid analyses[45-47] have revealed sirenian fat to contain little in the way of ω3 fatty acids. The major fatty acids are oleate, palmitate, and myristate. Apparently, dietary polyenoic acids are either biohydrogenated or converted to absorbable metabolites by the microorganisms in the hindgut[47,48] of these herbivores.

Sirenians are known to be extremely sensitive to sudden changes in water temperature, and they will perish from exposure.[49] Their inability to withstand cold ocean water may be related to their low metabolic rate.[49] However, their depot fat being so highly saturated could solidify at temperatures not much lower than 37°C. If this were to occur, a structural stress would be placed on the body and the solidified fat stores would not be available for energy utilization.

Polyenoic ω3 Fatty Acids: Possible Adaptive Mechanisms

Examination of the nature of marine mammalian fat suggests that the hydrolysis of tissue triglycerides and mobilization of fatty acids may be more complicated than in terrestrial mammals. Stereospecific analysis of triglycerides isolated from the blubber of marine mammals have shown that the polyenoic ω3 fatty acids are present primarily in the 1-position, rather than mostly in the 2-position as in fish fat.[39] Triglycerides from marine mammalian tissue are quite resistant to the action of pancreatic lipase.[39,50] When acted upon by this lipase, the digestion products are diglycerides enriched in these polyenoic acids rather than the expected monoglycerides and fatty acids. Patton et al.[51] have reported that in fish a second lipase, acting on these diglycerides, is available to carry this reaction to completion. It is very likely that marine mammals also have such a lipase. In a similar manner, hormone sensitive lipase,[52] the enzyme involved in the initial hydrolysis of stored fat, may either have unique properties in marine mammals or act in concert with other enzymes. In building up these stored fats, the synthetic machinery of these unusual triglycerides may also involve a distinct group of enzymes and co-factors.

The ability to maintain themselves on a diet rich in polyenoic fatty acids of the ω3 series raises several questions about other metabolic controls in these animals. For example, how do these animals retard the oxidation of their tissue fats? Do these animals have some way for rapidly metabolizing ketone bodies expecially during prolonged periods of fasting? With little or no ω6 fatty acids in the diet, what are the structure and function of prostaglandins in these animals? With a continued interest in the use of polyunsaturated fatty acids to lower cholesterol levels in human plasma, marine mammals may prove to be excellent models for exploring the metabolic consequences resulting from sustained ingestion of polyenoic acids.[53] Apropos to the topic of this chapter, what effect do ω3 fatty acids have on the structure, function and metabolism of circulating lipoproteins?

Lipoproteins: Salient Features

The physicochemical data presented in this chapter do not indicate any obvious characteristic which would be associated with the presence of polyenoic acids. Postprandial studies clearly have shown that these animals are able to secrete and to metabolize chylomicra, after the rapid ingestion of fat often equivalent to 5% of their body weight. Moreover, rapid clearance of triglycerides by postheparin lipase has been seen in both bottlenose dolphins and elephant seals following i.v. injection of heparin.[16] In terms of structural consideration at the molecular level, it would be interesting to determine if stereospecific analysis of circulating triglycerides would reveal a pattern similar to that in the depot fat of the ingested fish, i.e., the polyenoic acids being located on the 2-position.

In contrast to the triglyceride-rich lipoproteins, the study of marine mammalian LDL has revealed these lipoproteins to be distinctive in both their ultracentrifugal and electrophoretic distributions. Although both types of LDL were found in pinnipeds and cetaceans, α-LDL generally predominated in pinnipeds and β-LDL in cetaceans. Polar bear LDL reportedly contained only a single electrophoretic class.[2]

The low levels or absence of β-LDL seen in elephant seals may be due to genetic factors. The animals were slaughtered almost to extinction during the early part of this century.[54] Existing herds along the Pacific Coast are thought to be derived from less than 100 survivors which included just a few harem bulls.

The elevation of α-LDL, noted in three young animals (a dolphin, an elephant seal, and a walrus), may reflect a limited ability to tolerate diets rich in triglycerides. Although supportive electrophoretic and analytical ultracentrifugal data were lacking, the lipid concentrations in Table 6 suggest a similar phenomenon occurred in Steller sea lion pups, which were being fed a milk formula diet at the time. Interestingly, Leat et al.[55] reported that in young lambs lipoproteins having flotation rates corresponding to α-LDL disappeared once the animals began to ruminate. Possibly, the buildup in α-LDL resulted in each case from the metabolism of surface components, i.e., phospholipids and cholesterol, shed from triglyceride-rich lipoproteins as their core lipids were being hydrolyzed. Uptake of surface components by α-Lp and the formation of cholesteryl esters from these components are thought to be mediated by the serum enzyme, lecithin cholesterol acyltransferase (LCAT).[56,57] While this reaction would provide core lipids for α-Lp, the availability of apoproteins might limit the number of α-Lp which could actually transport these cholesteryl esters. If this were the case in a young animal, it would be more efficient to sequester the cholesteryl esters into the interior of α-LDL which, based on size difference, have a lower surface to volume ratio than α-HDL.

The serum concentration of β-LDL also depends on complicated control mechanisms. Nevertheless, considering that killer whales and polar bears eat a diet rich in polyunsaturated fats, lower levels of β-LDL might have been expected. It should be noted that while these animals do have a diet consisting of fish, as do all marine carnivores, polar bears and killer whales do eat other marine mammals. Whether these dietary differences actually influence the level of β-LDL remains to be determined.

The concentrations of LDL were comparable in polar bears and killer whales, but polar bear HDL attained the high levels found in both elephant seal and harbor seals. Moreover, the composition of HDL was very similar for polar bears and pinnipeds. Based on these high ratio of surface lipids to cholesteryl esters, it is very likely that polar bear HDL consist primarily of HDL_3, as is the case in pinnipeds. Canine HDL also have a high phospholipid content;[8] and are predominantly HDL_3.[1]

It is not clear why certain HDL subclasses are elevated in one animal and not in another. The serum enzyme LCAT is thought to be involved in the conversion of HDL_3 to HDL_2.[59] The author has attempted such a conversion in vitro. Although human LCAT was able to react with seal HDL_3, the animal's own enzyme produced no change in the HDL distribution.[10] In vitro studies also failed to demonstrate LCAT activity in a variety of marine mammalian sera.[1] However, presence of high levels of unesterified fatty acid, as was noted in the sera of these animals, could have inhibited the action of LCAT, as has been reported by Rutenberg et al.[60]

Additional studies still need to be done on marine mammalian lipoproteins. Data on apoproteins have been reported only for polar bears.[2] The results of Kaduce et al.[2] are quite remarkable, for they have found a dimeric form of polar bear apo A-II. In addition, apo A-I, apo B, and apo C were shown to be present in these animals. As apoprotein data are obtained on other marine mammals, it will be possible to look for phylogenetic patterns. Moreover, knowledge concerning the distribution of these apoproteins on lipoprotein sub-classes will provide a better basis for the understanding of the dynamics of lipid transport in these animals.

The ability to adapt both to prolonged fast and to dive to great depths has made the marine mammal the subject of extensive physiological studies. Unfortunately, none of these studies have focused on the enzymes involved in fat mobilization. In a recent review, Kooyman et al.[61,62] have proposed that fat metabolism is a major source of energy not only to the fasting animal but to the diving animal as well (particularly during short aerobic dives).

Based on the duration (10 to 15 hr) of alimentary lipemia in marine mammals, circulating triglyceride-rich lipoproteins are probably a major source of energy during this period. This would be particularly true when the fed animals swim for extended periods, often at high speeds, in the ocean. Even when less active, as in the case of fasting pinnipeds, triglyceride-rich lipoproteins are probably providing energy to these animals. In studying fasting elephant seal pups, Costa and Ortiz[32] found triglyceride concentrations greater than 100 mg/dℓ, levels higher than found in animals after an overnight fast (see Tables 1 and 8). Whether this demand for energy also influences the distribution of the other lipoprotein classes remains to be demonstrated. It is very likely that during these periods changes occur in the activity of enzymes involved in fat metabolism and in the serum concentration of unesterified fatty acids.

SUMMARY

α-Lp were the predominant lipid carriers in the blood of marine mammals, and they were isolated almost exclusively within the HDL ultracentrifugal class, d 1.063 to 1.21 g/mℓ. Analytical ultracentrifugal analyses revealed that these lipoproteins comprise a highly polydisperse class and that their concentrations were greater than human HDL. In many cases, the concentration of marine mammalian HDL exceeded the combined levels of all lipoprotein classes in normal human sera.[63] The content of cholesteryl esters in cetacean HDL was higher than in pinniped HDL. This difference in the content of core lipids was in agreement with analytical ultracentrifugal data which indicated that cetacean HDL were predominantly larger than pinniped HDL.

LDL, the other carrier of cholesteryl esters, comprised between 5 and 40% of the total lipoproteins. Electrophoretic analyses of cetacean and pinniped LDL revealed that the class was heterogeneous, containing both α- and β-LDL. Among pinnipeds, the α-LDL were the predominant class. In both killer whale and polar bear sera, levels of β-LDL were comparable, or exceeded, levels in normal human sera. On the other hand, elephant seal sera did not contain a readily demonstrable β-LDL.

In fasting (24 to 70 hr) sera, the triglyceride concentration in marine mammalian sera were 100 mg/dℓ or less. In most cases, the triglycerides were associated with lipoproteins with densities less than 1.006 g/mℓ. However, in both polar bears and manatees, lipoproteins with densities greater than 1.006 g/mℓ had an appreciable triglyceride content. During alimentary lipemia, the triglyceride concentrations in dolphins, elephant seals, and harbor seals rose to levels of 200 mg/dℓ or greater. This increase was due to an influx of chylomicra into the blood.

Finally, the herbivorous manatee, with more β-Lp than α-Lp, was found to be distinctly different from the other marine mammals listed in this chapter.

ACKNOWLEDGMENTS

The author wishes to express his thanks to Professor Alex V. Nichols and Ms. Elaine Coggiola Gong for their assistance during much of the work done at the Donner Laboratory. Studies at the Donner Laboratory were supported by Public Health Service Biophysics Training Grant, NIH Grants HE 12710-01 and HD 10878-04 and a Bay Area Heart Post-doctoral Fellowship. The author also wishes to acknowledge the cooperation of the following marine laboratories and oceanariums: U.S. Naval Station in San Diego; Marine World, Redwood City, Calif.; Marineland of the Pacific, Palos Verdes Estates, Calif.; Sea World, San Diego, Calif.; Sea World, Orlando, Fla.; Stanford Research Institute, Palo Alto, Calif. The author also thanks the following veterinarians for their assistance: L. H. Cornell, D. W. Kenney, R. C. Hubbard, S. H. Ridgway, and J. C. Sweeney. Preparation of this chapter was made possible through the support of an NIH Atherosclerosis Training Grant (HL 07386). The author expresses his gratitude to Ms. Elaine Stieglitz for her assistance in preparing the manuscript.

REFERENCES

1. **Puppione, D. L.,** Physical and Chemical Characterization of the Serum Lipoproteins of Marine Mammals, Rep. UCRL-18821, Lawrence Radiation Laboratory, Berkeley, 1969.
2. **Kaduce, T. L., Spector, A. A., and Folk, G. E., Jr.,** Characterization of the plasma lipids and lipoproteins of the polar bear, *Comp. Biochem. Physiol.,* 69B, 541, 1981.
3. **Jensen, L. C., Rich, T. H., and Lindgren, F. T.,** Graphic presentation of computer-derived schlieren lipoprotein data, *Lipids,* 5, 491, 1970.
4. **Hatch, F. T. and Lees, R. S.,** Practical methods for plasma lipoprotein analysis, in *Advances in Lipid Research,* Vol 6, Paoletti, R. and Kritchevsky, D., Eds., Academic Press, New York, 1968, chap. 1.
5. **Noble, R. P., Hatch, F. T., Mazrimas, J. A., Lindgren, F. T., Jensen, J. C., and Adamson, G. T.,** Comparison of lipoprotein analysis by agarose gel and paper electrophoresis with analytical ultracentrifugation, *Lipids,* 4, 55, 1969.
6. Lipid Research Clinics Program Manual Laboratory Organizations, Vol. 1, DHEW Publication (NIH) 75-628, Department of Health, Education and Welfare, Washington, D.C., 1974.
7. **Burstein, M. and Scholnick, H. R.,** Lipoprotein-polyanion-metal interaction, in *Advances in Lipid Research,* Vol. 11, Paoletti, R. and Kritchevsky, D., Eds., Academic Press, New York, 1973.
8. **Puppione, D. L. and Nichols, A. V.,** Characterization of the chemical and physical properties of the serum lipoproteins of certain marine mammals, *Physiol. Chem. Phys.,* 2, 49, 1970.
9. **Nelson, G. J.,** The lipid composition of the blood of marine mammals. II. Atlantic bottlenose dolphin, *Tursiops truncatus,* and two species of seals, *Halichoerus grypus* and *Phoca vitulina, Comp. Biochem. Physiol.,* 40B, 428, 1971.
10. **Puppione, D. L.,** unpublished data, 1970.
11. **Puppione, D. L., Forte, T., and Nichols, A. V.,** Serum lipoproteins of killer whales, *Comp. Biochem. Physiol.,* 39B, 673, 1971.
12. **Hashimoto, S., Dayton, S., and Roberts, J. C., Jr.,** Aliphatic wax alcohols and other lipids in atheromata and arterial tissues of cetaceans, *Comp. Biochem. Physiol.,* 20, 975, 1967.
13. **Puppione, D. L.,** Serum lipoproteins in two species of phocids (*Phoca vitulina* and *Mirounga angustirostris*) during alimentary lipemia, *Comp. Biochem. Physiol.,* 59A, 127, 1978.
14. **Puppione, D. L., Forte, G. M., Nichols, A. V., and Strisower, E. H.,** Partial characterization of serum lipoproteins in the density interval 1.04—1.06 g/mℓ, *Biochim. Biophys. Acta,* 202, 392, 1970.
15. **Enholm, C., Garoff, H., Simon, K., and Aro, H.,** Purification and quantification of the human plasma lipoprotein carrying the Lp(a) antigen, *Biochim. Biophys. Acta,* 236, 431, 1971.
16. **Puppione, D. L.,** unpublished data, 1979—1981.
17. **Dangerfield, W. G., Finlayson, R., Myatt, G., and Mead, M. G.,** Serum lipoproteins and atherosclerosis in animals, *Atherosclerosis,* 25, 95, 1976.
18. **Leat, W. M. F. and Baker, J.,** Distribution of fatty acids in the plasma lipids of herbivores grazing pasture: a species comparison, *Comp. Biochem. Physiol.,* 36, 153, 1970.

19. **Vague, J. and Fenase, R.,** Comparative anatomy of adipose tissue, in *Handbook of Physiology,* Section 5, Renold, A. E. and Cahill, G. F., Jr., Eds., American Physiological Society, Washington, D.C., 1965, chap. 5.

20. **Kenyon, K. W.,** *The Sea Otter in the Eastern Pacific Ocean,* Dover Publications, New York, 1975.

21. **Costa, D. P. and Kooyman, G. L.** Oxygen consumption and temperature effects of the sea otter, *Enhydra lutria,* in water and the effect of fur oiling and washing, *Can. J. Zool.,* in press.

22. **Hoffman, R. A.,** Terrestrial animals in cold: hibernators in adaption to the environment, in *Handbook of Physiology,* Section 4, Renold, A. E. and Cahill, G. F., Jr., Eds., American Physiological Society, Washington, D.C., 1965, chap. 24.

23. **LeBoeuf, B. J.,** Male-male competition and reproductive success in elephant seals, *Am. Zool.,* 14, 163, 1974.

24. **King, J. E.,** *Seals of the World,* British Museum, London, 1967.

25. **Peterson, R. S.,** Social behavior in pinnipeds, in *Behavior and Physiology of Pinnipeds,* Harrison, R., Hubbard, R., Peterson, R., Rice, C., and Schusterman, B., Eds., Appleton-Century-Crofts, New York, 1968, chap. 1.

26. **Pilson, M. E. Q.,** Absence of lactose from the milk of *Otarioidea* a superfamily of marine mammals, *Am. Zool.,* 5, 220, 1965.

27. **Lauer, B. H. and Baker, B. E.,** Whale milk. I. Fin whale *(Balaenoptera physalus)* and beluga whale *(Delphinapterus leucas)* milk: gross composition and fatty acid constitution, *Can. J. Zool.,* 47, 95, 1969.

28. **Cook, H. W. and Baker, B. E.,** Seal milk. I. Harp seal *(Pagophilus groenlandicus)* milk: composition and pesticide residue content, *Can. J. Zool.,* 47, 1129, 1969.

29. **VanHorn, D. R. and Baker, B. E.,** Seal milk. II. Harp seal *(Pagophilus groenlandicus)* milk; effect of stage of lactation on the composition of the milk, *Can. J. Zool.,* 49, 1085, 1971.

30. **Jenness, R. and Odell, D. K.,** Composition of the milk of the Pigmy Sperm Whale *(Kogia breviceps) Comp. Biochem. Physiol.,* 61A, 383, 1978.

31. **Frisch, J. and Oritsland, N. A.,** Insulative changes in the harp seal pup during moulting, *Acta Physiol. Scand.,* 74, 637, 1968.

32. **Costa, D. P. and Ortiz, C. L.,** Blood chemistry homeostasis during prolonged fasting in the northern elephant seal, *Am. J. Physiol.,* 242, R591, 1982.

33. **Slijper, E. J.,** Distribution and migration, in *Whales,* Hutchinson and Co., London, 1962, chap. 12.

34. **Rice, D. W. and Wolman, A. A.,** Life History and Ecology of the Gray Whale, Spec. Pub. No. 3, American Society of Mammologists., Pittsburgh, 1971, 1.

35. **Albritton, E. C.,** *Standard Values in Nutrition and Metabolism,* W. B. Saunders, Philadelphia, 1954.

36. **West, G. C., Burns, J. J., and Modafferi, M.,** Fatty acid composition of Pacific walrus skin and blubber fats, *Can. J. Zool.,* 57, 1249, 1979.

37. **West, G. C., Burns, J. J., and Modafferi, M.,** Fatty acid composition of blubber from the four species of Bering Sea phocid seals, *Can. J. Zool.,* 57, 189, 1979.

38. **Brockerhoff, H., Ackman, R. G., and Hoyle, R. J.,** Specific distribution of fatty acids in marine lipids, *Arch. Biochem. Biophys.,* 100, 9, 1966.

39. **Brockerhoff, H.,** Fatty acid distribution patterns of animal depot fats, *Comp. Biochem. Physiol.,* 19, 1, 1966.

40. **Sunshine, P. and Kretchmer, N.,** Intestinal disaccharidases: absence in two species of sea lions, *Science,* 144, 850, 1964.

41. **Kretchmer, N. and Sunshine, P.,** Intestinal disaccharidase deficiency in the sea lion, *Gastroenterology,* 53, 123, 1967.

42. **Kerry, K. R. and Messer, M.,** Intestinal glycosidases of three species of seals, *Comp. Biochem. Physiol.,* 25, 437, 1968.

43. **Fried, G. H., Ray, C., Hiller, J., Rabinow, S., and Antopol, W.,** Alpha glycerophosphate dehydrogenase and glucose-6-phosphate dehydrogenase in tissues of the Weddell seal, *Science,* 155, 1560, 1967.

44. **Murie, J.,** On the form and structure of the manatee *(Manatus americanus), Trans. Zool. Soc. London,* 8, 127, 1872.

45. **Tsuyuki, H. and Itoh, S.,** Fatty acid composition of the dugong oil, *Bull. Jpn. Soc. Sci. Fish.,* 33, 1035, 1967.

46. **Itoh, S. and Tsuyuki, H.,** Fatty acid component of Senegal manatee fats, *Sci. Rep. Whales Res. Inst.,* 26, 307, 1974.

47. **Marsh, H., Spain, A. V., and Heinsohn, G. E.,** Physiology of the dugong, *Comp. Biochem. Physiol.,* 61A, 159, 1978.

48. **Murray, R. M., Marsh, H., Heinsohn, G. E., and Spain, A. V.,** The role of the mid-gut caecum and the large intestine in the digestion of sea grasses by the dugong *(Mammalia sirenia), Comp. Biochem. Physiol.,* 56A, 7, 1977.

49. **Odell, D. K.,** West Indian manatee, in *Wild Mammals of North America: Biology, Management and Economics,* Chapman, J. A. and Feldhammer, G. A., Eds., The Johns Hopkins University Press, Baltimore, 1982, chap. 41.

50. **Bottino, N. R., Vandenberg, G. A., and Reiser, R.,** Resistance of certain long-chain polyunsaturated fatty acids of marine oils to pancreatic lipase hydrolysis, *Lipids,* 2, 489, 1967.

51. **Patton, J. S., Nevenzel, J. C., and Benson, A. A.,** Specificity of digestive lipases in hydrolysis of wax esters and triglycerides studied in anchovy and other selected fish, *Lipids,* 10, 575, 1975.

52. **Khoo, J. C., Steinberg, D., Huang, J. J., and Vagelos, P. R.,** Triglyceride, diglyceride, monoglyceride and cholesterol ester hydrolases in chicken adipose tissue activated by adenosine $3':5'$-monophosphate-dependent protein kinase, *J. Biol. Chem,* 251, 2882, 1976.

53. **Goodnight, S. H., Jr., Harris, W. S., Connor, W. E., and Illingworth, D. R.,** Polyunsaturated fatty acids, hyperlipidemia and thrombosis, *Arteriosclerosis,* 2, 87, 1982.

54. **Bonnell, M. L. and Selander, R. K.,** Elephant seals: genetic variation and near extinction, *Science,* 184, 905, 1974.

55. **Leat, W. M. F., Kubasek, F. O. T., and Buttress, N.,** Plasma lipoproteins of lambs and sheep, *Q. J. Exp. Physiol.,* 61, 193, 1976.

56. **Schumaker, V. N. and Adams, G. H.,** Circulating lipoproteins, in *Annual Review of Biochemistry,* Vol. 38, Snell E. E., Ed., Annual Revue Inc., Palo Alto, Calif., 1969, 113.

57. **Tall, A. R. and Small, D. M.,** Plasma high density lipoproteins, *New Engl. J. Med.,* 229, 1232, 1978.

58. **Hillyard, L. A., Chaikoff, I. L., Entenman, C., and Reinhardt, W. O.,** Composition and concentration of lymph and serum lipoproteins during fat and cholesterol absorption in dogs, *J. Biol. Chem.,* 223, 838, 1958.

59. **Glomset, J. A. and Norum, K. R.,** The metabolic role of lecithin: cholesterol acyl transferase: perspectives from pathology in *Advances in Lipid Research,* Vol. 2, Paoletti, R. and Kritchevsky, D., Eds., Academic Press, New York, 1968, 1.

60. **Rutenberg, H. L., Lacko, A. G., and Soroff, L. A.,** Inhibition of lecithin: cholesterol acyl transferase following intravenous injection of heparin in man, *Biochim. Biophys. Acta,* 326, 419, 1973.

61. **Kooyman, G. L., Wahrenbrock, E. A., Castellini, M. A., Davis, R. W., and Simmett, E. E.,** Aerobic and anaerobic metabolism during voluntary diving in Weddell seals: evidence of preferred pathways from blood chemistry and behavior, *J. Comp. Physiol.,* 138, 735, 1980.

62. **Kooyman, G. L., Castellini, M. A., and Davis, R. W.,** Physiology of diving in marine mammals, in *Annual Review of Physiol.,* Vol 43, Edelman, T. S., Ed., Annual Revue Inc., Palo Alto, Calif., 1981, 343.

63. **Nichols, A. V.,** Human serum lipoproteins and their interrelationships, in *Advances in Biological and Medical Physics,* Vol. 2, Academic Press, New York, 1967, 110.

*Lipoproteins of Common Small Laboratory Species
and of Some Large Animals*

CHARACTERIZATION OF RAT PLASMA LIPOPROTEINS

Karl H. Weisgraber and Robert W. Mahley

INTRODUCTION

The rat has been used for many years as a model for the study of lipoprotein metabolism. Rat lipoproteins have been studied under normal physiological conditions as well as under various experimentally induced perturbations of normal metabolism, such as cholesterol-fat feeding. In most instances, the lipoproteins in these studies have been isolated by ultracentrifugation and designated using the density ranges established for human lipoproteins: very low density lipoproteins (VLDL), d < 1.006; intermediate density lipoproteins (IDL), d = 1.006 to 1.019 or 1.020; low density lipoproteins (LDL), d = 1.019 or 1.020 to 1.063; high density lipoproteins-2 (HDL$_2$), d = 1.063 to 1.125; and high density lipoproteins-3 (HDL$_3$), d = 1.125 to 1.21. It is now clear that the application of the human density ranges to isolate and classify rat lipoproteins is inadequate. As will be discussed later, this is particularly true in the low density range (d = 1.02 to 1.063), where two distinctly different lipoproteins (LDL and HDL$_1$) occur.

The first part of this chapter will review the characterization of lipoproteins from normal rat plasma, with particular emphasis on the lower density range. Although it is not our objective here to treat rat lipoprotein metabolism in detail, selected examples of either in vitro or in vivo lipoprotein metabolic behavior will be discussed when they serve to emphasize or augment specific points. The second part of the chapter will review studies of perturbations in normal rat lipoprotein metabolism, both experimentally induced and genetically determined. Topics to be covered are cholesterol-fat feeding, hypothyroidism, and a form of genetic obesity. In addition, the effect of an experimental hypolipemic drug on the lipoprotein profile of the cholesterol-fed rat will be discussed.

NORMAL RAT PLASMA LIPOPROTEINS

Unlike cholesterol transport in humans, which is primarily by the LDL, most of the cholesterol in normal rats fed a commercial chow is transported by the HDL. Typical studies have shown that 45 to 65% of the plasma cholesterol in rats occurs above density 1.063.[1,2] As shown in Figure 1, paper electrophoresis of rat plasma, followed by lipid staining of the electrophoretogram, indicates that the α_1-migrating HDL are the main lipoprotein class.

Rat VLDL (d < 1.006), like the VLDL of many other species including humans, are triglyceride-rich lipoproteins with pre-β-mobility on paper electrophoretograms (Figure 1). Examples of the chemical compositions of VLDL isolated from normal rat plasma illustrate that there is some variation in VLDL composition (Table 1). Triglyceride is the most abundant component of the VLDL, ranging from 49.6 to 73.6% in these examples. Phospholipid is the next most abundant component (11.2 to 26.4%), and protein and total cholesterol comprise ~15 to 20% of the VLDL. The triglyceride and total cholesterol contents of the VLDL isolated by zonal ultracentrifugation (Table 1, Reference 5) differ from those of VLDL isolated in fixed angle rotors. The reason for this difference is not apparent. However, the variations observed in the other VLDL preparations (presented in Table 1) are probably due to differences in rat strain, in nutritional state (degree of fasting), and in individual animals within a strain.

As determined by negative staining electron microscopy, rat VLDL are spherical particles ranging in size from 200 to 1000 Å in diameter.[9] In a detailed study of size vs. chemical composition of rat chylomicrons and VLDL and their remnants, Mjøs et al. determined that

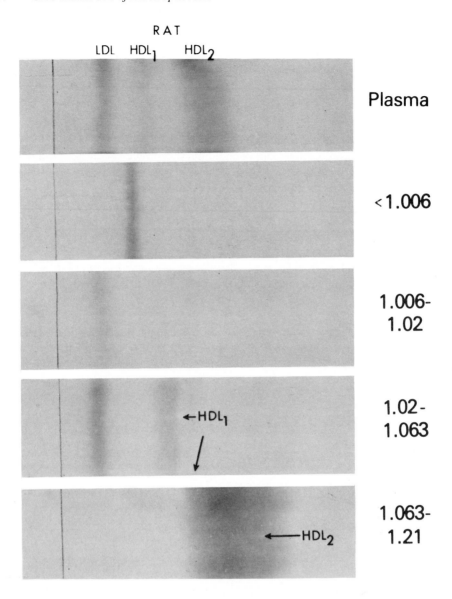

FIGURE 1.   Paper electrophoretograms of normal rat plasma and the isolated ultracentrifugal density fractions. The d = 1.02 to 1.063 fraction contains both the β-migrating LDL and the α_2-migrating HDL$_1$. (From Weisgraber, K. H., et al. *Atherosclerosis*, 28, 126, 1977. With permission.)

Table 1
PERCENT CHEMICAL COMPOSITION OF
RAT VLDL

	(a)[3]	(b)[4]	(c)[5]	(d)[6]	(e)[7]	(f)[8]
Triglyceride	53.3	73.6	49.6	61.0	58.7	70.1
Total cholesterol	5.8	4.6	22.2	13.1	9.7	7.4
Phospholipid	26.4	12.6	18.3	13.8	20.2	11.2
Protein	14.2	9.4	10.0	12.0	12.3	10.3

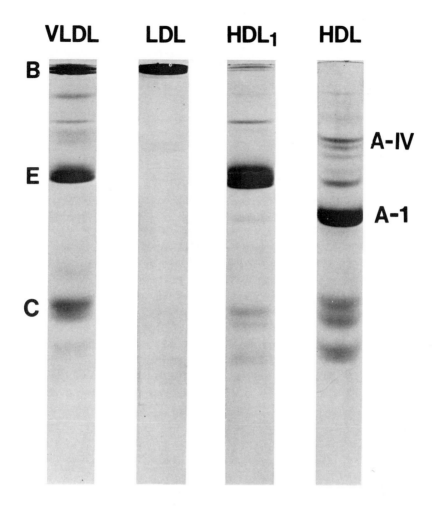

FIGURE 2.   SDS-PAGE of normal rat lipoproteins.

size correlated well with the percent volume of surface components (unesterified cholesterol, phospholipid, and protein).[4] This is consistent with the observations for different sized subfractions of human VLDL,[10] and also consistent with a ''pseudomicellar'' model of rat VLDL in which a core of nonpolar lipids, cholesteryl esters, and triglycerides is covered by a monolayer of polar lipids and protein.[4] Thus, it would appear that rat VLDL resemble human VLDL in the structural organization of their components. As shown in Figure 2, rat VLDL contain the B, E, and C apoproteins. The evidence supporting these apoprotein designations will be treated in a later section.

In fasted rats, unlike humans, the IDL fraction (d = 1.006 to 1.020) does not contain significant amounts of lipoprotein.[2] Presumably, this is due to the liver's efficient removal of VLDL remnants from circulation.[11,12,13] As a result, only 10% of the plasma VLDL passes through the lipolytic cascade via IDL to LDL.[12] This efficient removal of VLDL remnants would also account for the low LDL level in rats. This is in contrast to man, where 90% of the plasma VLDL are converted to LDL.[14]

In most studies of rat lipoproteins, the LDL density range (d = 1.02 to 1.063) established for human lipoproteins has been applied, even though this density range is known to be

Table 2
PERCENT CHEMICAL COMPOSITION AND SIZE
OF RAT LDL AND HDL$_1$

	HDL$_1$				LDL	
	(a)[16]	(b)[19]	(c)[20]	(d)[3]	(d)[3]	(c)[20]
Triglyceride	4.0	3.8	1.8	12.8	21.8	16.9
Total cholesterol	39.9	33.7	28.8	20.3	25.4	28.7
Phospholipid	26.0	26.7	37.0	34.4	22.2	27.5
Protein	30.1	30.0	32.6	32.5	30.6	27.0
Diameter, Å	192	140	200	—	—	28.0

heterogeneous in the rat (for summary see Reference 15). For example, immunochemical investigation of sequential density fractions has shown that most of the LDL occur between d = 1.04 and 1.05, with demonstrable HDL immunoreactivity below d = 1.060. The first attempt to clarify the heterogeneity of this density fraction was performed in our laboratory[16] using an approach identical to that developed for dealing with a similar situation in the dog.[17] Paper electrophoresis of the d = 1.02 to 1.063 fraction from normal rat plasma clearly demonstrated the presence of two lipoprotein classes (Figure 1), one with β-migration (LDL) and the other with α$_2$-migration, designated as HDL$_1$. Separation and purification of the LDL and HDL$_1$ were accomplished using preparative Geon-Pevikon electrophoresis[18] (see also chapter on canine lipoproteins by Mahley and Weisgraber). A lipoprotein with properties similar to those of HDL$_1$ has also been isolated by zonal ultracentrifugation in two other laboratories.[19,20] A summary of the chemical compositions and sizes of HDL$_1$ and LDL isolated by either Geon-Pevikon block electrophoresis or zonal ultracentrifugation is shown in Table 2.

As indicated, HDL$_1$ are cholesterol- and phospholipid-rich lipoproteins that contain approximately 30% protein. The amount of triglyceride contained in HDL can vary over a wide range, which is most likely reflective of the nutritional state of the animals. We have found that the level of HDL$_1$ in the d = 1.02 to 1.063 density fraction varies and ranges from 25 to 50% of the lipoprotein protein in this fraction. This is in agreement with HDL$_1$ levels determined by zonal ultracentrifugation.[20] Thus, HDL$_1$ can account for approximately one half of the cholesterol in the 1.02 to 1.063 density fraction.

Because HDL$_1$ and LDL are similar in size and composition, it is not possible to distinguish them or determine their relative levels either by negative staining electron microscopy or chemical composition. Thus, when a detailed characterization of lipoproteins in this density range is required, the question of HDL$_1$ levels must be considered. This is particularly important when comparing various metabolic perturbations in the rat and relating them to changes in control rats. As shown in Figure 2, HDL$_1$ are characterized by the presence of apo E as the major apoprotein component. As will be discussed, this has important metabolic consequences. By contrast, the LDL isolated by the Geon-Pevikon electrophoresis method contain only apo B (Figure 2).

The HDL (d = 1.063 to 1.21) of normal rats are protein-rich lipoproteins with α$_1$-migration on paper electrophoresis (Figure 1), similar to the HDL studied in other mammals. As with other lipoprotein classes in the rat, chemical compositions can vary (Table 3). Like human HDL, in which d = 1.063 to 1.125 and d = 1.125 to 1.21 define the HDL$_2$ and HDL$_3$ subclasses, rat HDL also appear to have two major subclasses defined approximately by these same densities, each with distinct chemical compositions (Table 3). The rat d = 1.125 to 1.21 HDL consist of spherical particles with diameters of 50 to 90 Å containing 46 to 51% protein.[9] The d = 1.063 to 1.125 HDL are spherical particles with an average diameter of 136 Å[9] and protein contents lower than those of the d = 1.125 to 1.21 HDL. The

Table 3
PERCENT CHEMICAL COMPOSITION OF RAT HDL

	1.063—1.21		1.063—1.125	1.125—1.21	
	(a)[21]	(b)[9]	(c)[6]	(d)[22]	(c)[6]
Triglyceride	0.6	2.8	1.8	3.2	7.6
Total cholesterol	23.3	29.8	30.0	28.6	20.6
Phospholipid	15.1	26.0	29.8	26.4	21.1
Protein	61.1	41.4	38.0	46.0	51.0

triglyceride in the HDL fraction appears to be carried by a subclass of HDL, which can be separated by heparin-Sepharose chromatography.[23] This triglyceride-rich subclass, like HDL$_1$, has apo E as a major apoprotein component.[23] From immunochemical and cholesterol analysis of sequential density fractions, it has been determined that most of the rat HDL occur at densities below d = 1.120.[2] This is consistent with earlier studies in which it was determined that rat HDL floats at a lower density than human HDL.[5,7] Rat HDL are characterized by the presence of the A-I, E, A-IV, and C apoproteins (Figure 2).

Another class of plasma lipoproteins is also present in rats. These are the large, triglyceride-rich chylomicron remnants. As in other species, these lipoproteins are synthesized in the intestine and have a very short residence time in plasma. As a consequence of their rapid turnover rate, under fasting conditions chylomicrons do not constitute a significant plasma lipoprotein class. They are usually isolated from thoracic or mesenteric duct lymph. Triglyceride is the major lipid class in chylomicrons, accounting for 90% of the mass. Approximately 5% of the mass is phospholipid. Cholesterol and protein each account for ~2.5% of the mass.[4,24] On paper electrophoretograms chylomicrons remain at the origin. Typically, chylomicrons are large, spherically shaped particles with diameters from 700 to 3000 Å.[4] The A-IV, A-I, E, and C apoproteins, as well as the low molecular weight intestinal form of apo B, are present in rat chylomicrons. Figure 3 compares the d < 1.02 lipoproteins isolated from human, swine, and dog thoracic duct lymph. These lipoproteins contain the B, A-IV, E, A-I, and C apoproteins. Note that the apo-B bands at the top of these 11% polyacrylamide gels penetrated the gels. As will be discussed later, this is a characteristic of intestinally derived apo B. The faint band migrating just above apo-A-IV in the ~60,000 M$_r$ position may represent the recently described apo A-V.[25]

A major function of chylomicrons is the transport of dietary fat. It has been determined that the triglyceride core of chylomicrons is rapidly depleted by the action of lipoprotein lipase located in the extrahepatic tissue, producing a chylomicron remnant particle. These remnants have been characterized in the rat. They are 400 to 600 Å diameter particles enriched in cholesteryl esters and apo B and depleted of phospholipids and the C apoproteins.[4] The cholesteryl esters of the remnants are rapidly removed from circulation by interaction with liver receptors.[26,27] It has been postulated that the E apoprotein associated with the remnant particle mediates the uptake of the remnant by interaction with hepatic receptors.[28,29]

SITES OF LIPOPROTEIN SYNTHESIS

In the rat, the liver and the intestine are the two major organs for lipoprotein synthesis and secretion. The evidence demonstrating the involvement of these organs in lipoprotein biosynthesis comes from several laboratories using a variety of techniques. These include electron microscopy, isolation of nascent lipoprotein particles from the Golgi apparatus or secretory vesicles, and isolated perfused organ studies. For example, it was demonstrated

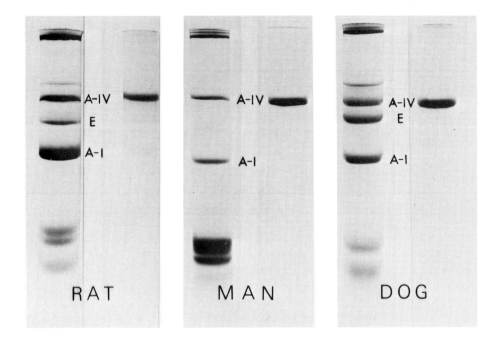

FIGURE 3.   SDS-PAGE of d < 1.02 lipoproteins isolated from rat, human, and dog thoracic duct lymph.

that the precursors of plasma VLDL could be isolated from the rat liver Golgi apparatus.[30,31] Examination of the properties of these nascent hepatic VLDL revealed that they closely resemble plasma VLDL. They float at d < 1.006 and range in size from 300 to 1000 Å in diameter. They are triglyceride rich (~60%) and their chemical compositions are essentially identical to plasma VLDL. In addition, as determined by immunochemical and polyacrylamide gel analysis, they appear to contain the same complement of apoproteins as plasma VLDL.[31] Similarly, the intestinal epithelial Golgi apparatus was shown to contain chylomicron-size particles (>800 Å).[32] A second size population of particles also has been demonstrated. These particles resembled plasma VLDL with respect to size (250 to 800 Å) and the presence of immunochemical determinants.

Studies with isolated perfused rat liver, in which LCAT is inhibited, have demonstrated that, unlike VLDL, the HDL that are secreted from the liver do not resemble their plasma counterparts.[8] The HDL secreted under these conditions are discoidal-shaped particles (46 by 190 Å) and are composed primarily of protein and polar lipids (phospholipids and unesterified cholesterol), with small amounts of cholesteryl esters. The major apoprotein component of these HDL is the E apoprotein. It has been suggested that these particles represent the nascent form of HDL secreted by the rat liver and that they are a preferred substrate for LCAT. HDL isolated from perfusion studies in which LCAT is not inhibited more closely resembled plasma HDL with regard to lipid composition, particle morphology, and particle size.[8]

RAT APOPROTEINS

In describing rat plasma apoproteins, investigators have applied the nomenclature developed for human apoproteins (i.e., apo A, apo B, apo C, and apo E). Designations are based on comparisons with human apoproteins, using one or more of the following criteria: (1) molecular weight, (2) amino acid composition, and (3) immunochemical cross-reactivity. In the earliest studies on rat apoproteins, it was noted that a major portion (>25%) of rat apo-VLDL was excluded from Sephadex gel permeation columns.[33,34] Based on behavior similar to that of apo B from human VLDL and LDL, the excluded rat protein was also designated as apo B. For many years, the fact that this protein (or proteins) did not penetrate polyacrylamide gels of greater than 10% served as the main criterion for its designation as rat apo B.

It was not until Kane et al.[35] described two forms of human apo B with different molecular weights that the similarities between rat and human apo B were more clearly demonstrated. On SDS-polyacrylamide gels, two forms of rat apo B, with molecular weights of 335,000 and 250,000, can be observed.[36,37] These forms co-migrate with their human counterparts. In addition, the amino acid compositions of the rat proteins are similar to those reported for the human proteins. However, in humans, the lower molecular weight form appears to be an exclusive product of the intestine, and thus a marker for apo-B-containing lipoproteins of intestinal origin.[35] By contrast, rat liver appears to secrete both forms.[36-38] Recent kinetic studies have indicated that the lower molecular weight form of the rat apo B is cleared from plasma faster than the higher molecular weight form, suggesting a metabolic heterogeneity of the apo B forms.[37-39]

In addition to apo B, rat VLDL has been demonstrated by Sephadex chromatography to contain other apoproteins. The early studies of Bersot et al.[34] and Koga et al.[33] indicated that two other apoprotein peaks are present, with molecular weights of ~35,000 and ~10,000, respectively. The first peak (~35,000 M_r), designated as VS-II[34] or P-II,[33] had an amino acid composition that was relatively high in arginine content.[33] The presence of another apoprotein in HDL with a similar amino acid composition was also demonstrated.[33] Later studies showed that this protein fraction contains the E, or arginine-rich, apoprotein. The ~10,000 M_r peak contains several peptides, which were later shown to be equivalent to the human C and A-II apoproteins.

One of the first detailed studies on rat apoproteins was performed by Swaney et al.[40,41] Using preparative SDS gels, they demonstrated that the major apoprotein (60% of total protein) is a 27,000 M_r peptide that appears to be homologous to human apo A-I in its molecular weight, amino acid composition, and amino terminal amino acid. In addition, the rat HDL contained 10 to 15% of a 46,000 M_r protein, designated apo-A-IV,[41] which is not normally detected in human HDL. The characterization of rat apo A-IV represents the only instance in which an apoprotein was first described in the rat before the homologous protein was described in humans, and other species.[42-44]

As it is in humans, rat apo A-IV is a major component of lymph chylomicrons (Figure 3).[45] However, in rats and humans the majority of plasma A-IV is not associated with a major plasma lipoprotein class and occurs in the d > 1.21 fraction.[46] Kinetic experiments in the rat using ^{125}I-A-IV associated with chylomicrons indicate that the A-IV is rapidly removed from the chylomicron particles as they circulate in plasma and that the A-IV passes into the d > 1.21 fraction before becoming associated with the HDL. Thus, the apo A-IV in rat HDL appears to be derived from chylomicrons and the d > 1.21 apo A-IV pool may play an important role in this process.[46] Recently, apo A-V has been described in rat chylomicrons.[25] This 59,000 M_r apoprotein has properties similar to β_2-glycoprotein-1 found in human HDL.[47]

Apolipoprotein E, formerly referred to as the arginine-rich apoprotein, also occurs in moderate concentrations (10 to 15%) in rat HDL.[41] This 35,000 M_r protein has an amino acid composition[16,41] similar to that described for human apo E. In the rat, apo E has been described in VLDL as well as HDL.[16] In addition, the E apoprotein is the predominant and sometimes exclusive apoprotein component of rat HDL$_1$.[3,16] The presence of apo E, along with low concentrations of apo A-I in HDL$_1$, would then account for the immunochemical detection of HDL antigens in the d = 1.02 to 1.063 fraction, which has been noted in previous investigations[2,15] and discussed earlier in this chapter. The presence of the apo-E-containing HDL_1 in the LDL density range has important consequences. The apo E of several species can bind to the same cell surface receptors as the apo B of LDL (for a review of the importance of this receptor system in cholesterol metabolism see Reference 48). This is also the case with rat apo E and apo B. Both the apo B of LDL isolated by Geon-Pevikon electrophoresis and the apo E of HDL_1 bind to the same receptor on cultured rat fibroblasts and smooth muscle cells.[3] By contrast, HDL subpopulations that are devoid of apo E do not interact with receptors on these cultured cell lines.[3] Analysis of normal and cholesterol-fed rats that have been injected with various ^{125}I-labeled lipoproteins containing apo E indicates that apo E freely redistributes among lipoprotein classes, with the tendency to redistribute to the lower density fractions in the hypercholesterolemic rats.[16]

As demonstrated by isoelectric focusing on polyacrylamide gels,[49] or by DEAE chromatography,[16] rat apo E is heterogeneous. It focuses into three to four isoforms with pHs between 5.31 and 5.46.[49] This heterogeneity appears to be a common feature of apo E, with multiple isoforms having been described for apo E from humans,[50,51,52] dogs,[53] and swine.[53] It has been shown that the heterogeneity of human apo E represents the result of a combination of genetic influence and the posttranslational addition of a variable number of sialic acid residues.[54] Although the possibility of genetic influence on rat apo E heterogeneity has not been explored, it appears likely that sialic acid differences contribute, at least in part, to this heterogeneity.[16] Rat apo E shows immunochemical cross-reactivity to human apo E, although the N-terminal sequence of rat apo E differs from that of human apo E.[53]

The low-molecular-weight apo C and apo A-II equivalents isolated from rat HDL have been characterized in a detailed study.[55] Rat apo A-II differs from human apo A-II in that it lacks cysteine and, as a result, exists in a monomeric form. In this regard, the rat apo A-II more closely resembles the A-II apoproteins of other species that also lack cysteine and cannot exist as disulfide-linked dimers (as does human apo A-II). In other respects, the rat apo A-II is homologous to the human protein. With the exception of cysteine content, the amino acid compositions of both are similar: each lacks histidine and tryptophan, and both have blocked amino-terminal residues.[55] The rat apo C-I and C-II also appear to be homologous to their human counterparts, based on similarities in amino acid composition and the lack of covalently linked carbohydrates. In addition, it appears that rat apo C-II activates lipoprotein lipase, as does its human equivalent.[55] Similar to human apo C-III, the rat counterpart exists in polymorphic forms, which contain varying amounts of carbohydrate. Rat C-III-0 and C-III-3, the two major forms, contain 0 and 3 mol of sialic acid per mole of protein, respectively. Swaney and Gidez have studied the rat apoproteins by isoelectric focusing on polyacrylamide gels and have developed a method by which the relative ratios of rat apo A-II, C-II, and C-III can be determined easily by the scanning of isoelectric focusing gels.[56]

Although the liver and intestine appear to be the major sources of lipoproteins in the rat,[57] the apoprotein profile that each organ secretes is quite different. The A-I[58-60,45] and A-IV[45] apoproteins are mainly synthesized by the intestine. Apoprotein E is a major product of the liver,[61,62] which also synthesizes apo A-I.[62] The intestine secretes only the lower-molecular-weight form of apo B, while the liver produces both forms.[36-38] Although apo E is associated with lymph chylomicrons and lymph VLDL, the intestine secretes little, if any, apo E.[57] Presumably, the apo E is transferred to newly secreted intestinal chylomicrons and VLDL from plasma lipoproteins that infiltrate into lymph.

CHOLESTEROL-FED RATS

Besides having a relatively low plasma cholesterol level (about 80 mg/dℓ), the rat is immune to the natural development of atherosclerosis, and particularly resistant to the induction of hypercholesterolemia by addition of dietary cholesterol alone. However, hy-

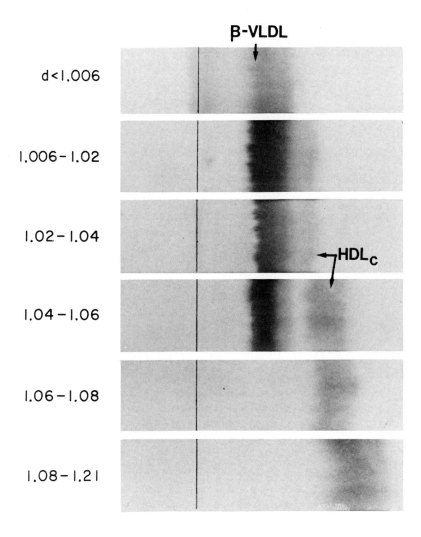

FIGURE 4.   Paper electrophoretograms of the lipoproteins in various ultracentrifugal density fractions for rats on the atherogenic diet. (From Mahley, R. W. and Holcombe, K. S., *J. Lipid Res.*, 18, 314, 1977. With permission.)

percholesterolemia sufficient to induce experimental atherosclerosis can be achieved by feeding rats diets containing cholesterol, saturated fat, bile acids, and the antithyroid drug, propylthiouracil.[63-65] Mahley and Holcombe's detailed characterization of the plasma lipoproteins of rats maintained on this atherogenic diet demonstrated that several changes in the plasma lipoprotein profile occur.[1] These changes are discussed in detail in the following paragraphs.

Plasma cholesterol levels in the range of 400 to 600 mg/dℓ can be reached after 2 to 4 weeks by feeding rats diets of laboratory chow supplemented to achieve final concentrations by weight of 5% lard, 1% cholesterol, 0.1% propylthiouracil, and 0.3% taurocholic acid. As determined by electrophoretic mobility on paper electrophoretograms, in cholesterol-fed rats several plasma lipoproteins are present in the various density fractions (Figure 4). In addition to the pre-β-migrating VLDL, the d < 1.006 fraction contains a β-migrating lipoprotein referred to as β-VLDL. Between d = 1.006 and 1.06, LDL (β-migrating) and

Table 4
PERCENT DISTRIBUTION OF LIPID AND PROTEIN
AMONG ULTRACENTRIFUGAL FRACTIONS

	Triglyceride	Total cholesterol	Phospholipid	Protein
Control[a]				
Plasma (mg/dℓ)	70	58	110	
d<1.006	97.5	14.6	29.8	9.8
1.006—1.04	1.2	5.2	4.4	5.4
1.04—1.08	1.1	26.1	16.7	15.0
1.08—1.21	0.2	54.0	49.0	69.8
Cholesterol-fed[b]				
Plasma (mg/dℓ)	194	680	488	
d<1.006	98.5	84.6	82.6	66.3
1.006—1.04	0.5	12.0	12.0	12.8
1.04—1.08	0.5	0.5	0.9	2.0
1.08—1.21	0.5	2.9	4.5	18.9
Cholesterol-fed[b]				
Plasma (mg/dℓ)	52	259	171	
d<1.006	95.8	54.2	47.8	18.0
1.006—1.02	1.8	24.8	24.2	20.6
1.04—1.08	0.1	3.7	4.6	9.2
1.08—1.21	0.1	6.9	12.1	39.3

[a] Data represent a composite from six pools of 12 to 20 rats each.
[b] Cholesterol diet. These two sets of data represent the extremes of plasma cholesterol.
    Each set of data was obtained from one pool of 12 rats.

HDL_c (α-migrating) are present. The designation HDL_c refers to a lipoprotein class that does not contain apo B but occurs in the lower density ranges in cholesterol-fed animals. The HDL_c have been described in cholesterol-fed dogs and swine.[66] In the case of the rat, the HDL_c are most likely equivalent to the HDL_1 of normal rats, described earlier in this chapter. The subscript ''c'' is used here, as it has been with other species, to designate that the plasma levels and density distribution of this lipoprotein class are affected by cholesterol feeding. The HDL_c in the d = 1.06 to 1.08 fraction migrate slightly more slowly than the HDL in the d = 1.08 to 1.21 fraction, and slightly faster than the HDL_c in the lower density fractions. Thus, the HDL_c class represents a lipoprotein fraction that spans a rather wide density range. The β-migrating lipoproteins are also prominent components of the d = 1.02 to 1.063 density range (Figure 3).

Examination of the density distribution of plasma lipids in normal and hypercholesterolemic rats reveals that several important changes occur (Table 4). Data are shown from two pools of rats, one group with plasma cholesterol levels of 680 mg/dℓ and the other with levels of 259 mg/dℓ. As shown, 54% of the plasma cholesterol and 69.8% of the lipoprotein protein are present in the HDL of control rats. These distributions are markedly altered in the cholesterol-fed animals. In the two hypercholesterolemic groups, more than 79% of the plasma cholesterol and 38.6% of the lipoprotein protein are in the d < 1.02 fraction. Comparison of the two hypercholesterolemic groups demonstrates that, as the plasma cholesterol increases, there is a further shift of cholesterol and protein to the d < 1.006 fraction. The phospholipid responds similarly. Comparable results have been obtained from euthyroid animals with a more moderate level of hypercholesterolemia.[66] The increased prominence of lower density lipoproteins as cholesterol carriers is associated with the appearance of β-VLDL in the d < 1.006 fraction. Another important change in the lipoprotein profile related to cholesterol feeding is the absolute decrease of protein in the HDL fraction (d = 1.08 to

Table 5
CHEMICAL PERCENT COMPOSITIONS OF PLASMA LIPOPROTEINS FROM CHOLESTEROL-FED RATS

Density	β-VLDL	VLDL	LDL	HDL$_c$	HDL
d<1.006					
Triglyceride	14.3				
Total cholesterol	58.0				
Phospholipid	22.2				
Protein	5.2				
1.006—1.02					
Triglyceride			0.8	0.9	
Total cholesterol			59.0	65.7	
Phospholipid			29.9	20.7	
Protein			10.2	12.6	
1.02—1.04					
Triglyceride			0.5	1.2	
Total cholesterol			63.4	42.8	
Phospholipid			21.1	41.5	
Protein			14.8	14.2	
1.04—1.08					
Triglyceride			1.0	0.9	
Total cholesterol			52.9	36.7	
Phospholipid			28.3	43.1	
Protein			17.3	19.2	
1.08—1.21					
Triglyceride					0.1
Total cholesterol					28.1
Phospholipid					28.2
Protein					43.7

1.21). Analysis of the various density fractions indicates a 1.5- to 2-fold increase in lipo-protein protein of the cholesterol-fed animals compared to control animals.

The various lipoproteins in the hypercholesterolemic rats have been isolated by a combination of ultracentrifugation and Geon-Pevikon block electrophoresis. Their chemical compositions, sizes, and apoprotein contents have been determined. As shown in Table 5, the d < 1.006 fraction, in this case almost exclusively β-VLDL, is a cholesterol-rich lipoprotein class with esterified ester making up approximately 76% of the cholesterol. This is in marked contrast to the triglyceride-rich d < 1.006 lipoproteins of control rats (Table 1). The β-migrating LDL in the d = 1.006 to 1.063 density range are also cholesteryl ester-rich lipoproteins and contain relatively little triglyceride even in the d = 1.006 to 1.02 density range. Like the HDL$_1$ of control rats, the HDL$_c$ are also rich in cholesterol, which accounts for 36.7 to 65.7% of their composition. Although there is a decrease in HDL concentration with the atherogenic diet, the composition of the HDL (d = 1.08 to 1.21) is similar to that of control HDL (Table 3). A summary of the sizes of the lipoproteins from hypercholesterolemic rats as determined by negative staining electron microscopy is presented in Table 6. The β-VLDL average diameter is 450 Å, ranging from 350 to 700 Å. The sizes of the LDL and HDL$_c$ are increased, compared with those of the LDL and HDL$_1$ from control rats (Table 2). The sizes of the hypercholesterolemic HDL are similar to those of control HDL.

Significant apoprotein changes also occur in the lipoproteins of cholesterol-fed vs. control, chow-fed rats. As determined by rocket immunoelectrophoresis, plasma apo E levels rise significantly with the atherogenic diet, increasing from the control level of 30.9 ± 6.0 mg/dℓ to 47.2 ± 13.0 mg/dℓ. In β-VLDL, the apo E content is markedly increased and the apo C content decreased relative to that of control rat VLDL (Figure 5 and Figure 2). Figure

Table 6
PARTICLE SIZE (Å) BY NEGATIVE STAINING[a]

Density	β-VLDL	VLDL	LDL	HDL$_c$	HDL
1.006	350—700 (450)	300—850 (500)			
1.02			300—400 (350)	250—450 (325)	
1.04			250—350 (310)	200—325 (275)	
1.08			185—300 (250)	125—200 (175)	
1.21					80—120 (100)

[a] The diameters of approximately 200 particles were measured. Data are presented as the range, with the mean diameter in parentheses.

5 also demonstrates that the E apoprotein is a prominent apoprotein component of the other cholesterol-rich lipoproteins, LDL and HDL$_c$, in the d < 1.06 density range. These results suggest a role for apo E in cholesterol metabolism. It is noteworthy that the hypercholesterolemic rat HDL (d = 1.08 to 1.21) do not contain apo E. Apoprotein A-I is also more prominent in the lower density region in the cholesterol-fed rat. The shift of apo E and A-I to the lower density regions most likely accounts for the immunochemical detection of HDL in the d = 1.006 to 1.030 fraction, as described in the cholesterol feeding studies of Lasser et al.[2]

The qualitative impression from SDS gels that the apo E distribution is shifted from the HDL to the lower density lipoproteins in cholesterol-fed rats was verified when the distribution of apo E was quantitated by rocket immunoelectrophoresis. Table 7 shows that in control rats less than 10% of the apo E is found in the d < 1.006 fraction, as compared to >50% in the d = 1.063 to 1.21 fraction (HDL), with the remainder in the d = 1.02 to 1.063 density fraction. The apo E represents 4 to 15% of the total d < 1.006 protein, 0 to 6% of the d = 1.006 to 1.02 protein, 7 to 30% of the d = 1.02 to 1.063 protein (LDL and HDL$_1$), and 8 to 15% of the d = 1.063 to 1.21 protein (HDL).

Both the distribution and concentration of the apo E changes markedly with cholesterol feeding (Table 7). In cholesterol-fed rats, greater than 95% of the apo E is associated with lipoproteins below d = 1.063, as contrasted to less than 50% in the controls. The apo E represents 21 to 46% of the total d < 1.006 protein (β-VLDL and VLDL), 29 to 58% of the total d = 1.006 to 1.02 protein, 20 to 51% of the total d = 1.02 to 1.063 protein (LDL and HDL$_c$), and only 1 to 5% of the total d = 1.063 to 1.21 protein (HDL).

However, there is an important consideration that must be mentioned when comparing the distribution of apo E among lipoprotein fractions. It is apparent from Table 7 that a significant portion of the plasma apo E was not recovered with the four centrifugally isolated lipoprotein fractions. Although the d > 1.21 fraction was not analyzed in these experiments, in subsequent experiments in which a 60-Ti rotor at 59,000 r/min was also used to isolate the lipoproteins, approximately 40% of the plasma apo E could be measured in the d > 1.21 fraction. Losses of apo E from plasma lipoproteins could be minimized by centrifugation at 39,000r/min in a 40 rotor. In these studies, approximately 23% of the plasma apo E from controls and 10% from the cholesterol-fed animals was recovered in the d > 1.21 fraction.[1] Loss of apo E from lipoproteins in centrifugation has been extensively studied by others.[67]

The β-VLDL of rats bear a striking resemblance to the β-VLDL associated with human Type III hyperlipoproteinemia. Both are cholesteryl ester-rich lipoproteins with the E apo-

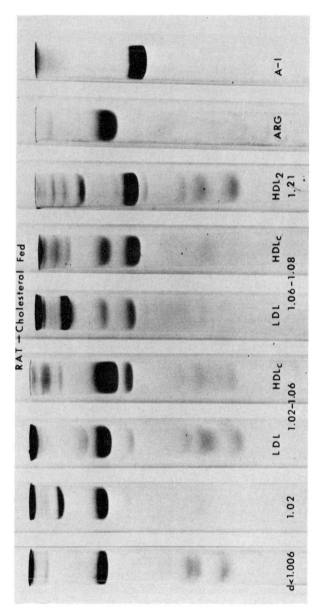

FIGURE 5.   SDS-polyacrylamide gels of the apolipoproteins in the isolated ultracentrifugal density fractions from rats on the atherogenic diet. The LDL and HDL$_c$ were purified by Geon-Pevikon electrophoresis prior to apoprotein analysis.

Table 7
MEAN CONCENTRATIONS[a] OF THE E
APOPROTEIN IN RAT PLASMA
LIPOPROTEINS

Density Fraction	Control[b] (mg/dℓ)	Cholesterol-fed[b] (mg/dℓ)
d<1.006	0.4 ± 2.6	5.2 ± 3.3
1.006—1.02	0.01 ± 0.02	4.8 ± 1.6
1.02—1.063	1.9 ± 1.4	17.2 ± 6.9
1.063—1.21	4.0 ± 1.6	0.77 ± 0.58
Total	6.2 ± 2.6	24.7 ± 7.3
Plasma	29.0	47.0

[a]   Data expressed as mean concentration ± SD.
[b]   Data determined from six pools of 12 rats each.

protein present as a prominent protein component.[68] The β-VLDL induced by cholesterol feeding in several species possess a very interesting and possibly very important metabolic property. When incubated with mouse peritoneal macrophages, the β-VLDL of rats, dogs, rabbits, and monkeys cause massive accumulation of cholesteryl esters in the macrophages.[69,70] The β-VLDL are the only naturally occurring lipoproteins to possess this ability.[70] The significance of this observation takes on added importance when it is considered that the appearance of β-VLDL, both in animal models and in Type III subjects, is associated with accelerated atherosclerosis. In addition, the lipid-laden macrophages bear a close resemblance to the foam cells of atherosclerotic lesions.

The increase in plasma apo-E levels with the atherogenic diet could be the result of increased synthesis, decreased catabolism, or a combination of both. To address this question, Wong and Rubenstein have studied the turnover of apo E in the d < 1.21 lipoprotein fraction of both normal and cholesterol-fed rats.[71] These types of studies are subject to several potential pitfalls, such as physical dissociation of apo-E from lipoproteins by ultracentrifugation or different kinetic behavior of the various apo-E-containing lipoproteins. Nevertheless, the data would indicate that the elevated apo E levels in hypercholesterolemic rats are the result of an increase in apo E secretion rather than a decrease in catabolism. This is consistent with the observation of other investigators using isolated perfused livers or liver slices from cholesterol-fed animals.[72-74] In a more direct approach at looking at apo E production, Lin-Lee et al. have compared the apo-E mRNA activities in normal and hypercholesterolemic rat livers. They have determined that the apo-E mRNA activity in cholesterol-fed rats is increased approximately twofold over normal levels.[75] Although these results suggest that increased apo-E mRNA activity is associated with increased cholesterol feeding, a conclusive answer to the question of the effect of cholesterol feeding on apo-E mRNA levels requires quantitative determination of the message for apo E.

There are several possibilities that might explain the origin of the abnormal lipoproteins that accumulate in the plasma of cholesterol-fed rats. The increased absorption of dietary lipids could overload the remnant removal mechanisms, resulting in an accumulation of remnants from intestinally synthesized lipoproteins. Alternatively, these lipoproteins may be directly synthesized by the liver and intestine. A combination of causes might also be possible. To address this question, isolated liver perfusion studies and studies on the nascent hepatic lipoproteins contained in the Golgi apparatus and secretory vesicles have been performed. Compared to isolated perfused livers from control rats, livers from cholesterol-fed rats secrete significantly more cholesterol than triglyceride.[74] This has a dramatic effect on the composition of the secreted VLDL and results in a VLDL particle that is cholesteryl ester rich rather than triglyceride rich.[74] Presumably, the cholesteryl ester is packed into the core of these particles in place of triglyceride. These results are consistent with the observation that atherogenic diets only moderately elevate plasma triglyceride levels, but have a marked effect on plasma cholesterol levels.

Studies of the composition of nascent lipoproteins isolated from the hepatic Golgi and secretory vesicles of rats have also demonstrated that the VLDL are enriched in cholesteryl ester.[76,77] It is noteworthy that the migration of these VLDL on agarose electrophoresis is retarded compared to that of normal VLDL. It has also been demonstrated that cholesterol-rich lipoproteins floating in the IDL and LDL range are secreted by the livers of cholesterol-fed rats.[74] Thus, it appears that at least a portion of the plasma β-VLDL arises as a result of a switch from synthesis and secretion of triglyceride-rich lipoproteins to cholesterol-rich lipoproteins. In addition, the direct secretion of abnormal lipoproteins in the density range d = 1.006 to 1.063 would also contribute to the increased prominence of these lipoproteins in the hypercholesterolemic rat. Cholesterol feeding also affects the distribution of cholesterol in the lymph, suggesting that the intestine may be contributing cholesterol directly to the d = 1.006 to 1.02 fraction and β-VLDL to the d < 1.006 fraction.[78]

Studies done by Ross and Zilversmit on cholesterol-fed rabbits, using retinol as a marker for intestinally produced lipoproteins, clearly indicate that the majority of the cholesterol in the d < 1.006 fraction arises from intestinal remnants.[79] This is at variance with retinol studies done by Melchior et al. in cholesterol-fed dogs, in which intestinally produced lipoproteins were rapidly cleared from the plasma in both normal and cholesterol-fed animals, with no accumulation in the d < 1.006 fraction.[80] While these differences might be explained by species differences, it is possible that the degree and duration of the hypercholesterolemia was also a factor. Central to this discussion is a study by Kris-Etherton and Cooper.[81] They demonstrated that although the plasma residence time of chylomicron remnants in cholesterol-fed animals was increased relative to that of control animals, the ability of isolated perfused rat livers to remove remnant from circulation was not impaired. These results suggest that the delayed clearance rate of remnants observed in vivo was the result of an overload of the clearance mechanism without a removal defect. In addition, as discussed above, Kris-Etherton and Cooper have demonstrated that the cholesterol-fed rat liver secretes a cholesteryl ester-rich β-VLDL. Based on their studies, they have postulated that the etiology of hyperlipoproteinemia in the rat is the result of a combination of several factors.[81]

By integrating the observations discussed in the preceding paragraphs, it is possible to speculate on the underlying factors involved in the development of hypercholesterolemia in the rat. With the initiation of a cholesterol-fat diet, the intestine would absorb and secrete more lipid into the plasma than it would with the control diet. Initially, the remnants of these particles would be cleared normally, but with time they would begin to accumulate as the removal system became saturated. Evidence of remnant saturation in cholesterol-fed animals comes from the studies of Redgrave et al. in rats,[82] as well as the Ross-Zilversmit studies in rabbits.[79] At this point the hypercholesterolemia would be moderate and most of the d < 1.006 cholesterol would result from intestinal remnants. This would be consistent with the studies of Ross and Zilversmit in the rabbit, since those studies showed that rabbits fed cholesterol for relatively short periods of time developed moderately elevated plasma cholesterol levels. With increased time on the diet, the liver would gradually become saturated with cholesterol as the normal pathways for cholesterol excretion, i.e., bile acid production, became saturated. The liver would then begin to secrete the abnormal cholesterol-rich β-VLDL and other cholesterol-rich lipoproteins. The lipoprotein protein and cholesteryl ester increase in the 1.063 density range then would be the result of a combination of effects: intestinal remnants, direct secretion of abnormal lipoproteins into this density range by both the liver and intestine, and lipolytic products of abnormal lipoproteins, i.e., β-VLDL conversion to cholesteryl ester-rich LDL. This scheme is speculative and remains to be proved.

Table 8
CHARACTERISTIC CHANGES IN RAT LIPOPROTEINS
WITH THE ATHEROGENIC DIET

Appearance of β-VLDL in the d<1.006 fraction
An increased prominence of the E apoprotein in the lower density lipoproteins
The presence of HDL_c in the d<1.08 fraction
A decrease in the typical HDL

At present, there is no direct experimental evidence that establishes the origin of the HDL_c. As discussed earlier, the HDL_c are analogous to the HDL_1 of control rats. Both lipoproteins are cholesterol rich, contain apo E as the major protein component, and are found in similar density ranges. They may originate from chylomicron catabolism. Studies in the dog, using retinol as a marker of intestinal core lipids, demonstrate that the core of HDL_c in the dog is not derived from chylomicrons making it unlikely that the HDL_1 arise from chylomicron metabolism.[80]

Another possible source of the HDL_1 and HDL_c is direct secretion by the liver or the modification of a secreted hepatic lipoprotein. Normal rat liver secretes large amounts of apo E into the HDL density range (d < 1.063) in the form of nascent lipoprotein particles that appear as discoidal particles composed of protein, phospholipid, and unesterified cholesterol, with only small amounts of polar core lipids.[8] It has been postulated that these discoidal HDL are capable of interacting with the plasma membranes of peripheral cells, acquiring cholesterol that is then esterified by LCAT and transferred into the core of the HDL.[8] These modified nascent lipoproteins may be the source of HDL_1.

Alternatively, HDL_1 might arise from the addition of unesterified cholesterol, derived from peripheral cells, to a typical HDL particle (for a detailed discussion of this mechanism see chapter on canine plasma lipoproteins by Mahley and Weisgraber). It is postulated that the newly acquired cholesterol is esterified by LCAT and transferred to the core of the particle. During this process, the lipoproteins become enriched with apo E. The addition of apo E targets the cholesterol for delivery to the liver for elimination. This process would then provide a route for cholesterol transport from peripheral cells to the liver and would give rise to HDL_1.

Presumably, the packing of cholesteryl esters into the core of the typical HDL would cause them to become less dense. The low concentrations of apo E in HDL above d > 1.063, and the absolute decrease in HDL in the cholesterol-fed rat, are consistent with the model explained in the previous paragraph. With cholesterol feeding there would be an increased influx of cholesterol into the system, generating an increase in demand for its elimination. In this situation, the apo-E-containing lipoproteins would be packed to capacity with esterified cholesterol, causing them to become less dense than they do in the normal rat. The distribution of apo E would gradually shift out of the d > 1.063 fractions. Another possibility for the shift of apo E to the lower densities is its transfer to β-VLDL. It has been demonstrated that when ^{125}I-HDL (containing apo E) from normal rats is injected into cholesterol-fed rats, the apo-E label redistributes to both β-VLDL and HDL_c.[16]

In summary, several changes occur in the plasma lipoproteins of rats fed an atherogenic diet (Table 8). The major changes include: (1) the appearance of the cholesteryl ester-rich β-migrating β-VLDL in the d < 1.006 fraction, (2) an absolute increase in the plasma concentration of the lower density lipoproteins, (3) the increased prominence of the E apoprotein in the lower density lipoproteins, (4) the presence of HDL_c in the d < 1.08 fractions, and (5) the decrease in the typical HDL (above d = 1.063). Several of these changes have been demonstrated in cholesterol feeding in other species as well,[65] including euthyroid dogs,[83] two species of swine,[66,84] rabbits,[85] and the Patas monkey.[66]

Table 9
PERCENT DISTRIBUTION OF TOTAL PLASMA
CHOLESTEROL IN CONTROL, UNTREATED-DIET,
AND TREATED-DIET RATS[a]

	Control	Atherogenic diet	Diet + drug
Experiment I			
Plasma (mg/dℓ)	54	327	140
d < 1.006	15.5	23.5	8.4
1.02	3.7	32.6	1.9
1.063	19.9	38.4	4.1
1.21	60.9	5.5	85.6
Experiment II			
Plasma (mg/dℓ)	74	326	327
d < 1.006	9.9	20.5	1.3
1.02	2.7	38.1	3.2
1.063	39.9	28.2	63.6
1.21	47.4	13.1	31.3

[a] 20 rats per group for each experiment.

EFFECT OF DRUG INTERVENTION ON LIPOPROTEIN AND APO-E LEVELS

As seen in the previous section, an atherogenic diet caused marked changes in rat plasma lipoproteins and apo E levels. Because of these changes, it was of interest to determine if an intervention regimen, other than simply returning the rats to normal laboratory chow, would reverse or alter any of the changes.*

In the experimental protocol, Osborne-Mendel rats were divided into three groups of 20 rats each. One group served as a control, and the two other groups were fed the atherogenic diet for 11 days. Drug treatment was then initiated and, in one of the diet groups, was continued for 10 days. The drug was administered by gastric intubation at a dose of 75 mg/ kg of body weight. During this treatment period, the atherogenic diet was continued in both diet groups. Plasma was obtained on the 21st day of the experiment from rats fasted 6 to 8 hr, and the lipoproteins were characterized. In two experiments, plasma cholesterol levels increased from an average of 64 mg/dℓ in control animals to a mean of 326 mg/dℓ in the diet group fed the atherogenic diet. This represents an increase of 250%.

The HDL (d > 1.063 to 1.21) of the control rat carried most of the plasma cholesterol. Following cholesterol feeding, the cholesterol distribution was shifted into the lower density fractions with less than 15% remaining in the HDL fraction (Table 9). Treatment of the cholesterol-fed animals with the Upjohn drug resulted in a remarkable redistribution of the cholesterol that more closely paralleled the distribution in control animals (Table 9), i.e., 1 to 8% of the total plasma cholesterol was in the VLDL and 1 to 3% in the IDL, with the bulk of the cholesterol redistributed to the d = 1.02 to 1.063 fraction and HDL. It is noteworthy that this redistribution from the d < 1.006 fraction to higher density lipoproteins occurred whether the plasma cholesterol was reduced or not (Experiment I vs. II, Table 9). In addition, the amount of cholesterol redistributed to the HDL fraction compared with the amounts present in the d = 1.02 to 1.063 fraction appeared to correlate with the degree to which the total plasma cholesterol had been reduced by drug treatment. For example, in

* In the studies to be described, rats on the atherogenic diet were treated with an experimental hypolipemic drug (Upjohn No. U792), which was kindly supplied by Dr. Charles Day of the Upjohn Company, Kalamazoo, Mich. These studies were performed in collaboration with Mrs. Kathleen Holcombe.

Experiment I, in which the level was reduced from 327 to 140 mg/dℓ, 85.6% of the plasma cholesterol was in the HDL and 4.1% in the d = 1.02 to 1.063 fraction. By comparison, in Experiment II the cholesterol level (327 mg/dℓ) was not reduced by drug treatment. In these studies, 31.3% of the cholesterol was in the HDL fraction while 63.6% was in the d = 1.02 to 1.063 fraction.

One study also revealed notable changes in the distribution of phospholipid after drug treatment of cholesterol-fed rats. Of the total plasma phospholipid concentration of 72 mg/ dℓ in the cholesterol-fed group, 38% was found in the VLDL, 19% in the IDL, 34% in the d = 1.02 to 1.063 fraction, and 9% in the HDL$_1$, as compared with 1.2, 0.8, 65, and 33% in the corresponding fractions from the drug-treated animals of this group. However, the drug redistributed rather than decreased phospholipid. In the example given above, the total plasma phospholipid was increased from 72 to 131 mg/dℓ.

In Figure 6, the paper electrophoretograms of control, cholesterol-fed, and drug-treated rats are shown for comparison. They reveal a number of striking changes in the lipoprotein profile of rats treated with drugs. The β-VLDL in the d < 1.006 fractions, characteristic of a cholesterol-fat diet, disappeared after drug treatment. In its place was a faint pre-β-migrating band whose migration pattern closely resembled that of normal rat VLDL (Figure 1). A faint β-migrating band was present in the IDL fraction of the drug-treated group, similar to that of the control group, but in marked contrast to the untreated, cholesterol-fed group, in which concentrations of both LDL and HDL$_c$ were high.

Drug treatment also altered the lipoprotein pattern of the d = 1.02 to 1.063 fraction. The α_1-migrating lipoproteins became a more prominent component. In examples shown in Figure 6, the HDL$_1$ or HDL$_c$ accounted for 30, 27, and 90%, respectively, of the lipoprotein protein in the d = 1.02 to 1.063 fraction of control, untreated-diet, and treated-diet rats. The intensity of the HDL band also increased with drug treatment, resembling that of normal HDL.

As we have seen, the E apoprotein plasma concentration and distribution are altered with cholesterol feeding. Thus, it was of interest in the light of the changes caused by the drug treatment to measure the apo E levels. The results, as measured by two-dimensional immunoelectrophoresis,[1] are shown in Table 10. Two groups of drug-treated rats were used, one with a 50% reduction in plasma cholesterol and the other with no reduction. Compared with control animals there was a significant increase in the lipoprotein-associated apo E with the untreated, cholesterol-fed group, the level rising from a mean of ~6 mg/dℓ to ~25 mg/ dℓ. In the control animals, the majority of the apo E was present above d = 1.02. By contrast, the apo-E distribution was shifted to the lower density fractions after cholesterol feeding, with most of the apo E now appearing below d = 1.063. The change in the apo-E levels with drug treatment appeared to be related to the magnitude of the change in plasma cholesterol levels. With no change in plasma cholesterol, the lipoprotein-associated apo E increased more than twofold, while with a 50% reduction in plasma cholesterol there was little or no change. Regardless of whether the plasma cholesterol levels changed or not, the distribution pattern of apo E among the various density fractions was markedly altered after drug treatment. There was a dramatic shift of the apo E to a higher density fraction (Table 10).

To illustrate the changes occurring with the E apoprotein, and to examine the other apoproteins, SDS gels comparing the control, cholesterol-fed, and drug-treated density fractions are presented in Figure 7. The E apoprotein labeled here as ARG increased significantly in the d = 1.21 fraction after drug treatment. It also increased in the d = 1.02 to 1.063 fraction. This was the result of the HDL$_1$/HDL$_c$ increase in this fraction. With regard to the other apoproteins, there were no marked changes in their relative concentrations in the various density fractions.

In summary, when rats fed an atherogenic diet were treated with the experimental hy-

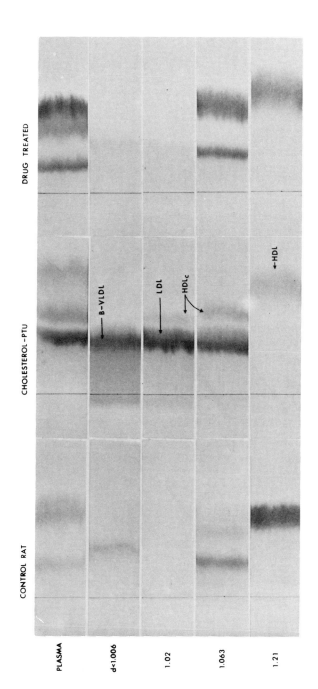

FIGURE 6. Paper electrophoretograms of plasma and the isolated ultracentrifugal density fractions from normal, cholesterol-fed, and drug-treated rats.

Table 10
CONCENTRATION OF THE E APOPROTEIN IN THE
ULTRACENTRIFUGAL FRACTIONS OF CONTROL,
UNTREATED-DIET, AND TREATED-DIET RATS

	Total	d<1.006	1.02	1.063	1.21
Control	6.3	0.4	<0.1	1.9	4.0
Cholesterol-fed	24.7	5.2	4.8	17.2	0.8
Cholesterol-fed + drug (50% cholesterol reduction)	25.1	0.1	0.6	4.4	20.0
Cholesterol-fed + drug (no cholesterol reduction)	63.3	<0.1	<0.1	38.4	24.9

polipemic drug, marked changes occurred in the lipoprotein profile. Whether the plasma cholesterol was reduced or not, the distributions of lipoprotein protein and cholesterol among the density fractions changed, coming to closely resemble those seen in control rats. In addition, the β-VLDL were no longer present in the d < 1.006 fraction; they were replaced by a normal-appearing VLDL. It is noteworthy that these changes can occur in the presence of high levels of plasma cholesterol. The distribution of the apo E associated with the various lipoprotein classes was also changed with drug treatment, more closely paralleling the distribution seen in control rats.

EFFECT OF HYPOTHYROIDISM

One factor that has not been treated in the cholesterol feeding studies is the effect of hypothyroidism. As discussed earlier, in order to induce atherogenic hypercholesterolemia in the rat, it is necessary to use antithyroid drugs, such as propylthiouracil (PTU). The question arises as to which effects are the result of cholesterol-fat feeding and which are the result of hypothyroidism. In an effort to determine the effects of hypothyroidism on plasma lipoproteins, Dory and Roheim have measured lipid and apoprotein levels in a control group of rats, and in two groups rendered hypothyroid by radiothyroidectomy or by treatment with PTU.[86] These investigations demonstrated a 50 to 100% increase in plasma cholesterol and a 20 to 40% reduction in plasma triglyceride. In a similar study, in which controls were compared with PTU-treated rats, Swift et al.[76] showed lower increases (~17%) in plasma cholesterol and no changes in triglyceride levels, although in these studies the diet of the PTU-treated rats included 5% lard.

Conflicting data exists with regard to the effect of hypothyroidism on plasma apo E levels. In the Dory and Roheim study,[86] the hypothyroid groups showed a 40 to 100% increase in apo E and apo B, and a 10 to 30% increase for apo A-I. Mahley and Holcombe have determined that apo-E levels remain unchanged in PTU-treated rats compared to normal controls.[1] The observation of increased levels of apo B in hypothyroid rats[86] would be consistent with decreased uptake of remnant VLDL.

With respect to the lipid composition of VLDL, Kris-Etherton and Cooper determined that perfusate VLDL from livers of PTU-treated rats fed a normal diet secreted triglyceride-rich VLDL, while livers from euthyroid rats fed cholesterol and fat secreted VLDL that was more cholesteryl ester rich than the hypothyroid group.[81] This has been confirmed by Swift et al.,[76] who examined the VLDL isolated from the Golgi apparatus of livers from PTU-treated rats and found that they were essentially triglyceride rich (>66%), although there was a slight increase in cholesterol content compared to the control group (13.9% vs. 9.3%). Dory and Roheim reported that the plasma VLDL of PTU-treated rats migrated intermediate between the pre-β and β positions.[86] Based on this retarded migration rate, they suggested

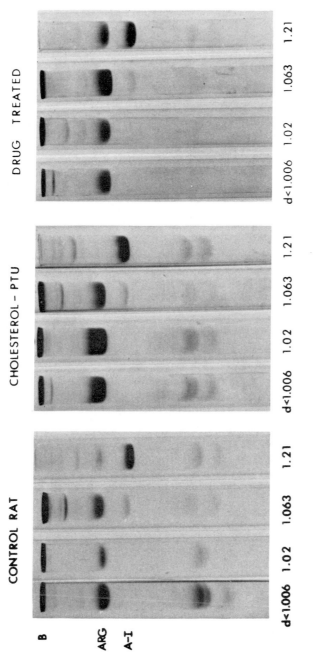

FIGURE 7. SDS-PAGE of lipoproteins from control, cholesterol-fed, and drug-treated rats.

that these lipoproteins were β-VLDL, and were the result of hypothyroidism. Although the apoprotein and lipid composition of VLDL were not presented in this study, it would appear, based on the percent of distribution of the plasma cholesterol among the various density fractions, that this lipoprotein class was not cholesteryl ester rich. This would also be consistent with the observation discussed above that PTU-treated livers secrete triglyceride-rich VLDL.[76,81] Swift et al. have reported pre-β migration for the plasma VLDL of their PTU-treated rats.[76]

Although it is clear that hypothyroidism causes secondary hyperlipoproteinemia in the rat, it does not appear to induce the presence of β-VLDL. The induction of β-VLDL is perhaps the most important characteristic change brought about by atherogenic hyperlipoproteinemia. In many species, it has been induced in euthyroid animals.[62] Thus, the hypothyroid, cholesterol-fed rat remains an important animal model in which the metabolic properties of β-VLDL can be further delineated. There is potential for such studies to provide insight into the role that β-VLDL plays in the development of atherosclerosis.

CORPULENT RATS

A strain of genetically obese rats that spontaneously develop an endogenous hyperlipidemia and hypertension has been developed by Koletsky.[87,88] The hyperlipidemia is characterized by marked hypertriglyceridemia and moderate hypercholesterolemia, and is reported to be associated with atherosclerosis.[87,88] Obesity in these rats is the result of increased food consumption of normal laboratory chow. Typically, between 2 and 5 months of age, these rats consume two to four times the amount of food that normal rats do.[89] In order to characterize the hyperlipidemia associated with this strain, a program was initiated to study the lipoproteins.*

The preliminary characterization of the corpulent rat lipoproteins was carried out with pooled plasma from both male and female rats. Plasma levels of triglyceride and cholesterol ranged from 169 to 332 and 130 to 220 mg/dℓ, respectively. Paper electrophoresis of the plasma (Figure 8) revealed that the corpulent rat lipoproteins qualitatively resembled those of normal rats, with the HDL (HDL$_2$) present as the major lipoprotein class (see Figure 1). However, the α$_2$-migrating HDL$_1$ were much more prominent than in normal plasma. The LDL were a minor lipoprotein class, as they are in normal rat plasma. Examination of the various density classes by paper electrophoresis showed that the d < 1.006 fraction contained only pre-β-migrating VLDL with no detectable β-VLDL. The intermediate density fraction (d = 1.006 to 1.02) contained two lipoprotein classes, one the β-migrating LDL and the other the α$_2$-migrating HDL$_1$. It is noteworthy that the d = 1.02 to 1.063 density fraction contained HDL$_1$ as the most prominent lipoprotein. This is in contrast to normal rats, where HDL$_1$ usually comprise 25 to 50% of the lipoprotein protein. The α$_1$-migrating HDL$_2$ were present above d = 1.063.

The protein distribution among the various lipoprotein fractions for two pools, and the cholesterol distribution for one pool, are shown in Table 11. The plasma triglyceride and cholesterol levels for pools 1 and 2 were 286 and 186, and 277 and 142 mg/dℓ, respectively. Although the plasma lipid levels were elevated, compared to controls, the protein and cholesterol distributions in the corpulent rats, unlike cholesterol-fed rats, were similar to those of normal rats. However, the corpulent rats had an increase in the total amount of lipoprotein protein as compared to normal rats. The average value for six pools of normal

*   These characterization studies were done in collaboration with Mrs. Kathleen Holcombe. The rats, referred to as "corpulent rats", were bred and raised by Dr. Carl T. Hansen, Division Research Services, Veterinary Resources Branch, National Institutes of Health, Bethesda. His essential contribution to these studies is gratefully acknowledged.

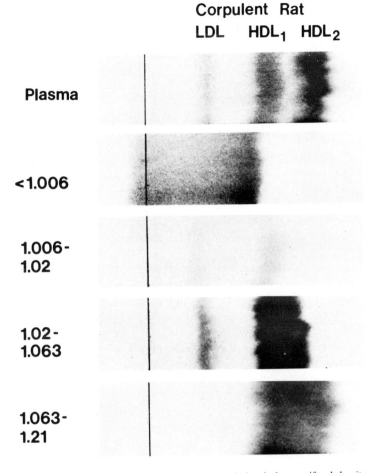

FIGURE 8. Paper electrophoresis of plasma and the isolated ultracentrifugal density fractions from corpulent rats.

Table 11
DISTRIBUTION OF PROTEIN AND CHOLESTEROL IN CORPULENT RATS[a]

	Protein (%)		Cholesterol (%)
Density	Pool 1	Pool 2	(Pool 2)
d<1.006	11.3	7.2	8.3
1.006—1.02	0.4	0.3	0.3
1.02—1.063	7.8	15.3	24.2
LDL	—	1.3	1.6
38 HDL$_1$	—	14.0	22.6
1.063—1.21	80.6	77.2	67.2

[a]   Six rats per group.

rats was 59.6 ± 10.5 mg/dℓ, compared to an average of 187.5 ± 39.9 mg/dℓ for seven pools of corpulent rats. This represents a 3.1-fold increase in the amount of lipoprotein protein in the corpulent rats.

As was suggested by paper electrophoresis of the d = 1.02 to 1.063 fraction, the HDL$_1$ were shown to account for the bulk of the lipoprotein protein (91.5%) in this density fraction

Table 12
PERCENT CHEMICAL COMPOSITION OF CORPULENT RAT LIPOPROTEINS

Lipoprotein	Triglyceride	Total cholesterol	Phospholipid	Protein
VLDL	74.7	3.5	16.2	5.6
LDL	33.0	16.5	25.6	24.8
HDL$_1$	0.8	28.0	39.1	32.0
HDL	0.7	20.6	35.2	43.5

when the LDL and HDL$_1$ were isolated by Geon-Pevikon block electrophoresis. This was clearly different than the case for control rats and underscores the need to isolate and characterize the rat LDL and HDL$_1$ separately. The d = 1.02 to 1.063 fraction cannot be treated as homogeneous for either detailed characterization or for metabolic studies. In addition, the HDL$_1$ in corpulent rats were major carriers of plasma cholesterol. For example, it was calculated for pool 2 that 22.6% of the plasma cholesterol was transported in the HDL$_1$.

The chemical compositions for the isolated and purified lipoproteins are presented in Table 12. The VLDL were triglyceride rich and contained slightly more triglyceride and one half to one third less protein than do normal VLDL (see Table 1). The LDL also contained more triglyceride and less protein, as well as less cholesterol, than do normal LDL. The HDL$_1$ composition was similar to that of normal HDL$_1$; they were cholesterol- and phospholipid-rich lipoproteins containing about one third of their mass as protein. The corpulent rat HDL were protein-rich lipoproteins (43.5%) with a low concentration of triglyceride. Overall, their composition resembled that of normal HDL.

As indicated above, the preceding data were obtained from pools of males and females. When the breeding colony was expanded, more rats became available and it became possible to examine the lipoproteins of the females and males separately. From four studies of females and five studies of males, the plasma triglyceride and cholesterol levels for females were determined to be 327.1 ± 22.8, and 176.7 ± 30.0 mg/dℓ, respectively, with the values for males being 137.7 ± 61.1, and 186.6 ± 8.1, respectively. As indicated, the males tended to have lower plasma triglyceride levels than the females, with both having comparable cholesterol levels. However, paper electrophoresis of the plasma and density fractions did not reveal any sex differences; the electrophoretograms were essentially identical to those shown in Figure 8. The distribution of lipid and lipoprotein protein in the various density classes was essentially identical in both males and females (Table 13). The majority (>96%) of the triglyceride was found in the d < 1.006 fraction and greater than 90% of the cholesterol occurred above d = 1.02. In all of the male and female pools, more than 90% of the lipoprotein protein in the d = 1.02 to 1.063 fraction was HDL$_1$, making this a major cholesterol carrying lipoprotein class in these rats. The compositions of the various lipoproteins did not show a sex difference and were essentially identical to those presented in Table 12 from the pooled male and female studies.

The sizes of the various lipoprotein classes were determined by negative staining electron microscopy and the average diameters are presented in Table 14. The VLDL were spherical particles ranging in size from 300 to 1100 Å with an average diameter of 600 Å. This diameter was significantly larger than that for VLDL from control rats.[9] The d = 1.006 to 1.02 fraction was more uniform in size, with the majority of the particles ranging in size from 250 to 400 Å, with an average diameter of 340 Å. The LDL isolated by Geon-Pevikon block electrophoresis ranged in size from 200 to 325 Å, and the average diameter of 250 Å was similar to that of LDL isolated from the control d = 1.02 to 1.063 fraction (Table 2). The HDL$_1$ from the corpulent rat were also similar in size to control rat HDL$_1$ (220 Å vs. 200 Å). The size of the HDL (d = 1.063 to 1.21) was also nearly identical with that reported for control rats (~140 Å)[8] but with a broader range.

Table 13
DISTRIBUTION OF TRIGLYCERIDE, CHOLESTEROL, AND
LIPOPROTEIN PROTEIN IN CORPULENT RAT LIPOPROTEINS

Sex/Density range	Cholesterol	Phospholipid	Protein	Female[a]
Plasma	308	158		
d<1.006	98.7	9.3	23.0	8.0
1.006—1.02	0.7	0.1	0.8	28.0
1.02—1.063	0.6	32.3	19.9	15.8
1.063—1.21	<.1	57.9	56.3	73.4
Male[b]				
Plasma	125	189		
d<1.006	96.3	4.9	18.3	4.9
1.006—1.02	1.3	1.5	0.9	3.7
1.02—1.063	2.5	38.1	31.3	20.2
1.063—1.21	<.1	48.9	49.5	71.2

[a] Pool of 25 rats.
[b] Pool of 30 rats.

Table 14
PARTICLE SIZE (Å) OF
CORPULENT RAT LIPOPROTEINS
BY NEGATIVE STAINING[a]

	Av diam	Range
d<1.006	600	300—1100
d = 1.006—1.02	340	250—550
LDL	250	200—325
HDL$_1$	220	150—325
HDL	140	100—180

[a] The diameters of approximately 200 particles were measured.

The apoprotein contents of the corpulent rat lipoproteins qualitatively resembled those of control rats. In addition, there were no differences between male and female pools. A representative set of SDS gels is shown in Figure 9. The VLDL contained the B, E, and C apoproteins. Both apo B and apo E were present in the d = 1.006 to 1.02 fraction. The d = 1.02 to 1.063 fraction contained the E apoprotein as the major component. This was the result of the high concentration of HDL$_1$ in this density fraction. As is the case with HDL$_1$ from control rats, the corpulent rat HDL$_1$ contained the E apoprotein as its major protein component (Figure 8). By contrast, the LDL isolated by Geon-Pevikon block electrophoresis contained primarily the B apoprotein, as is the case with normal rat HDL isolated by this method (Figure 2). The HDL (d = 1.063 to 1.21 fraction) contained the A-I, E, A-IV, and C apoproteins and the apoprotein pattern closely resembled the normal rat HDL pattern (for comparison, see Figure 2).

In summary, several features distinguish the lipoproteins of these genetically obese rats from those of normal rats. The elevated plasma triglyceride levels of the corpulent rats are associated with increased plasma concentrations of VLDL. There is about a threefold increase in lipoprotein protein, with the major increases occurring in VLDL, HDL$_1$, and HDL. In

Corpulent Rat Apoproteins

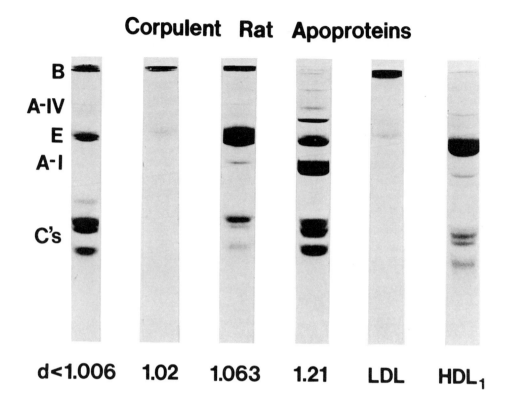

FIGURE 9.   SDS-PAGE of corpulent rat lipoproteins.

contrast to the normal rat d = 1.02 to 1.063 fraction, the HDL$_1$ in the corpulent rats are the predominant lipoproteins in this fraction. More than 20% of the plasma cholesterol in the corpulent rat is transported by HDL$_1$.

ACKNOWLEDGMENTS

We wish to acknowledge the important contributions of colleagues with whom it continues to be our pleasure to be associated: Drs. T. P. Bersot, M. Fainaru, T. L. Innerarity, R. E. Pitas, S. C. Rall, and Mrs. K. S. Holcombe. Also, thanks are expressed to Mrs. K. S. Holcombe, Mr. J. F. Andres, Mr. R. P. Levine, and Mr. R. A. Wolfe for editorial assistance and manuscript preparation.

REFERENCES

1. **Mahley, R. W. and Holcombe, K. S.,** Alterations of the plasma lipoproteins and apoproteins following cholesterol feeding in the rat, *J. Lipid Res.,* 18, 314, 1977.
2. **Lasser, N. L., Roheim, P. S., Edelstein, D., and Eder, H. A.,** Serum lipoproteins of normal and cholesterol-fed rats, *J. Lipid Res.,* 14, 1, 1973.
3. **Innerarity, T. L., Pitas, R. E., and Mahley, R. W.,** Disparities in the interaction of rat and human lipoproteins with cultured rat fibroblasts and smooth muscle cells, *J. Biol. Chem.,* 255, 11163, 1980.

4. **Mjøs, O. D., Faergeman, O., Hamilton, R. L., and Havel, R. J.,** Characterization of remnants produced during the metabolism of triglyceride-rich lipoproteins of blood plasma and intestinal lymph in the rat, *J. Clin. Invest.,* 56, 603, 1975.

5. **Wilcox, H. G. and Heimberg, M.,** Isolation of plasma lipoproteins by zonal ultracentrifugation in the B14 and B15 titanium rotors, *J. Lipid Res.,* 11, 7, 1970.

6. **Calandra, S., Pasquali-Ronchetti, I., Gherardi, E., Fornieri, C., and Tarugi, P.,** Chemical and morphological changes of rat plasma lipoproteins after a prolonged administration of diets containing olive oil and cholesterol, *Atherosclerosis,* 28, 369, 1977.

7. **Camejo, G.,** Structural studies of rat plasma lipoproteins, *Biochemistry,* 6, 3228, 1967.

8. **Hamilton, R. L., Williams, M. C., Fielding, C. J., and Havel, R. J.,** Discoidal bilayer structure of nascent high density lipoproteins from perfused rat liver, *J. Clin. Invest.,* 58, 667, 1976.

9. **Pasquali-Ronchetti, I., Calandra, S., Baccarani-Contri, M., and Montaguti, M.,** The ultrastructure of rat plasma lipoproteins, *J. Ultrastruct. Res.,* 53, 180, 1975.

10. **Sata, T., Havel, R. J., and Jones, A. L.,** Characterization of subfractions of triglyceride-rich lipoproteins separated by gel chromatography from blood plasma of normolipemic and hyperlipemic humans, *J. Lipid Res.,* 13, 757, 1972.

11. **Eisenberg, S. and Rachmilewitz, D.,** Metabolism of rat plasma very low density lipoprotein. I. Fate in circulation of the whole lipoprotein, *Biochim. Biophys. Acta,* 326, 378, 1973.

12. **Faergeman, O., Sata, T., Kane, J. P., and Havel, R. J.,** Metabolism of apolipoprotein B of plasma very low density lipoproteins in the rat, *J. Clin. Invest.,* 56, 1396, 1975.

13. **Faergeman, O. and Havel, R. J.,** Metabolism of cholesteryl esters of rat very low density lipoproteins, *J. Clin. Invest.,* 55, 1210, 1975.

14. **Eisenberg, S., Bilheimer, D. W., Levy, R. I., and Lindgren, F. I.,** On the metabolic conversion of human plasma very low density lipoprotein to low density lipoprotein, *Biochim. Biophys. Acta,* 326, 361, 1973.

15. **Mills, G. L. and Taylaur, C. E.,** The distribution and composition of serum lipoproteins in eighteen animals, *Comp. Biochem. Physiol.,* 40B, 489, 1971.

16. **Weisgraber, K. H., Mahley, R. W., and Assmann, G.,** Identification of the rat arginine-rich apoprotein and its redistribution following injection of iodinated lipoproteins into normal and hypercholesterolemic rats, *Atherosclerosis,* 28, 121, 1977.

17. **Mahley, R. W. and Weisgraber, K. H.,** Canine lipoproteins and atherosclerosis. I. Isolation and characterization of plasma lipoproteins from control dogs, *Circ. Res.,* 35, 713, 1974.

18. **Mahley, R. W. and Weisgraber, K. H.,** An electrophoretic method for the quantitative isolation of human and swine plasma lipoproteins, *Biochemistry,* 13, 1964, 1974.

19. **Danielsson, B., Ekman, R., Johansson, B. G., Nilsson-Ehle, P., and Petersson, B. G.,** Isolation of a high density lipoprotein with high contents of arginine-rich apoprotein (apoE) from rat plasma, *FEBS Lett.,* 86, 299, 1978.

20. **Lusk, L. T., Walker, L. F., DuBien, L. H., and Getz, G. S.,** Isolation and partial characterization of high density lipoprotein HDL₁ from rat plasma by gradient centrifugation, *Biochem. J.,* 183, 83, 1979.

21. **de Pury, G. G. and Collins, F. D.,** Composition and concentration of lipoproteins in the serum of normal rats and rats deficient in essential fatty acids, *Lipids,* 7, 225, 1971.

22. **Kuksis, A., Breckenridge, W. C., Myher, J. J., and Kakis, G.,** Replacement of endogenous phospholipids in rat plasma lipoproteins during intravenous infusion of an artificial lipid emulsion, *Can. J. Biochem.,* 56, 630, 1978.

23. **Quarfordt, S. H., Jain, R. S., Jakoi, L., Robinson, S., and Shelburne, F.,** The heterogeneity of rat high density lipoproteins, *Biochem. Biophys. Res. Commun.,* 83, 786, 1978.

24. **Zilversmit, D. B.,** Assembly of chylomicrons in the intestinal cell, in *Disturbances in Lipid and Lipoprotein Metabolism,* Dietschy, J. M., Gotto, A. M., Jr., and Ontko, J. A., Eds., American Physiological Society, Bethesda, Md., 1978, 169.

25. **Fidge, N. H. and McCullagh, P. J.,** Studies on the apoproteins cf rat lymph chylomicrons: characterization and metabolism of a new chylomicron-associated apoprotein, *J. Lipid Res.,* 22, 138, 1981.

26. **Sherrill, B. C. and Dietschy, J. M.,** Characterization of the sinusoidal transport process responsible for uptake of chylomicrons by the liver, *J. Biol. Chem.,* 253, 1859, 1978.

27. **Carella, M. and Cooper, D. D.,** High affinity binding of chylomicron remnants to rat liver plasma membranes, *Proc. Natl. Acad. Sci. U.S.A.,* 76, 338, 1979.

28. **Sherrill, B. C., Innerarity, T. L., and Mahley, R. W.,** Rapid hepatic clearance of the canine lipoproteins containing only the E apoprotein by a high affinity receptor. Identity with the chylomicron remnant transport process, *J. Biol. Chem.,* 255, 1804, 1980.

29. **Mahley, R. W., Hui, D. Y., Innerarity, T. L., and Weisgraber, K. H.,** Two independent lipoprotein receptors on hepatic membranes of the dog, swine, and man, the apo-B,E and apo-E receptors, *J. Clin. Invest.,* 68, 1197, 1981.

30. **Mahley, R. W., Hamilton, R. L., and LeQuire, V. S.,** Characterization of lipoprotein particles isolated from the Golgi apparatus of rat liver, *J. Lipid Res.,* 10, 433, 1969.

31. **Mahley, R. W., Bersot, T. P., LeQuire, V. S., Levy, R. I., Windmuellar, H. G., and Brown, W. V.,** Identity of very low density lipoprotein apoproteins of plasma and liver Gogli apparatus, *Science,* 168, 380, 1970.

32. **Mahley, R. W., Bennett, B. D., Morré, D. J., Gray, M. E., Thistlethwaite, W., and LeQuire, V. S.,** Lipoproteins associated with Golgi apparatus isolated from epithelial cells from rat small intestine, *Lab. Invest.,* 25, 435, 1971.

33. **Koga, S., Bolis, L., and Scanu, A. M.,** Isolation and characterization of subunit polypeptides from apoproteins of rat serum lipoproteins, *Biochim. Biophys. Acta,* 236, 416, 1971.

34. **Bersot, T. P., Brown, W. V., Levy, R. I., Windmueller, H. G., Fredrickson, D. S., and LeQuire, V. S.,** Further characterization of the apolipoproteins of rat plasma lipoproteins, *Biochemistry,* 9, 3427, 1970.

35. **Kane, J. P., Hardman, D. A., Paulus, H. E.,** Heterogeneity of apolipoprotein B: isolation of a new species from human chylomicrons, *Proc. Natl. Acad. Sci. U.S.A.,* 77, 2465, 1980.

36. **Krishnaiah, K. V., Walker, L. F., Borensztajn, J., Schonfeld, G., and Getz, G. S.,** Apolipoprotein B variant derived from rat intestine, *Proc. Natl. Acad. Sci. U.S.A.,* 77, 3806, 1980.

37. **Elovson, J., Huang, Y. O., Baker, N., and Kannan, R.,** Apolipoprotein B is structurally and metabolically heterogeneous in the rat, *Proc. Natl. Acad. Sci. U.S.A.,* 78, 157, 1981.

38. **Wu, A.-L. and Windmueller, H. G.,** Variant forms of plasma apolipoprotein B: hepatic and intestinal biosynthesis and heterogeneous metabolism in the rat, *J. Biol. Chem.,* 256, 3615, 1981.

39. **Sparks, C. E. and Marsh, J. B.,** Metabolic heterogeneity of apolipoprotein B in the rat, *J. Lipid Res.,* 22, 519, 1981.

40. **Swaney, J. B., Reese, H., and Eder, H. A.,** Polyeptide composition of rat high density lipoprotein: characterization by SDS-gel electrophoresis, *Biochem. Biophys. Res. Commun.,* 59, 513, 1974.

41. **Swaney, J. B., Braithwaite, F., and Eder, H. A.,** Characterization of the apoliproproteins of rat plasma lipoproteins, *Biochemistry,* 16, 271, 1977.

42. **Weisgraber, K. H., Bersot, T. P., and Mahley, R. W.,** Isolation and characterization of an apoprotein from the d<1.006 lipoproteins of human and canine lymph homologous with the rat A-IV apoprotein. *Biochem. Biophys. Res. Commun.,* 85, 292, 1978.

43. **Beisiegel, U. and Utermann, G.,** An apolipoprotein homolog of rat apolipoprotein A-IV in human plasma, *Eur. J. Biochem.,* 93, 601, 1979.

44. **Green, P. H., Glickman, R. R., Saudek, C. D., Blum, C. B., and Tall, A. R.,** Human intestinal lipoproteins. Studies in chyluric subjects, *J. Clin. Invest.,* 64, 233, 1979.

45. **Wu, A.-L. and Windmueller, H. G.,** Identification of circulating apolipoproteins synthesized by rat small intestine *in vivo, J. Biol. Chem.,* 253, 2525, 1978.

46. **Fidge, N. H.,** The redistribution and metabolism of iodinated apolipoprotein A-IV in rats, *Biochim. Biophys. Acta,* 619, 129, 1980.

47. **Polz, E. and Kostner, G. M.,** The binding of β_2-glycoprotein-1 to human serum lipoproteins. Distribution among density fractions, *FEBS Lett.,* 102, 183, 1979.

48. **Brown, M. S. and Goldstein, J. L.,** Receptor-mediated control of cholesterol metabolism, *Science,* 191, 150, 1976.

49. **Gidez, L. I., Swaney, J. B., and Murnane, S.,** Analysis of rat serum apolipoproteins by isoelectric focusing. I. Studies on the middle molecular weight subunits, *J. Lipid. Res.,* 18, 59, 1977.

50. **Shore, V. G. and Shore, B.,** Heterogeneity of human plasma very low density lipoproteins. Separation of species differing in protein components, *Biochemistry,* 12, 502, 1973.

51. **Utermann, G., Hess, M., and Steinmetz, A.,** Polymorphism of apolipoprotein E and occurrence of dysbetalipoproteinemia in man, *Nature (London),* 269, 604, 1977.

52. **Pagnan, A., Havel, R. J., Kane, J. P., and Kotite, L.,** Characterization of human very low density lipoproteins containing two electrophoretic populations: double pre-beta lipoproteinemia and primary dys-betalipoproteinemia, *J. Lipid Res.,* 18, 613, 1977.

53. **Weisgraber, K. H., Troxler, R. F., Rall, S. C., and Mahley, R. W.,** Comparison of the human, canine, and swine E apoproteins, *Biochem. Biophys. Res. Commun.,* 95, 374, 1980.

54. **Zannis, V. I. and Breslow, J. L.,** Human VLDL and apo-E isoprotein polymorphism is explained by genetic variation and post translational modification, *Biochemistry,* 20, 1033, 1981.

55. **Herbert, P. N., Windmueller, H. G., Bersot, T. P., and Shulman, R. S.,** Characterization of the rat apolipoproteins. I. The low molecular weight proteins of rat plasma high density lipoproteins, *J. Biol. Chem.,* 249, 5718, 1974.

56. **Swaney, J. B. and Gidez, L. I.,** Analysis of rat serum apolipoproteins by isoelectric focusing. II. Studies on the low molecular weight subunits, *J. Lipid. Res.,* 18, 69, 1977.

57. **Wu, A.-L. and Windmueller, H. G.,** Relative contributions by liver and intestine to individual plasma apolipoproteins in the rat, *J. Biol. Chem.,* 254, 7316, 1979.

58. **Windmueller, H. G., Herbert, P. N., and Levy, R. I.,** Biosynthesis of lymph and plasma lipoprotein apoproteins by isolated perfused rat liver and intestine, *J. Lipid Res.,* 14, 215, 1973.
59. **Glickman, R. M. and Green, P. H. R.,** The intestine as a source of apolipoprotein A, *Proc. Natl. Acad. Sci. U.S.A.,* 74, 2569, 1977.
60. **Imaizumi, K., Havel, R. J., Fainaru, M., and Vigne, J.-L.,** Origin and transport of the A-I and arginine-rich apolipoprotein in mesenteric lymph of rats, *J. Lipid Res.,* 19, 1038, 1978.
61. **Marsh, J. B.,** Lipoproteins in a nonrecirculating perfusate of rat liver, *J. Lipid Res.,* 15, 544, 1974.
62. **Felker, T. F., Fainaru, M., Hamilton, R. I., and Havel, R. J.,** Secretion of the arginine-rich and A-I apolipoproteins by the isolated perfused rat liver, *J. Lipid Res.,* 18, 465, 1977.
63. **Wissler, R. W., Eilert, M. L., Schroeder, M. A., and Cohen, L.,** Production of lipomatous and atheromatous arterial lesions in the albino rat, *Am. Med. Assoc. Arch. Pathol.,* 57, 333, 1954.
64. **Fillias, L. C., Andrus, S. B., Mann, G. V., and Stare, F. J.,** Experimental production of gross atherosclerosis in the rat, *J. Exp. Med.,* 104, 539, 1956.
65. **Malinow, M. R., Hojman, R. D., and Pellegrino, A.,** Different methods for the experimental production of generalized atherosclerosis in the rat, *Acta Cardiol.,* 9, 480, 1954.
66. **Mahley, R. W.,** Alterations in plasma lipoproteins induced by cholesterol feeding in animals including man, in *Disturbances in Lipid and Lipoprotein Metabolism,* Dietschy, J. M., Gotto, A. M., Jr., and Ontko, J. A., Eds., American Physiological Society, Bethesda, 1978, 181.
67. **Fainaru, M., Havel, R. J., and Imaizumi, K.,** Apoprotein content of plasma lipoproteins of the rat separated by gel chromatography or ultracentrifugation, *Biochem. Med.,* 17, 347, 1977.
68. **Havel, R. J. and Kane, J. P.,** Primary dysbetalipoproteinemia: predominance of a specific apoprotein species in triglyceride-rich lipoproteins, *Proc. Natl. Acad. Sci. U.S.A.,* 70, 2015, 1973.
69. **Goldstein, J. L., Ho, Y. K., Brown, M. S., Innerarity, T. L., and Mahley, R. W.,** Cholesteryl ester accumulation in macrophages resulting from receptor-mediated uptake and degradation of hypercholesterolemic canine β-very low density lipoproteins, *J. Biol. Chem.,* 255, 1839, 1980.
70. **Mahley, R. W., Innerarity, T. L., Brown, M. S., Ho, Y. K., and Goldstein, J. L.,** Cholesteryl ester synthesis in macrophages: stimulation by β-very low density lipoproteins from cholesterol-fed animals of several species, *J. Lipid Res.,* 21, 970, 1980.
71. **Wong, L. and Rubinstein, D.,** Turnover of apo-E in normal and hypercholesterolemic rats, *Atherosclerosis,* 34, 249, 1979.
72. **Roheim, P. S., Haft, D. E., Gidez, L. I., White, A., and Eder, H. A.,** Plasma lipoprotein metabolism in perfused rat livers. II. Transfer of free and esterified cholesterol into the plasma, *J. Clin. Invest.,* 42, 1277, 1963.
73. **Camejo, G., Bosch, V., Arreaza, C., and Mendez, H. C.,** Early changes in plasma lipoproteins and biosynthesis in cholesterol-fed rabbits, *J. Lipid Res.,* 14, 61, 1973.
74. **Noel, S.-P., Wong, L., Dolphin, P. J., Dory, L., and Rubinstein, D.,** Secretion of cholesterol-rich lipoproteins by perfused livers of hypercholesterolemic rats, *J. Clin. Invest.,* 64, 674, 1979.
75. **Lin-Lee, Y.-C., Tanaka, Y., Lin, C.-T., and Chan, L.,** Effects of an atherogenic diet on apolipoprotein E biosynthesis in the rat, *Biochemistry,* 20, 6474, 1981.
76. **Swift, L. L., Manowitz, N. R., Dewey Dunn, G., and LeQuire, V. S.,** Isolation and characterization of hepatic Golgi lipoproteins from hypercholesterolemic rats, *J. Clin. Invest.,* 66, 415, 1980.
77. **Dolphin, P. J.,** Serum and hepatic nascent lipoproteins in normal and hypercholesterolemic rats, *J. Lipid Res.,* 22, 971, 1981.
78. **Riley, J. W., Glickman, R. M., Green, P. H. R., and Tall, A. R.,** The effect of chronic cholesterol feeding on intestinal lipoproteins in the rat, *J. Lipid Res.,* 21, 942, 1980.
79. **Ross, A. C. and Zilversmit, D. B.,** Chylomicron remnant cholesteryl esters as the major constituent of very low density lipoproteins in plasma of cholesterol-fed rabbits, *J. Lipid Res.,* 18, 169, 1977.
80. **Melchior, G. W., Mahley, R. W., and Buckhold, D. K.,** Chylomicron metabolism during dietary-induced hypercholesterolemia in dogs, *J. Lipid Res.,* 22, 598, 1981.
81. **Kris-Etherton, P. M. and Cooper, A. D.,** Studies on the etiology of the hyperlipemia in rats fed an atherogenic diet, *J. Lipid Res.,* 21, 435, 1980.
82. **Redgrave, T. G. and Snibson, D. A.,** Clearance of chylomicron triacylglycerol and cholesteryl ester from the plasma of streptozotocin-induced diabetic and hypercholesterolemic hypothyroid rats, *Metabolism,* 26, 493, 1977.
83. **Mahley, R. W., Innerarity, T. L., Weisgraber, K. H., and Fry, D. L.,** Canine hyperlipoproteinemia and atherosclerosis: accumulation of lipid by aortic medial cells *in vivo* and *in vitro, Am. J. Pathol.,* 87, 205, 1977.
84. **Reitman, J. S. and Mahley, R. W.,** Yucatan miniature swine lipoproteins: changes induced by cholesterol feeding, *Biochim. Biophys. Acta,* 575, 446, 1979.
85. **Shore, V. G., Shore, B., and Hart, R. G.,** Changes in apolipoproteins and properties of rabbit very low density lipoproteins on induction of cholesterolemia, *Biochemistry,* 13, 1579, 1974.

86. **Dory, L. and Roheim, P. S.,** Rat plasma lipoproteins and apolipoproteins in experimental hypothyroidism, *J. Lipid Res.*, 22, 287, 1981.
87. **Koletsky, S.,** Obese spontaneously hypertensive rats — a model for study of atherosclerosis, *Exp. Mol. Pathol.*, 19, 53, 1973.
88. **Koletsky, S.,** Pathologic findings and laboratory data in a new strain of obese hypertensive rats, *Am. J. Pathol.*, 80, 129, 1975.
89. **Butkus, A., Tan, E., and Koletsky, S.,** Tissue lipid distribution in genetically obese and spontaneously hypertensive rat. II. Liver and adipose tissue lipids, *Artery*, 2, 208, 1976.

GUINEA PIG LIPOPROTEINS AND THEIR CHANGES IN RESPONSE TO DIETARY CHOLESTEROL

Rosemarie Ostwald and Mark Fitch*

THE NORMAL GUINEA PIG

Density Classes

The guinea pig differs from other laboratory rodents in many aspects of lipid composition and lipid metabolism. Arachidonic acid is practically absent in guinea pig tissues, including plasma, with the exception of red blood cell phospholipids.[1] Guinea pigs, unlike most other animals studied, possess very little HDL (d 1.07 to 1.21).[2,3] The content of HDL in normal human plasma, for instance, ranges from 140 to 340 mg/100 mℓ plasma,[4] while that of guinea pigs is about 2 mg/100 mℓ plasma.[2] The principal serum lipoprotein of the guinea pig is the low density family of lipoproteins (LDL).[5] The hydrated density of this lipoprotein was found to be as high as 1.10 g/mℓ.[6] Unlike the plasma LDL in man in which apo B constitutes about 95% of the total lipoprotein protein,[7] we found the guinea pig LDL to contain approximately 75% of apo B and a relatively high concentration of soluble proteins with molecular weights between 8000 and 66,000 daltons.[8]

Apolipoproteins

Analytical SDS gel electrophoresis of serum lipoproteins of guinea pigs indicated the presence of four major apoproteins that are homologous to those in man and other species.[8] The molecular weights, electrophoretic mobilities, and amino acid compositions showed the presence of apo B, apo E, apo A-I, and some of the apo C groups corresponding to those of humans and rats.

Apo B is predominant in serum VLDL and LDL of both normal and cholesterol-fed guinea pigs.[8] This apoprotein, isolated by tetramethylurea extraction of LDL from normal animals, shows two major components with molecular weights of 422,000 and 402,000 daltons (comparable to the B-100 and B-95 of humans).[6]

Apoprotein A-I is present mainly in the very small HDL fraction of normal guinea pigs. This apoprotein has a molecular weight of 25,000 and exists in six polymorphic forms (pI 5.75 to 5.40). Unlike the apo A-1 of humans, which lacks cysteine and isoleucine, apo A-I of guinea pigs contains 20 and 8 mol/10³ of isoleucine and cysteine, respectively. Despite this and other small differences in amino acid composition, isolated guinea pig apo A-I is as potent as human apo A-I as co-factor for activating purified LCAT from human plasma. Among mammals, the normal guinea pig has the lowest reported plasma concentration of apo A-I (6.2 ± 2.0 mg/dℓ), whereas plasma level of this apoprotein is doubled in 8- to 10-week cholesterol-fed animals.[9]

THE CHOLESTEROL-FED GUINEA PIG

Lipoproteins

In the guinea pig, dietary cholesterol at the level of 1% causes large increases in the concentration of unesterified cholesterol in all density classes and an increase of lipoproteins floating at a hydrated density of 1.07 to 1.09 g/mℓ and 1.09 to 1.21 g/mℓ, but unchanged or decreased amounts of VLDL (d < 1.006 g/mℓ).[2,5] This is in contrast to the large increase in VLDL containing a tenfold increase of cholesterol ester in cholesterol-fed rabbits.[10,11]

* The experimental part of this work has been supported in part by USPHS NIH Grant #AM08480-15.

A most unusual observation is the appearance in plasma of an α-migrating, discoidal HDL in response to dietary cholesterol. These discoidal HDL are enriched in unesterified cholesterol[2] and apo E.[8] It is particularly interesting that this apo E-rich discoidal HDL is remarkably similar to discs accumulating in the plasma of humans with the genetic disease of LCAT deficiency.[12,15] Cholesterol-induced hyperlipoproteinemia in other animal models, on the other hand, is characterized by the appearance of an α-migrating cholesteryl ester-rich particle floating with LDL and IDL, designated as HDL_c concomitant with a markedly reduced amount of typical HDL_2.[16]

Apoproteins

Guinea pigs respond to dietary cholesterol with (1) a significant increase of apo E, which is a minor component of VLDL in normal guinea pigs and (2) the appearance of this protein in the other density classes.[8] In chow-fed guinea pigs the plasma concentration of apo E measured by immunoelectrophoresis, is 2.2 ± 0.5 mg/dℓ.

In animals fed a 1% cholesterol diet, plasma levels of this apoprotein increased by 10-fold in 1 week and up to 22-fold in 8 to 10 weeks.[9] This apoprotein, isolated by column chromatography from serum HDL of cholesterol-fed animals is similar to the apo E reported in other mammalian species in its molecular weight (34,000), electrophoretic mobility, amino acid composition, and N-terminal amino acid (lysine).[8] This protein and the other major polypeptides present in guinea pig plasma are immunologically distinguishable.[8,9,17] The amino acid composition of this protein is very similar to that of apo E reported in the VLDL fraction of patients with Type III hyperlipoproteinemia.[18] It has been suggested that this apo E is directly or indirectly involved with cholesterol transport and metabolism although no role as a co-factor in enzyme systems of lipid metabolism has as yet been identified.[8,17,19] The physiological function and the binding properties of this protein to cholesterol or phospholipids are not yet understood. Because of immunological differences, the very large increase in response to exogenous cholesterol, and its sequential appearance in VLDL, HDL, and LDL, apo E may be of special importance and lend itself as a marker for the study of apoprotein biosynthesis and lipoprotein metabolism.

Recent studies by Guo et al. have shown that apo E is present in both nascent VLDL and HDL from perfused livers of normal guinea pigs and contains three isoforms (pI 5.42 to 5.34). Following cholesterol feeding, the numbers of apo E isoforms are increased from three to five or more by shifting the major component (pI 5.42) to more acidic isoforms (pI 5.28 to 5.17). This shift is mostly reversible when apo E is treated with neuraminidase, suggesting that cholesterol feeding leads to a modification of apo E by increasing its content of sialic acid.[9]

In addition, the normal guinea pig has an apoprotein which has the same electrophoretic mobility as apo E on SDS polyacrylamide gel. This apoprotein, however, has a different elution volume on Sephadex® gel chromatography and a different electrophoretic mobility on alkaline polyacrylamide gels.[17] This apoprotein has been designated "co-migrating protein" and is also different from apo E in amino acid composition and is not cross-reactive with antiserum to apo E. Furthermore, this apoprotein contains 34% carbohydrate by weight compared to 3.5% present in apo E.[17] Human serum apo D, a cholesteryl ester transfer protein is also rich in carbohydrate, but its amino acid composition is quite different from that of the apo E "co-migrating protein".[20] The biological function of this co-migrating apoprotein as well as the other apoproteins in the guinea pig remains to be elucidated.

Lipids

Dietary cholesterol changes not only the lipoprotein density classes and their apoproteins but also their lipid composition. They are characterized by a high concentration of unesterified cholesterol in all lipoprotein density classes reaching the unusually high molar ratio of two

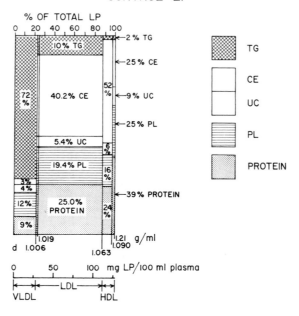

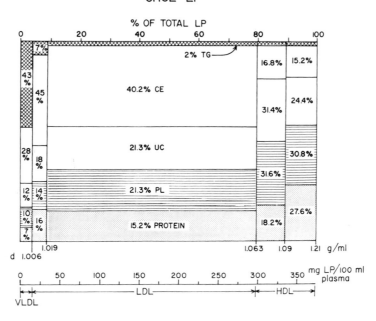

FIGURE 1. Chemical composition of five plasma lipoprotein classes of control and chol guinea pigs: VLDL d < 1.006 g/mℓ; LDL d 1.006—1.019 g/mℓ and d 1.019 to 1.063 g/mℓ; HDL d 1.063 to 1.090 g/mℓ and d 1.090 to 1.21 g/mℓ. Data are means of analyses of five chol and three control guinea pigs. Column width shows the lipoprotein (LP) concentration (mg LP/100 mℓ plasma) in each density fraction. (Total control LP = 137.8 ± 26.8 mg/100 mℓ; total chol LP = 372.2 ± 14.2 mg/100 mℓ.) Numbers in areas indicate wt % composition of each lipoprotein. (SD of the data were less than 15% of the means for chol HDL and chol LDL constituents and were 3 to 30% of the means for VLDL and control HDL constituents.) Percent TG for HDL and LDL are data from earlier experiments.[3] Mg CE = 1.67 × mg cholesterol. To convert percentages of UC, CE, and PL to mole ratios: moles CE per moles UC = 0.6 × %CE/%UC; moles UC per moles PL = 2 × %UC/%PL. Recovery of lipids averaged 80 to 90% of total lipid; data were corrected to 100% recovery of lipid. (From Sardet, C., Hansma, H., and Ostwald, R., *J. Lipid Res.*, 13, 624, 1972. With permission.)

Table 1
PERCENT COMPOSITION OF THE NON-APO B
FRACTION OF VLDL, LDL, AND HDL FROM GUINEA
PIGS FED CHOLESTEROL-FREE DIETS CONTAINING
COTTONSEED OIL (CSO) OR HYDROGENATED
COCONUT OIL (HCNO).[25]

Density class	Diet	Apo E	Apo A-I	ACP[a]	Apo C
1.006 g/mℓ	CSO	22	.7	4.4	73
	HCNO	44	.2	3.8	52
1.01—1.07 g/mℓ	CSO	12	4.2	9.3	75
	HCNO	15	8.8	28	48
1.09—1.20 g/mℓ	CSO	1.8	26	0	73
	HCNO	1.5	45	0	53

Note: The animals were fed semisynthetic diets containing 10% total fat and were fasted for 16 to 18 hr prior to blood sampling.

[a] Apo E co-migrating polypeptide.[17]

cholesterols for each phospholipid without major changes in total phospholipid.[17] Although total phospholipids were not changed significantly, there were changes in the individual phospholipid classes.[21] The increase in free cholesterol and in LDL in the guinea pig is similar to certain conditions seen in humans. Unusually high concentrations of free cholesterol are found in patients with obstructive liver disease[22] and with familial LCAT deficiency.[12,23,24] For details of the lipoprotein composition in the normal and cholesterol-fed guinea pig see Figure 1, Sardet et al.,[2] and Puppione et al.[3]

The apoproteins of the guinea pig are not only changed in response to dietary cholesterol but also in response to changes in the saturation of dietary fat (see Table 1). When the dietary fat was changed from cottonseed oil (polyunsaturated fat) to hydrogenated coconut oil (saturated fat), the percentage of apo C in the non-apo B fraction decreased in all lipoprotein density classes while the percentages of apo E and the apo E co-migrating apoprotein increased several-fold.

ACKNOWLEDGMENT

The authors wish to thank Dr. Luke Guo for reading the manuscript and for his helpful suggestions for changes and additions.

REFERENCES

1. **Ostwald, R. and Shannon, A.,** Composition of tissue lipids and anaemia of guinea pigs in response to dietary cholesterol, *Biochem. J.,* 91, 146, 1964.
2. **Sardet, C., Hansma, H., and Ostwald, R.,** Characterization of guinea pig plasma lipoproteins: the appearance of new lipoproteins in response to dietary cholesterol, *J. Lipid Res.,* 13, 624, 1972.
3. **Puppione, D. L., Sardet, C., and Yamanaka, W.,** Plasma lipoproteins of cholesterol-fed guinea pigs, *Biochim. Biophys. Acta,* 231, 295, 1971.
4. **Barclay, M.,** Lipoprotein class distribution in normal and diseased states, in *Blood, Lipids and Lipoproteins,* Nelson, G. J., Ed., Interscience, New York, 1972, chap. 12.
5. **Mills, G. L., Chapman, M. J., and McTaggart, F.,** Some effects of diet on guinea pig serum lipoproteins, *Biochim. Biophys. Acta,* 260, 401, 1972.

6. **Guo, L., Hamilton, R. L., Ostwald, R., and Havel R. J.,** Secretion of nascent lipoproteins and apoproteins by perfused livers of normal and cholesterol-fed guinea pigs, *J. Lipid Res.,* 23, 543, 1982.

7. **Shonfeld, G., Lees, R. S., George, P. K., and Pfleger, B.,** Assay of total plasma apolipoprotein B concentration in human subjects, *J. Clin. Invest.,* 53, 1458, 1974.

8. **Guo, L. S. S., Meng, M., Hamilton, R. L., and Ostwald, R.,** Changes in plasma lipoprotein-apoproteins of guinea pigs in response to dietary cholesterol, *Biochemistry,* 16, 5805, 1977.

9. **Guo, L., Hamilton, R. L., Kane, J. P., Fielding, C. J., and Chen, G. C.,** Characterization and quantitation of apolipoproteins A-1 and E of normal and cholesterol-fed guinea pigs, *J. Lipid Res.,* 23, 531, 1982.

10. **Shore, V. G., Shore, B., and Hart, R. G.,** Changes in apolipoproteins and properties of rabbit very low density lipoproteins on induction of cholesterolemia, *Biochemistry,* 13, 1579, 1974.

11. **Camejo, G., Bosch, V., and Arreaza, C.,** Early changes in plasma lipoprotein structure and biosynthesis in cholesterol-fed rabbits, *J. Lipid Res.,* 14, 61, 1973.

12. **Norum, K. R., Glomset, J. A., Nichols, A. V., and Forte, T.,** Plasma lipoproteins in familial lecithin: cholesterol acyltransferase deficiency: physical and chemical studies of low and high density lipoproteins, *J. Clin. Invest.,* 50, 1131, 1971.

13. **Forte, T., Norum, K. R., Glomset, J. A., and Nichols, A. V.,** Plasma lipoproteins in familial lecithin: cholesterol acyltransferase deficiency: structure of low and high density lipoproteins as revealed by electron microscopy, *J. Clin. Invest.,* 50, 1141, 1971.

14. **Glomset, J. A. and Norum, K. R.,** The metabolic role of lecithin: cholesterol acyltransferase: perspectives from pathology, in *Advances in Lipid Research,* Vol. 11, Paoletti, R. and Kritchevsky, D., Eds., Academic Press, New York, 1973, 1—65.

15. **Mitchell, C. D., King, W. C., Applegate, K. R., Forte, T., Glomset, J. A., Norum, K. R., and Gjone, E.,** Characterization of apoprotein acyltransferase deficiency, *J. Lipid Res.,* 21, 625, 1980.

16. **Mahley, R. W.,** Alterations in plasma lipoproteins induced by cholesterol feeding in animals including man, in *Disturbances in Lipid and Lipoprotein Metabolism,* Deitschy, J. M., Grotto, A. M., Jr., and Ontko, J. A., Eds., American Physiological Society, Bethesda, Md., 1978, 181—197.

17. **Meng, M., Guo, L., and Ostwald, R.,** Isolation and partial characterization of a guinea pig serum apolipoprotein comigrating with apo E on sodium dodecyl sulfate-polyacrylamide electropherograms, *Biochim. Biophys. Acta,* 576, 134, 1979.

18. **Havel, R. J. and Kane, J. P.,** Primary dysbetalipoproteinemia: predominance of a specific apoprotein species in triglyceride - rich lipoproteins, *Proc. Natl. Acad. Sci. U.S.A.,* 70, 2015, 1973.

19. **Fielding, C. J., Shore, V. G., and Fielding, P. E.,** A protein cofactor of lecithin: cholesterol acyltransferase, *Biochem. Biophys. Res. Commun.,* 46, 1493, 1972.

20. **McConathy, W. J. and Alaupovic, P.,** Studies on the isolation and partial characterization of apolipoprotein D and lipoprotein D of human plasma, *Biochemistry,* 15, 515, 1976.

21. **Ostwald, R., Yamanaka, W., and Light, M.,** The phospholipids of liver, plasma and red cells in normal and cholesterol-fed anemic guinea pigs, *Proc. Soc. Exp. Biol. Med.,* 134, 814, 1970.

22. **Seidel, D., Alaupovic, P., and Furman, R. H.,** A lipoprotein characterizing obstructive jaundice. I. Method for quantitative separation and identification of lipoproteins in jaundiced subjects, *J. Clin. Invest.,* 48, 1211, 1969.

23. **Norum, K. R. and Gjone, E.,** Familial serum cholesterol esterification failure. A new inborn error of metabolism, *Biochim. Biophys. Acta,* 114, 698, 1967.

24. **Glomset, J. A., Norum, K. R., and King, W.,** Plasma lipoproteins in familial lecithin: cholesterol acyltransferease deficiency: lipid composition and reactivity, *in vitro, J. Clin. Invest.,* 49, 1827, 1970.

25. **Fitch, M. D., Guo, L., and Ostwald, R.,** unpublished manuscript, 1982.

LIPOPROTEINS IN RABBITS ON USUAL AND SPECIAL DIETS

D. C. K. Roberts

INTRODUCTION

The rabbit has been used as a model for the study of cholesterol metabolism for many years. While the lipoprotein changes seen under conditions of cholesterol feeding have been well documented, less research has been undertaken into the normal pattern seen in chow-fed animals. One of the problems has been, and remains, the definition of "chow". Even though the overall composition remains similar (that is, the percent protein, fat, carbohydrate, and ash), chow can vary in components from country to country and, more importantly, from batch to batch, depending on the available supply of cereals, meals, oil, etc. This problem has been noted in the past.[22,35] Other problems beset data obtained from cholesterol supplemented, chow-fed rabbits, since the vehicle in which the cholesterol is dispensed can affect the cholesterolemic response[23] and hence the lipoprotein pattern. Other variables include different strains responding in different ways[19,30,36,45] and even variations in response within strains.[43]

COMMERCIAL CHOW

Most electrophoretic techniques have been used for separation of serum lipoproteins including paper, cellulose acetate, starch, agarose, and polyacrylamide. Recent work has involved ultracentrifugal separation of lipoproteins followed by polyacrylamide gel separation of the apoproteins. Early work was essentially comparative, with some authors suggesting similarities with man[1,21] and others not.[4,18,34] The chemical composition of the lipoproteins has been reviewed for New Zealand White, Dutch belted, and Fauve de Bourgogne strains by Chapman.[15] Other strains noted are Swedish Landrace[18] and Japanese White.[19]

Some authors report HDL (d = 1.063 to 1.210) to be present in the highest concentration[30,31,36] with others doing so only when IDL (d = 1.006 to 1.019) and LDL (d = 1.019 to 1.063) have been separately estimated.[15] In those reports where LDL has been estimated as β-lipoprotein (β-Lp) or as d = 1.006 to 1.063, it is present in the greater concentration (Table 1). Two reports mention sex differences. In one, α-lipoprotein (α-Lp) was higher in males than females as a result of β-Lp decreasing with age in males and increasing in females.[30,31] This may be a reflection of the decrease in total cholesterol seen with increasing age in male rabbits.[43] In contrast, Borresen[7-9] reported males have less α-Lp than females, but in a different strain from that reported by Lozsa.[30,31]

Lipoprotein polymorphism has been reported for both LDL and HDL (Table 1) and has been reviewed by Rapascz.[38] Allotyping of HDL has proceeded to the level of identifying polypeptides within which the allotypes reside. Hℓ-1 is present in a polypeptide chain of 40,000 daltons (which may be a dimer) and R-67 resides in a polypeptide of 17,000 daltons which may be within the apo A family.[7-10]

Of the apoproteins, A-I, B, and Cs, and E have all been shown to be present in rabbit lipoproteins (Table 1, 2). Apo A-I has been further characterized as having a molecular weight of 25,000 and being similar to dog[10] and human.[10,24] Rabbit Apo A-I has also been reported to bind a bacterial lipopolysaccharide (LPS)[54] and to complex with and act as a carrier for the amyloid-related protein SAA produced in response to injected LPS.[49]

CHOLESTEROL AND COMMERCIAL CHOW

Cholesterol feeding produces rapid, early changes in the lipoproteins of rabbits. There is

Table 1
SUMMARY OF RABBIT LIPOPROTEINS AS AFFECTED BY DIET

Natural or Commercial Chow Diets

Author	Ref. no.	Year	Pattern	Normal values[a]	± SD	Characteristics of subject (age, sex, strain, diet)	Summary of results
Adlersberg, D. et al.	(1)	1955	β predominates	60[b] (15) α = 32,[c] β = 50 origin = 18		Strain unspecified, fasting males, commercial chow	Paper electrophoresis
Kirkeby, K.	(21)	1966	β predominates	α = 127[d] (17) β = 281 (17)	29 65	Strain unspecified, fasting males and females, "natural habits" diet	Paper electrophoresis
Alexander, C. and Day, C. E.	(4)	1973	β predominates	57[b] (6) α = 46[c] β = 54		New Zealand White, nonfasting, sex unspecified, commercial chow	Agarose electrophoresis. β value the sum of zone 2 + 3
Johansson, M. B. and Karlsson, B. W.,	(18)	1976	β predominates	α = 25[c] β = 65 origin = 1		Swedish Landrace, nonfasting males, commercial chow	Agarose electrophoresis; albumin region was lipid stained to 10%. 2 peaks in β region
Lozsa, A.	(30)	1972	α predominates in males	Gray 92[b] (12) Brown 83 (61)	15 11	White, gray, and brown Hungarian, fasting males "natural habits" diet	Paper electrophoresis; females have lower cholesterol than males
Lozsa, A. et al.	(31)	1972	α,β equal in females	White 80 (70) α/β[c] ratio 100:0 to 80:20 Gray 45[b] (63) Brown 44 (55) White 42 (65) α/β[c] ratio 60:40 to 30:70	12 10 10 9	Fasting females	β decreased with age (2—10 months) in males, gray > brown > white; β increased in females with age and S-band (VLDL) decreased; β reappeared in castrated males

Author	Ref	Year	Marker	Value	Ratio	Strain/Diet	Comments
Pescador, R.	(36)	1978	HDL^f predominates	47^b (4) VLDL^f = 5^g, LDL^f = 25, HDL^f = 53; 65^b (4) VLDL^f = 3^g, LDL^f = 15, HDL^f = 63	1.2^e; 3.5^e	New Zealand White males, commercial chow; Red Burgundy males, commercial chow	New Zealand White had less LDL$_2$ (d = 1.019—1.063) than Red Burgundy
Borresen, A. L. and Kindt, T. J.	(10)	1978	Apo A-I	25000^i(5)		Albino females, commercial chow	SDS polyacrylamide gels; rabbit Apo A-I had a difference index of 7.4 with human A-I and 6.5 with dog A-I
Kawai, K. et al.	(19)	1980		50^b(5) β = 100 250—etc.; $250—500^{b,h}$(8) β = $500—1200^{d,h}$		Japanese White, fasting, sex unspecified Heritable etc.; Heritable hyperlipemic (HLR), fasting males and females, commercial chow	On diet for up to 36 months. α decreased and pre-β and β increased in HLR compared to controls

Antigenic Markers

Author	Ref	Year	Marker	Value	Strain/Diet	Comments
Kelus, A. S.	(20)		Lpl	(18)	Strain unspecified, sex unspecified	Antigenic marker for β-lipoprotein
Albers, J. J. and Dray, S.	(2) (3)	1968 1969	Lpq 1-4		Flemish giant × New Zealand White in closed colony, males and females, commercial chow	Antigenic markers for LDL. 78 families × 327 offspring
Berg, K., et al. Boman, H., et al.	(5) (6)	1971 1972	Hℓ-1 R-67	(90) (128)	Albino, commercial chow Albino, commercial chow	Antigenic marker for HDL Antigenic marker, distinct from Hℓ-1 for HDL
Gilman-Sachs, A. and Knight, K. K.	(17)	1972	Lhj-1 Lhj-2		Flemish Giant × New Zealand White in closed colony, commercial chow	Genetic markers for HDL not linked to 1pq locus; 83 families × 312 offspring

Table 1 (continued)
SUMMARY OF RABBIT LIPOPROTEINS AS AFFECTED BY DIET

Author	Ref. no.	Year	Pattern	Normal values[a]	± SD	Characteristics of subject (age, sex, strain, diet)	Summary of results
Borresen, A. L.	(7-9)	1976	R-67 Hℓ-1	17,000[i] 40,000[i]		Albino, males and females, Dutch Black and White, males and females, fasting and non-fasting, commercial chow	Females had more α than males. Hℓ-1 may be a dimer; R-67 antigen showed partial identity with Apo A-I

a   Number of animals in parentheses.
b   Cholesterol concentration mg/dℓ.
c   Percent area under peaks.
d   Lipoprotein concentration mg/dℓ.
e   SEM
f   Separated by preparative ultracentrifugation:
VLDL, d < 1.006; LDL, d = 1.006—1.063; HDL, d = 1.063—1.210.
g   Percent area under OD$_{280}$ curve from ultracentrifugal fractions.
h   Range of observations.
i   Apparent molecular weight.

Table 2
SUMMARY OF RABBIT LIPOPROTEINS AS AFFECTED BY DIET

Commercial Chow + Crystalline Cholesterol

Author	Ref. no.	Year	Pattern	Normal values[a]	± SD	Pathology[a]	± SD	Characteristics of subject; age, sex, strain, diet	Summary of results
Adlersberg, D. et al.	1	1955	α reduced			1210[b](1) α = 2[c] β = 53 origin = 45		Strain unspecified, fasting males, 1 g cholesterol/day	Paper electrophoresis
Shore, V. G. et al.	46	1974	α reduced, β increased					New Zealand White, nonfasting females, 1% cholesterol	2—24 days on diet. VLDL[d] of β mobility on agarose. On acrylamide (10% 8 M urea) large increase in bands R2/R3 (Apo E) in VLDL[d]; present in decreasing amounts in IDL[e] and LDL[e] Band R1, found in VLDL[d] corresponds to human Apo CI
Shore, B. and Shore V.	44	1974	Apo E increased in VLDL[d]					New Zealand White nonfasting, sex unspecified, 1% cholesterol	24 days on diet; VLDL[d] has β mobility on agarose; apo E increased to 50% of soluble apoprotein
Shore, B. and Shore, V.	45	1976		80[b](12), 70—100[f] VLDL[d]=25[g] IDL[e]=100 LDL[e]=115 HDL[d]=130 75[b](8), 65—90[f] VLDL[d] = 25[g] IDL[e] = 85 LDL[e] = 105 HDL[d] = 140	5 7 8 9 5 8 9 8	1640[b](12) 1268—2200[f] 1880[g] 820 380 75 880[b](8), 657—1200[f] 1110[g] 270 170 95		New Zealand White nonfasting females, 1% cholesterol Dutch Belted non-fasting females, 1% cholesterol	10—32 days on diet from 10 weeks of age; data from 10 days on diet; Dutch Belted are hyporesponder strain; apo E is major component of VLDL in both strains; the E:B ratio decreases with increasing density; apo A-1 is present in HDL, with a apo E a minor component but in greater quantity in HDL_3 than HDL_2
Sirtori, C. R. et al.	47	1977	Apo E increased			1085[b](6) 1409[b](6)	212 297	New Zealand White males, 2 g cholesterol/day + metformin (135 mg/kg)-Metformin	60 days on diet; cholesterol decreased on metformin; on PAGE (8 M urea, 10%) bands R2 > R3 (both apo E) in VLDL[h]. With metformin treatment apo E decreased as apo Cs increased and band R3 > R2
Bosanquet, A. G. et al.	11	1978	α reduced, broad β					New Zealand White fasting females, 0.8 g cholesterol/day	7 days on diet; cellulose acetate electrophoresis; broad β component included particles of S_f 20-150 and S_f > 150

Table 2 (continued)
SUMMARY OF RABBIT LIPOPROTEINS AS AFFECTED BY DIET

Author	Ref. no.	Year	Pattern	Normal values[a]	± SD	Pathology[a]	± SD	Characteristics of subject; age, sex, strain, diet	Summary of results
Pescador, R.	36	1978				1408[b](4) VLDL[d] = 14[j] LDL[d] = 37 HDL[d] = 26 666[b](4) VLDL[d] = 9 LDL[d] = 29 HDL[d] = 33	108[i] 50[j]	New Zealand White males, 2% cholesterol Red Burgundy males, 2% cholesterol	3 weeks on diet; Red Burgundy are hypo-responder strain; New Zealand White had more IDL[c] than Red Burgundy strain
Ploplis, V. A. et al.	37	1979	α reduced, pre-β/β ratio increased	95[b](3)		650[b]—1665[f](3)		New Zealand White, fasting, sex unspecified, 2% cholesterol	VLDL[h] still had pre-β mobility on PAGE; α mobility decreased
Sirtori, C. R. et al.	48	1981	Apo B increased Apo E increased	76[b](6) 15 Apo B VLDL[h] = 37[k] LDL[e] = 72		1102[b](6) 73[k] 95 35[b](6) Apo B VLDL[h] = 30[k] LDL[e] = 55	103 7	New Zealand White males, 2 g cholesterol/day 8 weeks post PIB partial ileal bypass	8 weeks on diet; one group then had PIB and fed for further 8 weeks; VLDL[h] enriched in apo E; content decreased after PIB with increase in apo A-I and apo Cs
Lovati, M. R. and DeMarchi, G.	20	1981		83[b](6)	10	2184(6) 2415(6)	411 408	Fawn Burgundy males, 2 g cholesterol/day + Pirinixil (50 mg/kg) – Pirinixil	60 days on diet; no change in cholesterol on pirinixil treatment; apo E increased in VLDL[h] on cholesterol feeding but less so in pirinixil group; apo A-I increased in VLDL[h] of pirinixil group

Commercial Chow and Cholesterol Dispersed in Oil

Author	Ref. no.	Year	Pattern	Normal values[a]	± SD	Pathology[a]	± SD	Characteristics of subject; age, sex, strain, diet	Summary of results
Camejo, G. et al.	13	1973	VLDL[h] of β mobility	VLDL[h] + 23[g](8) LDL[e] = HDL[d] = 134 (8)	3 8	848[g](6) 122(6)	200 21	Strain and sex unspecified, 1% cholesterol + 10% sesame oil	21 days on diet; appearance of a slow migrating component in VLDL[h] on SDS acrylamide gels
	14	1974						Inbred colony, 1% cholesterol + 10% sesame oil	On diet 2—21 days. VLDL[h] separated by gel filtration into two fractions showed similarities in apoproteins but different lipid composition
Stange, E. et al.	50 51	1975 1977	α reduced VLDL[d] of β mobility	128[b](5) 75[b](4) 85[b](6)	29 16 34	1444[b] 2629[b] 2002b2b	340 119 333	New Zealand White males, nonfasting 1% cholesterol 1% cholesterol + 5% coconut oil 1% cholesterol + 5% corn oil	18 weeks on diet; on agarose, LDL[d] is β-migrating but slower when cholesterol present; both VLDL[d] and LDL[d] showed broad β when corn oil in diet; on acrylamide (15%, TMU) up to 12 bands identified in VLDL and LDL

Author	Ref.	Year	Apo/notes	Component				Subjects	Comments
Kushwaha, R. S. and Hazzard, W. E.	24	1978	Apo E increased, apo B increased	$70^b(4)$	9	1515^b	177	New Zealand White females, nonfasting, 0.5% cholesterol + 5% sesame oil	10—12 weeks on diet
				$VLDL^d=5^b$	4	790^b	62		
				$IDL^c=13$	18	608	68		
				$LDL^c=31$	1	89	6		
				$HDL^d=15$	6	25	3		
				$VLDL^d$ Apo C-I$=11^c$		7^c			
				Apo E$=36$		77			
				Apo C-II + C-III $=54$		16			
Kushwaha, R. S. et al.	26	1978		$HDL^d=19^b(3)$	6	20^b	6	New Zealand White females, nonfasting, 0.5% cholesterol + 5% sesame oil	Apo A-I$_1$ and A-I$_2$ on polyacrylamide (10%, 8 M urea) were homologous with human A-I.
				C-I + E$=6^c$	1	7^c	1		
				A-I$^1=22$	1	27	1		
				A-I$_2=46$		35			
				C-III$=27$	3	31	3		
Kushwaha, R. S. and Hazzard, W. R.	25	1981		$84^b(4)$	45	825^b	259	New Zealand White females, fasting, 0.2% cholesterol + 2% sesame oil + estradiol(0.2—0.5 mg/kg)	Up to 6 months on diet; after estrogen treatment the cholesterol fell and LDL^e became the major carrier; the ratio of apo C/E increased in VLDL; atherosclerosis was much less at 6 months in those on estrogen
				$VLDL^d$ apo C/E ratio $=1.73(3)$	0.5	0.22	0.03		
				$313^b (4)$			16		
				$VLDL^d$ apo C/E ratio$=0.91 (3)$			0.08		

Semipurified diets

Author	Ref.	Year	Apo/notes	Component				Subjects	Comments
Mettler, L. et al.	33	1971	α decreased, β increased	$178^b (7)$	14	801^b	186	Strain unspecified, females, nonfasting ovariectomized + estrogen (0.05 mg/day) – Estrogen	27 days on semipurified diet containing casein (17%) and cholesterol (3%); estrogen treatment slowed the development of hypercholesterolemia
				α$=31^c$	8	12^c	4		
				pre-β + β$=40$	1513	52	8		
				$198^b (7)$	10	958^b	80		
				α$=42^c$	7	10^c	3		
				pre-β + β$=28$	24	55	7	Males, nonfasting	
				$129^b (8)$	8	936^b	180		
				α$=29^c$	4	10^c	3		
				β$=27$		47	13		
Lacombe, C. and Nibbelink, C. M.	27	1980		$VLDL^d = 5^b(3)$	1^i	$8^b (3)$	0.3^i	Fauve de Burgogne males, fasting; low cholesterol (0.04%), high fat (20%) diet containing casein (20%)	12 weeks on diet; both casein and soy-fed developed a moderate hypercholesterolemia but casein more marked; LDL and HDL predominated
				$IDL^c = 7$	3	11	1.6		
				$LDL^e = 4$	0.5	23	27		
				$HDL^d = 10$	1	23	1.8	Soya (20%)	
						$VLDL^d = 8^b (3)$	1^i		
						$IDL^e = 7$	0.2		
						$LDL^e = 14$	0.1		
						$HDL^d = 19$	5		

Table 2 (continued)
SUMMARY OF RABBIT LIPOPROTEINS AS AFFECTED BY DIET

Author	Ref. no.	Year	Pattern	Normal values[a]	± SD	Pathology[a]	± SD	Characteristics of subject; age, sex, strain, diet	Summary of results
Semipurified diets									
Lacombe, C. and Nibbelink, C. M.	28	1980		VLDL[d] = 19[f] (10) LDL[d] = 27 HDL[d] = 55 VLDL[d] = 22[f] LDL[d] = 35 HDL[d] = 43	1[i] 4 3 6[i] 4 3	302[b] (10) 33[f] 55 11 68[b] 8[f] 49 43	43[i] 7 6 3 5[i] 1[i] 1 2	New Zealand White males, fasting, low cholesterol (0.04%), high fat (20%) diet containing casein (0.04%), high fat (20%) diet containing casein (20%), Fauve de Burgogne males, fasting	Fauve de Burgogne are hyporesponders; VLDL[d] and LDL[d] major carriers of cholesterol in New Zealand White, whereas LDL[d] and HDL[d] are major carriers in Fauve de Burgogne
Roberts, D. C. K. et al.	42	1981	Apo E					New Zealand White fasting males, cholesterol-free, low fat (1%) diet containing casein (25%) or soy protein isolate (25%)	Several months on diet acrylamide urea gels, isoelectric-focusing; apo E increased in VLDL[d] and IDL[e]; apo C increased in VLDL[d], casein vs. soy
Roberts, D. C. K.	39	1981	Apo B increased, apo E increased						
Roberts, D. C. K. and Cohn, J. S.,	40	1981		VLDL[d] = 4[b] (4) IDL[e] = 7 LDL[e] = 15 HDL[d] = 42	1[i] 0.2 3 7	118[b](4) 281 67 11 VLDL[d] = 12[b](4) IDL[e] = 15 LDL[e] = 10 HDL[d] = 11	11[i] 45 15 2 1[i] 1 1 0.2	Castle Hill laboratory white, fasting males, cholesterol-free, low far (1%) diet containing casein (25%), Soy protein isolate (25%)	16 weeks on diet apo B and E increased in VLDL[d] and IDL[e] casein vs. soy; IDL[e] major carrier of cholesterol in casein-fed animals

a   Number of animals in parenthesis.
b   Cholesterol concentration mg/dℓ.
c   Percent area under peaks.
d   VLDL, d < 1.006; LDL, d = 1.006—1.063; HDL, d = 1.063—1.210.
e   IDL, d = 1.006—1.019; LDL, d = 1.019—1.063.
f   Range of observations.

g   Lipoprotein concentration mg/dℓ.
h   VLDL, d < 1.019; HDL_2 d = 1.065—1.125; HDL_3 d = 1.125—1.200.
i   SEM
j   Percent area under OD_{280} curve from ultracentrifugal fractions.
k   Percent of total protein, TMU soluble + TMU insoluble.
l   Percent of total cholesterol.

Table 3
MEAN[a] PLASMA CHOLESTEROL
CONCENTRATIONS IN THE LIPOPROTEINS[b]
OF HYPER- AND HYPO-RESPONDER RABBITS[c]

Lipoprotein	Hyper-responders	Hypo-responders
Noncholesterol-Fed		
VLDL < 1.006	16 ± 1	13 ± 1
IDL = 1.006—1.019	28 ± 1	30 ± 2
LDL = 1.019—1.063	19 ± 1	19 ± 1
HDL = 1.063—1.210	11 ± 1	10 ± 2
Cholesterol-Fed[d]		
VLDL < 1.006	602 ± 80	274 ± 22[e]
IDL = 1.006—1.019	320 ± 33	315 ± 17
LDL = 1.019—1.063	78 ± 15	60 ± 7
HDL = 1.063—1.210	20 ± 3	20 ± 1

[a] Results are mean ± SE mg/dℓ for six animals per group.
[b] Unpublished data, Roberts, D. C. K. and West, C. E.
[c] Semi-lop strain bred for hyper- or hypo-response to cholesterol feeding.[43]
[d] 200 mg cholesterol in 70 g commercial chow per day, fed for 3 weeks.
[e] $p < 0.01$, comparison by Student's t-test, hyper- vs. hypo-responders.

a marked increase in the cholesteryl ester content of VLDL and IDL at the expense of triglyceride and an increase in the apo E content of these fractions. Continued feeding produces increases in the cholesterol content of all lipoprotein fractions (Table 3). Variations in response have been noted between strains[36,45] and within strains [43] (Table 3).

When VLDL is electrophoresed most authors report a change in mobility from pre-β to β upon cholesterol feeding, whether the cholesterol is fed as such[11,44-46] or dispersed in oil[13,14,50,51] (Table 2). Ploplis et al.,[37] however, report no change in VLDL mobility although the pre-β/β ratio still increases with cholesterol feeding. In this case the lipoproteins were separated on polyacrylamide rather than cellulose acetate or agarose as reported by the other authors and this may account for the apparent difference.

The proportion of apo E in VLDL (defined as d < 1.019) increases to as much as 70% of the soluble apoproteins[24,29,44,48] at the expense of the C apoproteins.[24] There is a concomitant increase in apo B with the E/B ratio decreasing with increasing density.[24,44,45,50] Kushwaha and Hazzard[24] suggest the presence of apo E in LDL (d = 1.019 to 1.063) may be HDL$_c$ as reported by Mahley et al.[32] for cholesterol-fed dogs.

Only minor changes are seen in HDL on cholesterol feeding. On agarose electrophoresis the mobility of HDL decreases[37,44,45] and it is present in a reduced proportion.[44,45,50,51] Apo E is present as a minor component of HDL[26,45] with a higher proportion in HDL$_2$ (d = 1.081 to 1.125) than HDL$_3$ (d = 1.125 to 1.200).[45] As in the chow-fed animals, apo A-I is the major soluble apoprotein of HDL.[26,45]

SEMIPURIFIED DIETS

Feeding of low cholesterol or cholesterol-free semipurified diets containing animal protein produces a hypercholesterolemia.[12,16,39,52,53] This results in an increase in apo B and apo E of VLDL and IDL[40,42] with the extra cholesterol being carried in either IDL[41] or LDL[12,52,53] (Table 2). Strain variations in response to casein-feeding are also noted.[27,28]

In summary, the results are reasonably consistent, despite the variables of diet, time on diet, and strain. On induction of hypercholesterolemia in the rabbit, whether by cholesterol or animal protein-feeding, there is an increase in the lipid (principally cholesteryl ester) of VLDL (d < 1.006) and IDL (d = 1.006 to 1.019) with the proportion in each lipoprotein class varying with the degree of hypercholesterolemia. There is an increase in the apo B content of these lipoproteins and the appearance of apo E while HDL remains relatively unchanged.

REFERENCES

1. **Adlersberg, D., Bossak, E. T., Sher, I. H., and Sobotka, M.,** Electrophoresis and monomolecular layer studies with serum lipoproteins, *Clin. Chem.,* 1, 18, 1955.
2. **Albers, J. J. and Dray, S.,** Identification and genetic control of two rabbit low-density lipoprotein allotypes, *Biochem. Genet.,* 2, 24—25, 1968.
3. **Albers, J. J. and Dray, S.,** Identification and genetic control of two new low density lipoprotein allotypes: phenogroups at the lpq locus, *J. Immunol.,* 103, 155—162, 1969.
4. **Alexander, C. and Day, C. E.,** Distribution of serum lipoproteins of selected vertebrates, *Comp. Biochem. Physiol.,* 46, 295—312, 1973.
5. **Berg, K., Boman, H., Torsvik, H., and Walker, S. M.,** Allotypy of high density lipoprotein of rabbit serum, *Proc. Natl. Acad. Sci. U.S.A.,* 68, 905—908, 1971.
6. **Boman, H., Torsvik, H., and Berg, K.,** A second polymorphic trait of rabbit serum high density lipoprotein, *Clin. Exp. Immunol.,* 11, 297—303, 1972.
7. **Borresen, A. L.,** High density lipoprotein (HDL) polymorphisms in rabbit. I. A comparative study of rabbit and human serum high density lipoprotein, *J. Immunogenet.,* 3, 73—81, 1976.
8. **Borresen, A. L.,** High density lipoprotein (HDL) polymorphisms in rabbit. II. A study of the inherited H*l* 1 and R 67 antigens in relation to HDL polypeptides, *J. Immunogenet.,* 3, 83—89, 1976.
9. **Borresen, A. L.,** High density lipoprotein (HDL) polymorphisms in rabbit. III. Quantitative determination of HDL and the inherited H*l* 1 and R 67 antigens, *J. Immunogenet.,* 3, 91—103, 1976.
10. **Borresen, A. L. and Kindt, T. J.,** Purification and partial characterization of the apoA-1 of rabbit high density lipoprotein, *J. Immunogenet.,* 5, 5—12, 1978.
11. **Bosanquet, A. G., Bickerstaffe, R., and Fraser, R.,** A new large chylomicron remnant from cholesterol-fed rabbits, *Aust. J. Exp. Biol. Med. Sci.,* 56, 157—169, 1978.
12. **Brattsand, R.,** Distribution of cholesterol and triglyceride among lipoprotein fractions in fat-fed rabbits at different levels of serum cholesterol, *Atherosclerosis,* 23, 97—110, 1976.
13. **Camejo, G., Bosch, V., Arreaza, C., and Mendez, H. C.,** Early changes in plasma lipoprotein structure and biosynthesis in cholesterol-fed rabbits, *J. Lipid Res.,* 14, 61—68, 1973.
14. **Camejo, G., Bosch, V., and Lopez, A.,** The very low density lipoproteins of cholesterol-fed rabbits. A study of their structure and *in vivo* changes in plasma, *Atherosclerosis,* 19, 139—152, 1974.
15. **Chapman, M. J.,** Animal lipoproteins: chemistry, structure comparative aspects, *J. Lipid Res.,* 21, 789—852, 1980.
16. **Carroll, K. K.,** Dietary protein in relation to plasma cholesterol levels and atherosclerosis, *Nutr. Rev.,* 36, 1—5, 1978.
17. **Gilman-Sachs, A. and Knight, K. L.,** Identification and genetic control of two rabbit high-density lipoprotein allotypes, *Biochem. Genet.,* 7, 177—191, 1972.
18. **Johansson, M. B. and Karlsson, B. W.,** Lipoproteins in serum of rat, mouse, gerbil, rabbit, pig and man studied by electrophoretical and immunological methods, *Comp. Biochem. Physiol.,* 54, 495—500, 1976.
19. **Kawai, K., Maruno, H., Watanabe, Y., and Hirohata, K.,** Fat necrosis of osteocytes as a causative factor in idiopathic osteonecrosis in heritable hyperlipemic rabbits. *Clin. Orthop.,* 153, 273—282, 1980.
20. **Kelus, A. S.,** Lpl allotypic determinant of rabbit beta-lipoprotein, *Nature (London),* 218, 595—6, 1968.
21. **Kirkeby, K.,** Total lipids and lipoproteins in animal species, *Scand. J. Clin. Lab. Invest.,* 18, 437—442, 1966.
22. **Kritchevsky, D.,** Experimental atherosclerosis in rabbits fed cholesterol-free diets, *Atherosclerosis,* 4, 103—105, 1964.
23. **Kritchevsky, D., Marcucci, A. M., Sallata, P., and Tepper, P. A.,** Comparison of amorphous and crystalline cholesterol in establishment of atherosclerosis in rabbits, *Med. Exp.,* 19, 185—193, 1969.

24. **Kushwaha, R. S. and Hazzard, W. R.,** Catabolism of very low density lipoproteins in the rabbit, effect of changing composition and pool size, *Biochim. Biophys. Acta,* 528, 176—189, 1978.
25. **Kushwaha, R. S. and Hazzard, W. R.,** Exogenous estrogens attenuate dietary hypercholesterolemia and atherosclerosis in the rabbit, *Metabolism,* 30, 359—366, 1981.
26. **Kushwaha, R. S., Hazzard, W. R., and Engblom, J.,** High density lipoprotein metabolism in normolipidemic and cholesterol-fed rabbits, *Biochim. Biophys. Acta,* 530, 132—143, 1978.
27. **Lacombe, C. and Nibbelink, M.,** Lipoprotein modifications with changing dietary proteins in rabbits on a high fat diet, *Artery,* 6, 280—289, 1980.
28. **Lacombe, C. and Nibbelink, M.,** Breed differences in nutritionally induced hyperlipoproteinaemia in the rabbit, *Artery,* 7, 419—427, 1980.
29. **Lovati, M. R. and De Marchi, G.,** Plasma lipoprotein composition in cholesterol-fed rabbits treated with pirinixil (BR 931), a new lipid lowering agent, *Pharmacol. Res. Commun.,* 13, 133—139, 1981.
30. **Lozsa, A.,** A considerable sex difference detectable by paper electrophoresis in the serum lipoprotein spectrum of rabbits, *Acta Physiol. Acad. Sci. Hung.,* 42, 137—149, 1972.
31. **Lozsa, A., Kereszti, Z., and Berencsi, G.,** Sex difference and seasonal variations of the serum total lipid and cholesterol levels of normal rabbits, *Acta. Physiol. Acad. Sci. Hung.,* 42, 1—11, 1972.
32. **Mahley, R. W., Weisgraber, K. G., Innerarity, T. L., and Windmueller, H. G.,** Accelerated clearance of low-density and high-density lipoproteins and retarded clearance of E apoprotein-containing lipoproteins from the serum of rats after modification of lysine residues, *Proc. Natl. Acad. Sci. U.S.A.,* 76, 1746—1750, 1979.
33. **Mettler, L., Parwaresch, M. R., and Greggersen, C.,** Säulenelektrophoretische Analyse der Lipoproteinfraktionen an Kaninchen unter dem Einfluss von Östrogenen bei experimenteller Arteriosklerose, *Z. Geburtshilfe Perinatol.,* 175, 277—291, 1971.
34. **Mills, G. L. and Taylaur, C. E.,** The distribution and composition of serum lipoproteins in eighteen animals, *Comp. Biochem. Physiol.,* 43, 489—501, 1971.
35. **Pollak, O. J.,** Purina rabbit chow as a variant of experimental atherosclerosis, *Atherosclerosis,* 5, 524—525, 1965.
36. **Pescador, R.,** Rabbit strain difference in plasma lipoprotein pattern and in responsiveness to hyper-cholesterolaemia, *Life Sci.,* 23, 1851—1862, 1978.
37. **Ploplis, V. A., Thomas, J. R., and Castellino, F. J.,** Comparative studies of the physical state of the lipid phase of normal and hypercholesterolemic very low density lipoprotein, *Chem. Phys. Lipids,* 23, 49—62, 1979.
38. **Rapacz, J.,** Lipoprotein immunogenetics and atherosclerosis, *Am. J. Med. Genet.,* 1, 377—405, 1978.
39. **Roberts, D. C. K.,** Diet and plasma cholesterol: effect of dietary protein, in *Festschrift for F.C. Courtice,* Garlick, D., Ed., University of New South Wales, Sydney, 1981, 130—138.
40. **Roberts, D. C. K. and Cohn, J. S.,** Rabbits with endogenous hypercholesterolaemia have Apoprotein E and Apoprotein AI proteins in their low density fractions, in Aust. Atherosclerosis Group 8th Annu. Meet., Adelaide, May 8, 1981, 24.
41. **Roberts, D. C. K., Huff, M. W., and Carroll, K. K.,** Plasma lipoprotein changes in suckling and weanling rabbits fed semi-purified diets, *Lipids,* 14, 566—571, 1979.
42. **Roberts, D. C. K., Stalmach, M. E., Khalil, M. W., Hutchinson, J. C., and Carroll, K. K.,** Effects of dietary protein on composition and turnover of apoproteins in plasma lipoproteins of rabbits, *Can. J. Biochem.,* 59, 642—647, 1981.
43. **Roberts, D. C. K., West, C. E., Redgrave, T. G., and Smith, J. B.,** Plasma cholesterol concentration in normal and cholesterol-fed rabbits. Its variation and hereditability, *Atherosclerosis,* 19, 369—380, 1974.
44. **Shore, B. and Shore, V.,** An apoliprotein preferentially enriched in cholesterylester-rich very low density lipoproteins, *Biochem. Biophys. Res. Commun.,* 58, 1—7, 1974.
45. **Shore, B. and Shore, V.,** Rabbits as a model for the study of hyperlipoproteinemia and atherosclerosis, *Adv. Exp. Med. Biol.,* 67, 123—141, 1976.
46. **Shore, V. G., Shore, B., and Hart, R. G.,** Changes in apolipoproteins and properties of rabbit very low density lipoproteins on induction of cholesteremia, *Biochemistry,* 13, 1579—1585, 1974.
47. **Sirtori, C. R., Catapalo, A., Ghiselli, G. C., Innocenti, A. L., and Rodriguez, J.,** Metaformin: an antiatherosclerotic agent modifying very low density lipoproteins in rabbits, *Atherosclerosis,* 26, 79—87, 1977.
48. **Sirtori, C. R., Ghiselli, G. C., Catapano, A. L., Lovati, M. R., Fragiacomo, C., Fox, U., Majone, G., and Buchwald, H.,** Reduced apoprotein B and increased lipoprotein turnover in cholesterol-fed rabbits after partial ileal bypass, *Surgery,* 89, 243—251, 1981.
49. **Skogen, B., Borresen, A. L., Matvig, J. B., Berg, K., and Michaelsen, T. A.,** High density lipoprotein as carrier for amyloid-related protein SAA in rabbit serum, *Scand. J. Immunol.,* 10, 39—45, 1979.
50. **Stange, E., Agostini, B., and Papenberg, J.,** Changes in rabbit lipoprotein properties by dietary cholesterol, and saturated and polyunsaturated fats, *Atherosclerosis,* 22, 125—148, 1975.

51. **Stange, E., Alavi, M., and Papenberg, J.,** Changes in metabolic properties of rabbit very low density lipoproteins by dietary cholesterol, and saturated and polyunsaturated fat, *Atherosclerosis,* 28, 1—14, 1977.
52. **Terpstra, A. H. M., Harkes, L., and van der Veen, F. H.,** The effect of different proportions of casein in semi-purified diets on the concentration of serum cholesterol and the lipoprotein composition of rabbits, *Lipids,* 16, 114—119, 1981.
53. **Terpstra, A. H. M. and Sanchez-Muniz, F. J.,** Time course of the development of hypercholesterolaemia in rabbits fed semipurified diets containing casein or soybean protein, *Atherosclerosis,* 39, 217—227, 1981.
54. **Ulevitch, R. J., Johnston, A. R., and Weinstein, D. B.,** New function for high density lipoproteins. Isolation and characterization of a bacterial lipopolysaccharide-high density lipoprotein complex formed in rabbit plasma, *J. Clin. Invest.,* 67, 827—837, 1981.

CANINE PLASMA LIPOPROTEINS: CHARACTERIZATION OF LIPOPROTEINS FROM CONTROL AND CHOLESTEROL-FED DOGS

Robert W. Mahley and Karl H. Weisgraber

INTRODUCTION

The dog is an extremely useful model for the study of lipoprotein metabolism and diet-induced atherosclerosis for several reasons: (1) its large size provides adequate quantities of blood for lipoprotein and apoprotein isolation and characterization, and the opportunity for repeated blood sampling that is required in turnover studies; its size also provides adequate quantities of tissues for various histochemical and biochemical analyses; (2) its docile nature and trainability make handling relatively easy; (3) the general availability of random source animals or specific breeds of dogs gives the investigator maximum flexibility; and (4) both the initial cost and the cost of maintenance are relatively low compared with costs for other large animals. In addition, as will be shown, canine plasma lipoproteins are similar to human lipoproteins, and the apoproteins of the major canine lipoproteins are homologous to the corresponding human apoproteins (for various reviews concerning canine lipoproteins see References 1 to 4).

For many years the dog was considered an inadequate model for studies of experimental atherosclerosis because its response to dietary manipulations varied and because it is naturally resistant to the disease. Development of procedures for metabolic and dietary perturbation, as well as a better understanding of the necessary degree of hypercholesterolemia required to induce atherosclerosis, now make the dog an excellent animal model. The absence of naturally occurring atherosclerosis provides a stable, no-background starting point for studies of diet-induced atherogenesis. The characteristics of the atherosclerosis and conditions necessary for its development have been described previously[2,5,6,7] and will be reviewed in the discussion to follow.

NORMAL CANINE PLASMA LIPOPROTEINS

The levels of the various plasma lipids found in normal dogs fed standard laboratory chow are shown in Table 1. These lipids are distributed among plasma lipoproteins which are equivalent to the major lipoprotein classes described in man. Canine lipoproteins include chylomicrons, very low-density lipoproteins (VLDL), low-density lipoproteins (LDL), and high-density lipoproteins (HDL). The densities at which these lipoproteins can be isolated by preparative ultracentrifugation differ from those of their human counterparts.[8] The typical density distribution for the major canine lipoproteins is shown in Figure 1. An interesting difference in canine plasma lipoproteins is the occurrence of lipoproteins that in many respects resemble HDL, but that float in the density range $d = 1.025$ to 1.10. This lipoprotein class, referred to as HDL_1, overlaps with LDL and cannot be isolated by ultracentrifugation alone.[1,8] Isolation and purification of canine LDL and HDL_1 required the development of a technique in which two distinctly different lipoproteins that float at similar densities could be separated. The methodology, which combines preparative ultracentrifugation and preparative Geon-Pevikon block electrophoresis, has proven highly useful in separating not only canine LDL and HDL_1, but also a variety of normal and cholesterol-induced lipoproteins (e.g., LDL and HDL_c).[5,6,8,9] The Geon-Pevikon electrophoretic technique is technically simple, provides almost quantitative recovery, and is highly reproducible.[8,9] The method is described in detail in the Appendix.

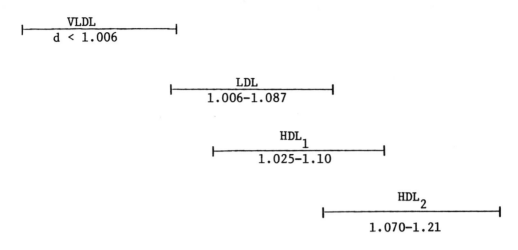

FIGURE 1.   Density distribution of normal canine lipoproteins as isolated by ultracentrifugation (From Mahley, R. W. and Weisgraber, K. H., *Circ. Res.*, 35, 713, 1974. With permission.)

Table 1
CANINE PLASMA LIPID LEVELS

	Triglycerides (mg/dℓ)	Total cholesterol (mg/dℓ)	Cholesteryl esters (mg/dℓ)	Phospholipids (mg/dℓ)
Mean	36	127	97	302
Range	15—70	60—185	40—150	198—437

Note: Results are from duplicate determinations on plasma from 25 control dogs.[8]

The density distribution and the relative mobilities of the canine lipoproteins are illustrated by the paper electrophoretograms shown in Figure 2. Chylomicrons remain at the origin (not seen in fasting plasma), whereas VLDL isolated at d<1.006 have a pre-β electrophoretic mobility. The canine LDL have β-mobility very similar to that of human LDL. The HDL$_1$ and HDL$_2$ have α-mobility but are clearly separable from each other by lipoprotein electrophoresis (Figure 2). The HDL$_1$ have an α$_2$-mobility, whereas HDL$_2$ have α$_1$-mobility.[1,8]

The HDL, specifically the subclass referred to as HDL$_2$, represent the major class of plasma lipoproteins in the normal dog. The HDL$_2$ carry approximately 70% of the total plasma cholesterol. The LDL are a minor subclass of plasma lipoproteins. The typical distribution of total cholesterol and lipoprotein protein within the different ultracentrifugal density fractions is shown in Table 2.

In many respects the chemical compositions of the various canine lipoproteins resemble those of their human counterparts (Table 3). It is of interest that canine LDL contain more triglyceride and less cholesterol than human LDL (Table 3). In fact, canine LDL are composed of approximately equal proportions, one-quarter each, of triglyceride, cholesterol, phospholipid, and protein. However, as will be shown, canine LDL resemble human LDL in most other properties, including their apoprotein content, particle size, and ability to interact with apo B,E receptors.[1,4,6] The HDL$_1$ and HDL$_2$ are protein- and phospholipid-rich lipoproteins with compositions similar to that of human HDL. It should be noted that HDL$_1$ contain more cholesterol than HDL$_2$.[1,8]

Canine plasma lipoproteins are spherical particles as observed by negative staining electron microscopy. As shown in Figure 3, the VLDL, LDL, and HDL$_2$ are approximately equivalent

153

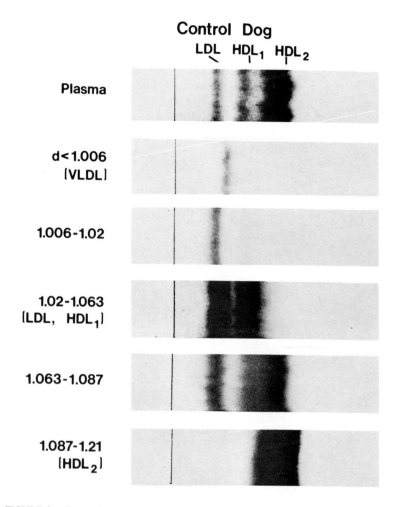

FIGURE 2. Paper electrophoretograms of the lipoproteins in the plasma, and of the lipoproteins obtained at different densities in the plasma of a normal dog.

in size to the corresponding human lipoproteins. The HDL$_1$ overlap the LDL in particle size
and range from ~120 to 300 Å in diameter.[1,8]

The apoprotein constituents of canine lipoproteins on the 11% SDS-polyacrylamide gels
are shown in Figure 4. The principal apoproteins of canine lipoproteins are the B, E, A-I,
and A-IV apoproteins, and a group of lower-molecular-weight constituents representing the
C and A-II apoproteins.[6,8] The properties of each individual canine apoprotein will be
discussed separately. Canine chylomicrons contain the B, E, A-I, A-IV, and C apoproteins.
The apoproteins of VLDL include predominantly apo B, apo E, and apo C. (In most cases,
lower-molecular-weight C apoproteins of VLDL are more abundant than what is shown in
Figure 4.) The HDL$_1$ lack apo B and contain predominantly the E and A-I apoproteins.
Lesser quantities of A-IV and the lower-molecular-weight apoproteins are present. The HDL$_2$
(d = 1.10 to 1.21) contain the A-I as a major constituent along with smaller amounts of
the C and A-II apoproteins.

Table 2
**PERCENT DISTRIBUTION OF
CHOLESTEROL AND LIPOPROTEIN
PROTEIN AMONG THE
ULTRACENTRIFUGAL DENSITY
FRACTIONS FOR A CONTROL AND TWO
CHOLESTEROL-FED DOGS**

Plasma and density fractions	Total cholesterol	Protein
Control dog		
Plasma (mg/dℓ)	102	—
d—<1.006 (%)	1.6	0.8
1.006—1.063	10.2	6.9
1.063—1.087	17.7	14.4
1.087—1.21	70.6	77.9
Cholesterol-fed (I)[a]		
Plasma (mg/dℓ)	250	—
d<1.006 (%)	4.6	1.6
1.006— 1.063	56.1	22.9
1.063—1.087	19.8	29.3
1.087—1.21	19.5	46.1
Cholesterol-fed (II)[a]		
Plasma (mg/dℓ)	750	—
d<1.006 (%)	61.5	30.1
1.006—1.063	28.5	26.0
1.063—1.087	4.2	12.9
1.087—1.21	5.8	30.9

[a]   Hypothyroid dogs fed a high-fat and high-cholesterol diet
(see Reference 5 for details).

Table 3
PROTEIN AND LIPID COMPOSITION OF CANINE LIPOPROTEINS

Fraction	Triglyceride (mg/dℓ plasma)	Cholesterol (mg/dℓ plasma)	Cholesteryl esters (mg/dℓ plasma)	Phospholipid (mg/dℓ plasma)	Protein (mg/dℓ plasma)
VLDL	9.0 ± 1.7 (58.8)	2.3 ± 0.5 (15.0)	1.8 ± 0.5	2.5 ± 1.0 (16.3)	1.5 ± 0.1 (9.8)
LDL	7.6 ± 1.8 (29.4)	5.7 ± 1.9 (22.1)	4.2 ± 0.2	6.4 ± 1.8 (24.8)	6.1 ± 1.9 (23.6)
HDL₁	0.3 ± 0.05 (2.2)	4.6 ± 1.4 (34.6)	3.4 ± 0.2	5.4 ± 1.3 (40.6)	3.0 ± 0.6 (22.6)
HDL₂	2.2 ± 0.3 (0.6)	70.0 ± 12.0 (20.4)	51.5 ± 1.3	124.6 ± 10.3 (36.4)	145.8 ± 8.0 (42.6)

Note: Results are means ± SE from duplicate determinations on the plasma lipoproteins from four dogs. The percent composition is given in parentheses.[8]

CHARACTERIZATION OF CANINE APOPROTEINS

Apoprotein B

Detailed comparative studies of canine and human apo B have not been conducted. As with several other animal species, the basis for the identification of a canine protein as an apo B equivalent is the observation that this protein constituent of LDL and VLDL does not

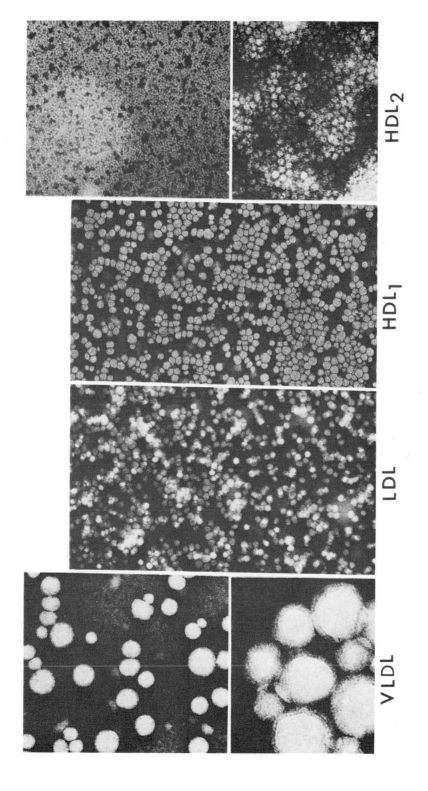

FIGURE 3. Negative stain electron microscopy of normal dog lipoproteins (From Mahley, R. W. and Weisgraber, K. H., *Circ. Res.*, 35, 713, 1974. With permission.)

Canine Lipoproteins

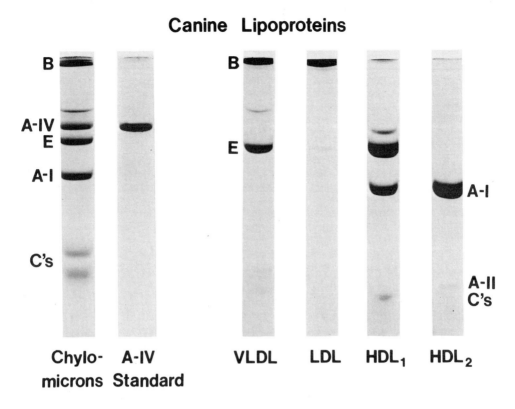

FIGURE 4.   SDS-PAGE (11% gel) of normal canine lipoproteins.

enter 10 to 15% polyacrylamide gels using either SDS or urea as a detergent. However, we have observed that the major protein constituent of canine LDL cross-reacts with antisera prepared against the human B apoprotein. The presence of this protein almost exclusively in VLDL and LDL is further evidence that this protein is equivalent to human apo B. Furthermore, it can be shown on 4% SDS-polyacrylamide gels that canine apo B exists predominantly in two forms, a high-molecular-weight form (B-100) and a lower-molecular-weight form (B-48) (Figure 5). It is also possible to demonstrate that, even with 10% SDS gels, chylomicron apo B penetrates the gel (Figure 4), whereas the apo B of VLDL and LDL does not (Figure 4). This suggests that the intestinal lipoproteins (chylomicrons) possess the lower-molecular-weight form of apo B (B-48) as described for humans and rats.[10,11]

Apoprotein E

Canine apo E is present principally in VLDL and HDL_1. It is a 37,000 mol wt (M_r) protein that is larger than human apo E ($\sim$34,000 M_r) (Figure 6). Canine apo E cross-reacts with immunochemical identity with human apo E when antisera prepared to human E apoprotein is used. Amino acid analyses comparing dog and human apo E reveal striking similarities (Table 4). However, the amino terminal residues as determined by automated Edman degradation are not homologous.[12] Amino acid sequence analysis of canine and human apo E does, however, reveal a remarkable homology between certain segments of these proteins (unpublished data). It is noteworthy that canine apo E lacks cysteine.

Canine apo E appears to have two major isoforms on isoelectric focusing gel electrophoresis. The pI values for the canine E isoforms (pI $\sim$ 5.0 to 5.4) are distinctly different from those of the human E isoforms (pI $\sim$ 5.4 to 6.1).

Apoprotein A-I

Canine apo A-I, the major constituent protein of HDL_2, coelectrophoreses with the human,

Canine Lipoproteins

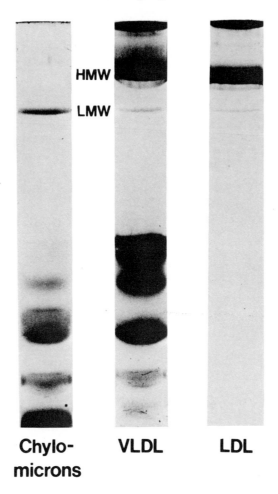

FIGURE 5.　PAGE of canine lipoproteins on 4% gels. The high-molecular-weight (HMW) and low-molecular-weight (LMW) forms of the B apoproteins are indicated.

swine, rat, and patas monkey A-I apoproteins and has an apparent molecular weight of 28,000, as determined by SDS-PAGE (Figure 7). Likewise, the amino acid analyses of dog and human apo A-I are remarkably similar. The first ten amino-terminal amino acid residues of canine apo A-I are very similar to those found in the human, swine, and patas monkey A-I apoprotein.[13]

Apoprotein A-II

A low-molecular-weight protein ($\sim$8500 M_r) has been isolated from canine HDL. This apoprotein appears to co-electrophorese with rat and patas monkey apo A-II (Figure 8) and has a lower molecular weight than the dimeric human A-II apoprotein ($\sim$17,000 M_r).

DIET-INDUCED CHANGES IN THE LIPOPROTEINS AND THE INDUCTION OF ATHEROSCLEROSIS

Dogs are highly resistant to the development of hypercholesterolemia and atherosclerosis,

Purified E Apoprotein

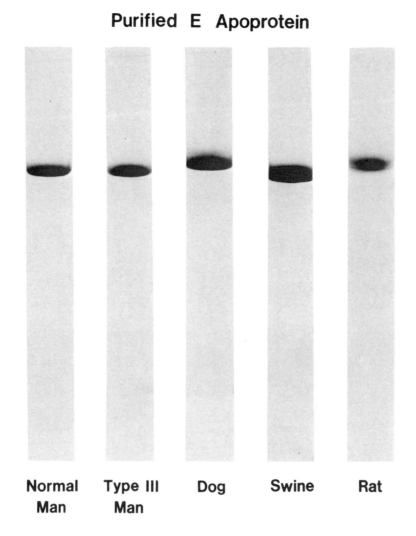

| Normal Man | Type III Man | Dog | Swine | Rat |

FIGURE 6. Comparison of the electrophoretic mobility (11% SDS gels) of the E apoprotein isolated from various species.

and more than a simple addition of fat and cholesterol to their diet is required to induce experimental atherosclerosis. A plasma cholesterol level in excess of 750 mg/dℓ must be maintained for more than 6 months before significant atherosclerosis is observed.[4,5,6] To accomplish this, the dog must be rendered hypothyroid and fed a diet that contains 1 to 3% cholesterol, 15 to 20% fat, and 0.75% taurocholic acid.[5,7] Alternatively, hypercholesterolemia can be induced in euthyroid dogs by feeding them a semisynthetic diet that contains 16% hydrogenated coconut oil as the only fat and 5% cholesterol.[6] The main characteristics of the hyperlipoproteinemia induced by both of these dietary protocols are essentially identical.[1,2,5,6] Cholesterol feeding alone results in only a minimal increase in the plasma cholesterol of dogs. Several investigators have used one or the other of these protocols to induce canine atherosclerosis and hypercholesterolemia (for review see Reference 2). When a diet containing 5% lard, 1% cholesterol, 0.75% taurocholic acid, and 0.1% propylthiouracil is fed to rats, plasma cholesterol levels of 400 to 600 mg/dℓ are observed[14,15] and plasma lipoprotein changes occur which are similar to those seen in dogs.

The cholesterol-induced alterations in canine plasma lipoproteins accomplished by either

Table 4
AMINO ACID COMPOSITION OF
THE E APOPROTEIN (MOL %)

Amino acids	Canine apo E[a]	Human apo E[b]
Arg	10.5	11.2
Lys	4.5	4.1
His	0.6	0.7
Asp	6.5	4.3
Thr	4.1	3.6
Ser	4.2	4.7
Glu	26.1	24.3
Pro	3.8	3.0
Gly	4.8	6.1
Ala	11.8	12.3
Val	5.3	7.5
Met	1.9	2.1
Ile	0.9	0.6
Leu	12.3	12.6
Tyr	0.8	1.2
Phe	1.1	1.0
Cys	0	0.4

[a]   See Reference 12.
[b]   Apo E isolated from an E3 homozygous subject.

of these dietary protocols can be seen in a comparison of electrophoretograms (Figure 9). The electrophoretic pattern of control dog plasma lipoproteins is contrasted with those of three different hypercholesterolemic dogs with increasing plasma cholesterol levels. As described above, in control animals, the lipoproteins seen are LDL, HDL_1, and HDL_2. (The VLDL are too low in concentration to be visualized.) Initially, as the plasma cholesterol level increases, the cholesterol is carried principally by the LDL and by α_2-migrating lipoproteins similar to HDL_1. We have called these α_2-migrating, cholesterol-rich lipoproteins "HDL_c" to indicate by the subscript "c" that they are cholesterol induced, and to distinguish them from control HDL_1.[1,4,5,6] The HDL_c represent an increased concentration of HDL_1, and their concentration rises with further increases in plasma cholesterol.[2,4,5] For dogs that develop significant atherosclerosis (those with cholesterol levels greater than 750 mg/dℓ; bottom two patterns in Figure 9), the electrophoretic patterns are more complicated. There is a broad β band that corresponds to the presence of β-VLDL and LDL. There is also a prominent HDL_c band and a markedly reduced HDL_2 band. In addition, the atherogenic hyperlipoproteinemia is often characterized by the presence of chylomicrons and chylomicron remnants, which stay at the origin of the electrophoretic pattern or trail from the origin to the β position.

As suggested by electrophoresis of the whole plasma, there are marked changes in the distribution of the various lipoprotein classes following cholesterol feeding. To study these changes, the lipoprotein classes were isolated by preparative ultracentrifugation (Figure 10). When a fraction contained more than one lipoprotein class, Geon-Pevikon block electrophoresis was used to purify the individual lipoproteins.[5,6,8] The altered distribution induced by these diets can be appreciated by comparing the distribution of the total cholesterol and lipoprotein protein in the different ultracentrifugal fractions (Table 2). In the cholesterol-fed dogs, plasma cholesterol is transported principally by the low and very low density lipoproteins. It is noteworthy that in the cholesterol-fed, hypothyroid dog (II), ~60% of the plasma cholesterol appears in the d<1.006 fraction (Table 2). Such high levels of d<1.006 lipoproteins do not develop in euthyroid dogs fed the coconut oil/cholesterol diet.[6]

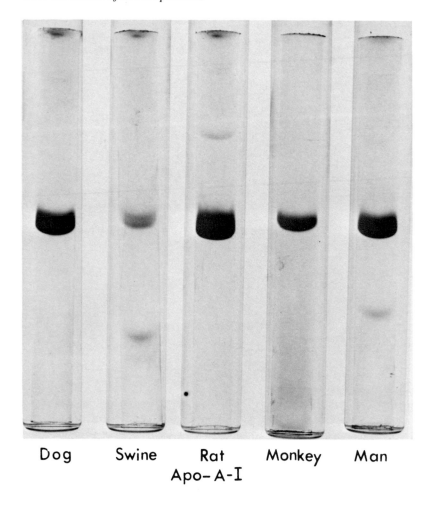

Dog   Swine   Rat   Monkey   Man

Apo-A-I

FIGURE 7.   Comparison of the electrophoretic mobility (11% SDS gels) of the A-I apoprotein isolated from various species.

In contrast to cholesterol-fed dogs with plasma cholesterol levels in excess of 750 mg/ dℓ, the d<1.006 fraction from control dogs fasted 15 to 16 hr contains only the pre-β-migrating VLDL (Figure 2). However, following cholesterol feeding, β-migrating lipoproteins, referred to as β-VLDL, appear in the d<1.006 fraction (Figure 10).[1,4-6] In addition, the d<1.006 fraction contains lipoproteins that remain at or near the origin. These have been shown to represent chylomicrons and chylomicron remnants that are present even after 15 to 24 hr of fasting.[16] The β-VLDL of cholesterol-fed dogs are similar to the β-VLDL described in patients with Type III hyperlipoproteinemia and contain 30 to 40% triglyceride and 30 to 40% cholesterol (Table 5). However, the cholesterol content of the β-VLDL rises with increasing concentrations of plasma cholesterol. In addition to the appearance of β-VLDL, cholesterol feeding also produces an increase in the d = 1.006 to 1.02 intermediate lipoprotein fractions (Figure 10). These lipoproteins are also cholesterol rich. Other changes visible in Figure 10 include an increase in LDL and the appearance of HDL_c. The LDL and HDL_c are both cholesterol rich, but they are quite different in their apoprotein contents (see below). There is both a relative and an absolute decrease in HDL_2, but the percent composition of the HDL_2 of control and cholesterol-fed animals is essentially unchanged. The chemical compositions and particle sizes of the control and cholesterol-induced lipoproteins are compared in Table 5.

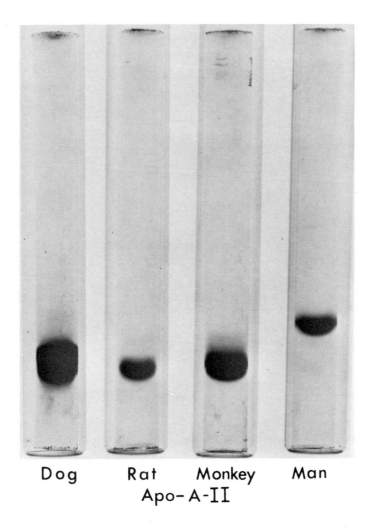

FIGURE 8.   Comparison of the electrophoretic mobility (11% SDS gels) of the A-II apoprotein from various species.

The apoproteins of the various cholesterol-induced canine lipoproteins are compared in Figure 11. The d<1.006 lipoproteins (β-VLDL) contain primarily the B and E apoproteins (Figure 11). The E apoprotein is a major constituent of the β-VLDL, as it is in the apoprotein patterns of β-VLDL from patients with Type III hyperlipoproteinemia and from other cholesterol-fed animals (for review see References 1,2,4). In HDL$_c$ isolated from various ultracentrifugal fractions, the E apoprotein is the major protein constituent. These HDL$_c$ contain no B apoprotein. The HDL$_c$ isolated from the d = 1.006 to 1.02 fraction of cholesterol-fed dogs contain apo E as the exclusive protein constituent and are referred to as apo E HDL$_c$.[6,17] The HDL$_c$ isolated at higher densities contain both E and A-I apoproteins (Figure 11). The presence of apo E as a major apoprotein is a consistent feature of the various cholesterol-induced lipoproteins.[1,2,4] For a description of the procedure used for isolating HDL$_c$, see the Appendix.

Control HDL$_1$ and cholesterol-induced HDL$_c$ are not unique to the dog. We have shown that there are HDL in normal human subjects that contain the E apoprotein (so-called HDL$_1$ or HDL-with apo E).[3,4,18-20] The HDL$_2$ of the dog lack the E apoprotein, as do the majority of HDL that occur in humans (so-called HDL-without apo E). In dogs, these two distinct HDL subclasses are easily and clearly distinguishable by electrophoresis.

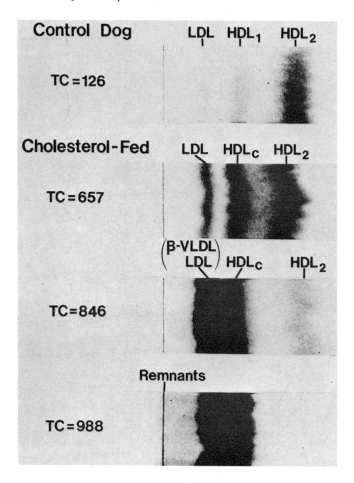

FIGURE 9.   Electrophoretic patterns of plasma lipoproteins from a normal dog and three cholesterol-fed dogs. The total plasma cholesterol levels (TC) of the cholesterol-fed dogs are indicated. The dogs fed fat and cholesterol were thyroidectomized (see Reference 5 for description of the diet).

The metabolism of the various lipoproteins has not been entirely elucidated. However, in the sections to follow, our current understanding of the metabolism will be reviewed.

ALTERATIONS IN THE HDL

As stated above, diet-induced hypercholesterolemia is accompanied by an increase in HDL-with-apo-E (HDL$_1$ or HDL$_c$) and a decrease in the typical HDL, HDL-without-apo-E (HDL$_2$). These changes can be observed qualitatively, as described, and quantitatively by using heparin-Sepharose affinity chromatography.[19,20] In fact, the HDL-with-apo-E have been observed to increase three- to fourfold within the first 4 days of cholesterol feeding in dogs.[3,4,19] Over the past several years, we have referred to these cholesterol-induced lipoproteins as HDL$_c$, indicating that they are HDL, increased in concentration by cholesterol feeding.[4] The HDL$_c$ become enriched in cholesteryl ester as the plasma cholesterol level is elevated by the diet. As the cholesteryl ester content of these HDL is increased, the particles become larger (160 to 300 Å in diameter), float at lower densities (as low as d = 1.006), and are progressively enriched in the E apoprotein at the expense of A-I apoprotein. In dogs, it is possible to obtain large, cholesteryl ester-rich HDL$_c$ that contain only the E apoprotein

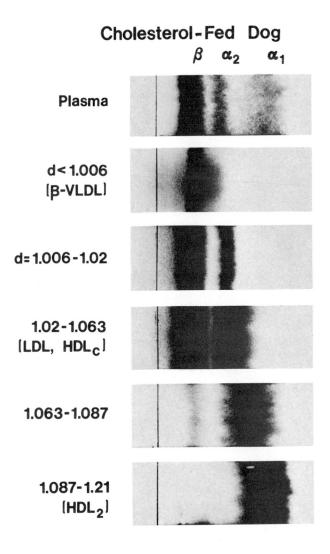

Cholesterol-Fed Dog

β α₂ α₁

Plasma

d< 1.006
[β-VLDL]

d= 1.006-1.02

1.02-1.063
[LDL, HDL꜀]

1.063-1.087

1.087-1.21
[HDL₂]

FIGURE 10. Electrophoretic mobility and ultracentrifugal density
distribution of plasma lipoproteins of a coconut oil/cholesterol-fed dog
(see Reference 6 for a description of the diet).

(referred to as apo E HDL꜀).[6,17] It should be emphasized that apo E HDL꜀ represent an
extreme in the spectrum of changes that occur with cholesterol loading of HDL. It has also
been possible to demonstrate that the concentration of HDL-with apo-E increases after
cholesterol feeding in humans,[3,4,21] swine,[22] and rats.[14]

ORIGIN OF HDL꜀

We have postulated that HDL꜀ are formed in the plasma or extracellular space as a result
of typical HDL acquiring cholesterol from peripheral tissues.[1,3,4] Furthermore, it has been
suggested that the free cholesterol is esterified by the action of lecithinolesterol acyltransferase
(LCAT) and that, as the particles are enriched in cholesteryl ester, the E apoprotein is
redistributed from other plasma lipoproteins to HDL꜀.[4] Many aspects of this scheme, shown
in Figure 12, remain speculative. However, results from several studies support the concept
(as reviewed elsewhere[1,3,4]).

Table 5

PERCENT CHEMICAL COMPOSITION AND PARTICLE SIZE OF CANINE LIPOPROTEINS

	Control diet				Cholesterol-fed			
	VLDL	LDL	HDL$_1$	HDL$_2$	β-VLDL	LDL	HDL$_c$	HDL$_2$
Triglyceride	58.8	29.4	2.2	0.6	40.1	8.3	0.9	1.2
Total cholesterol	15.0	22.1	34.6	20.4	35.0	43.8	51.5	21.0
Phospholipid	16.3	24.8	40.6	36.4	18.2	26.3	32.0	36.3
Protein	9.8	23.6	22.6	42.6	6.7	21.6	15.6	41.5
Size (A)	260—900	160—250	120—350	60—90	220—600	160—300	120—340	70—90

Note: The lipoproteins were obtained from the following ultracentrifugal fractions: VLDL or β-VLDL (d <1.006), LDL, HDL$_1$ or HDL$_c$ (1.02—1.063), and HDL$_2$ (1.10—1.21). The cholesterol-fed dog was surgically thyroidectomized and fed a diet containing 16% beef tallow, 1 to 3% cholesterol, and 0.75% taurocholate. In addition, the dog received 500 mg/day of propylthiouracil by capsule. The plasma cholesterol level of the cholesterol-fed dog was 1120 mg/dℓ.

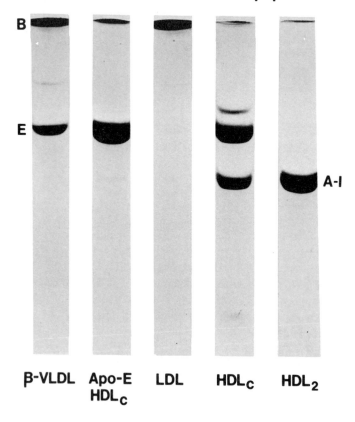

FIGURE 11.   SDS-PAGE (11% gels) of various cholesterol-induced canine plasma lipoproteins. The diet used was the coconut oil/cholesterol protocol. The apo E HDL$_c$ were isolated at d = 1.006 to 1.02 and purified by Geon-Pevikon block electrophoresis. The LDL and HDL$_c$ were separated from each other by Geon-Pevikon electrophoresis after isolation by ultracentrifugation at d = 1.02 to 1.063. The HDL$_2$ (HDL-without apo E) were obtained at d = 1.10 to 1.21.

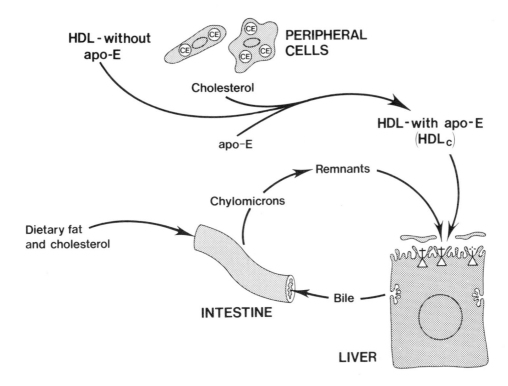

FIGURE 12.   Diagram illustrating the conversion of HDL-without apo E to HDL-with apo E (HDL_c) as they acquire cholesterol from peripheral cells. The HDL-with apo E are recognized by hepatic lipoprotein receptors (apo B,E and apo E). Chylomicron remnants are cleared from the plasma by the liver via the same receptor-mediated process responsible for HDL_c uptake. (From Mahley, R. W., in *Medical Clinics of North America: Lipid Disorders,* Vol. 66, Havel, R. J., Ed., W. B. Saunders, Philadelphia, 1982, 375. With permission.)

ROLE OF HDL-WITH APO-E IN CHOLESTEROL METABOLISM

HDL-with apo-E interact with the apo B,E receptors and regulate cholesterol metabolism.[6,17,23,24] In this regard, they resemble LDL. The question arises as to whether it would be reasonable for the HDL to have taken up cholesterol from tissue and be converted to the HDL-with apo-E, which could in turn be taken up again by those same tissues via the apo B,E receptors. It should be remembered that cells possessing the apo B,E receptors express these receptors only when the cells lack the requisite amount of cholesterol for cholesterol homeostasis. Therefore, the HDL-with-apo E would redistribute cholesterol to other cells that require it, transporting cholesterol from tissues or cells with elevated levels of cholesterol to cells requiring cholesterol for cell growth, steroidogenesis, etc. (depicted schematically in Figure 13).

Another major role for HDL-with apo E appears to be the delivery of cholesterol to hepatocytes for excretion from the body (see Figures 12 and 13). The HDL-with apo E would then function in the role of reverse cholesteryl ester transport. Recent studies have indicated that the E apoprotein may serve as the major determinant for lipoprotein recognition by hepatic receptors, especially in adult human, swine, and dog livers.[25,26] It has been shown that HDL_c are rapidly and efficiently cleared from the plasma by the liver; more than 90% of the injected dose is cleared in 20 min.[27,28] This uptake process has been shown to be receptor mediated. The kinetic parameters describing the uptake of apo E HDL_c are identical with those observed for chylomicron remnants, suggesting the involvement of the same hepatic uptake process for both lipoproteins.[29] Various studies emphasize the importance of the E apoprotein in mediating the uptake of lipoproteins by the liver.[25,28-33]

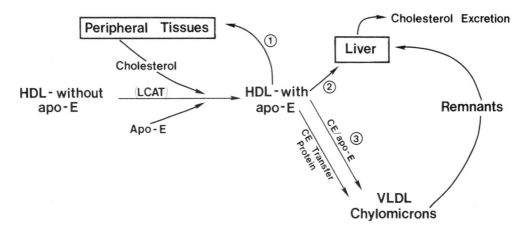

FIGURE 13.   Scheme illustrating the metabolic role of HDL-with apo E (HDL$_1$ and HDL$_c$) in cholesterol metabolism. (1) The cholesteryl ester-enriched HDL-with apo E (HDL$_c$) may function to redistribute cholesterol to peripheral tissues possessing the apo B,E receptor. (2) Alternatively, the HDL may deliver cholesterol directly to the liver for excretion. (3) In addition, under certain metabolic conditions HDL-with apo E may serve as intermediaries for the transfer of cholesteryl ester to other plasma lipoproteins, which in turn are taken up by the liver. (From Mahley, R. W., in *Medical Clinics of North America: Lipid Disorders,* Vol. 66, Havel, R. J., Ed., W. B. Saunders, Philadelphia, 1982, 375. With permission.)

Recently, the importance of the E apoprotein and the potential importance of HDL-with apo-E in hepatic cholesterol metabolism has been further emphasized because of the observation that the liver possesses a unique receptor (the apo E receptor) that interacts specifically with apo E-containing lipoproteins, and does not interact with LDL.[25,26] This apo E receptor, which binds HDL-with apo-E and chylomicron remnants, but not HDL-without apo-E or chylomicrons, is present on the hepatic membranes of adult dogs, swine, and humans as the only detectable receptor.[25,26] However, young, developing dogs or swine possess typical apo B,E receptors in addition to the apo E receptors. Furthermore, the apo B,E receptors in immature dog livers are completely repressed by cholesterol feeding, but the apo E receptors are not significantly reduced in number.[25,26] The expression of the hepatic apo E receptors does not appear to be strictly regulated by dietary perturbations and may serve a primary role in the hepatic uptake of cholesterol-rich chylomicron remnants and HDL-with apo-E in the delivery of cholesterol to the liver for excretion.

The scheme shown in Figure 13 indicates that cholesterol may be delivered to the liver by one or more alternate pathways. The importance of cholesteryl ester transfer protein(s) in the transfer of cholesteryl esters to other lipoproteins, e.g., chylomicrons or VLDL, which can then transport the cholesterol to the liver, has been described (for review see Reference 4).

ROLE OF HDL IN ATHEROGENESIS

Epidemiologic studies have shown a negative correlation between HDL levels and coronary artery disease (for review see Reference 34). One of the characteristics of atherogenic hypercholesterolemia in dogs is a marked reduction in typical HDL (α_1-migrating HDL-without apo-E).[5] Where there are high levels of lipoproteins capable of delivering cholesterol to cells and low levels of typical HDL, cholesterol may be deposited in cells of the arterial wall, resulting in atherosclerosis. This scheme suggests that HDL-with apo-E (HDL$_c$) have an anti-atherogenic role and are produced to mediate the removal of excess cholesterol from tissues and from the body.

β-VERY LOW DENSITY LIPOPROTEINS (β-VLDL)

Cholesterol-induced hypercholesterolemia and accelerated atherosclerosis are associated with the production of cholesteryl ester-rich lipoproteins that float at d<1.006 and have β-electrophoretic mobility. They are referred to as β-VLDL. These lipoproteins have been observed in a variety of species fed high levels of cholesterol or cholesterol and fat, including dogs, rats, rabbits, swine, and monkeys.[1,5,6,14,22,35]

In recent studies from our laboratory it has become apparent that the β-VLDL (the β-migrating, d<1.006 lipoproteins isolated by Geon-Pevikon block electrophoresis) are heterogeneous.[16] There are two distinct subclasses of lipoproteins present, which can be separated by agarose column chromatography. One subclass of the β-VLDL (Fraction I) elutes from the column in the void volume. These lipoproteins resemble chylomicron remnants and appear to originate in the intestine. They are approximately 1500 Å in diameter and possess the lower-molecular-weight form of apo B (B-48) as a major protein constituent. The kinetics of their clearance from the plasma resemble those of chylomicron remnants. The second subclass of β-VLDL (Fraction II) elutes from the column as a clearly distinct population of particles (~350 Å in diameter). The Fraction II lipoproteins possess the higher-molecular-weight form of apo B (B-100) and may originate in the liver. Both β-VLDL subclasses contain the E apoprotein as a major constituent. In addition, they both interact with the so-called β-VLDL receptors on the surface of macrophages and cause a marked accumulation of cholesteryl esters within these cells. Fraction I lipoproteins are 3- to 15-fold more active, however, in stimulating cholesteryl ester accumulation in macrophages than Fraction II lipoproteins.[16]

Cholesterol-induced β-VLDL from monkeys, dogs, rabbits, and swine are the only naturally occurring normal or cholesterol-induced lipoproteins that will cause a marked (20- to 160-fold) increase in the cholesteryl ester content of macrophages (Figure 14).[36,37] This observation has focused attention on the macrophage as a possible progenitor of foam cells in the atherosclerotic lesion and on β-VLDL as a potentially important atherogenic lipoprotein. None of the other normal or cholesterol-induced lipoproteins will cause a similar accumulation of cholesterol in these cells. Studies from various laboratories, using a variety of techniques, strongly implicate the macrophage as a key cell in the atherosclerotic lesion (for review see Reference 4).

β-VLDL AND ATHEROSCLEROSIS IN DOGS

A detailed examination of the changes in the lipoproteins of cholesterol-fed dogs provides insight into the role of specific lipoproteins in the development of atherosclerosis. Characterization of both atherogenic and nonatherogenic hypercholesterolemia reveals two major differences (see Figures 9 and 10). One is a marked reduction in the typical HDL in the atherogenic hyperlipoproteinemia; the other is the production of β-VLDL. These two changes appear to combine to accelerate atherosclerosis in dogs. As suggested above, the mechanism would involve the delivery of cholesterol to the arterial wall and eventual deposition of cholesterol into the macrophages.

The synergistic effects of high levels of d<1.006 lipoproteins (β-VLDL) and low levels of HDL in the production of atherosclerosis in the canine model are clearly emphasized in studies comparing the ability of the d<1.006 lipoproteins of nonatherogenic hyporesponders and atherogenic hyperresponders to deliver cholesterol to macrophages. It has been shown that the d<1.006 lipoproteins of both hyporesponders and hyperresponders effectively stimulate [^{14}C]oleate incorporation into the cholesteryl esters of macrophages.[38] The d<1.006 lipoproteins from normal dogs are very ineffective by comparison. The d<1.006 lipoproteins from hyporesponders generally do not have β-electrophoretic mobility, but they are enriched

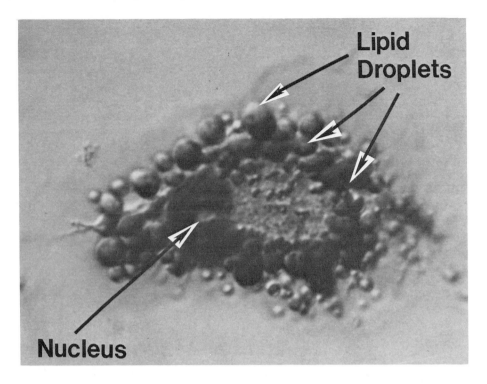

FIGURE 14.   Macrophages from the peritoneum of mice were incubated with canine β-VLDL for 16 hr (100 μg of β-VLDL cholesterol per milliliter in the medium). This cell, which is loaded with oil red O-positive lipid droplets (cholesteryl esters), resembles a lipid-laden foam cell. (From Mahley, R. W., in *Medical Clinic of North America: Lipid Disorders,* Vol. 66, Havel, R. J., Ed., W. B. Saunders, Philadelphia, 1982, 375. With permission.)

in cholesterol. Although the hyporesponder d<1.006 lipoproteins can deliver cholesterol to macrophages, hyporesponder dogs (plasma cholesterol less than 750 mg/dℓ; Figure 9) do not develop atherosclerosis. This observation has focused our attention on the possible protective effect of high (normal) levels of typical HDL in the face of abnormal d<1.006 lipoproteins. In fact, it has been shown that when typical HDL are added at various concentrations along with very high levels of β-VLDL cholesterol, there is a marked inhibition in cholesteryl ester synthesis and accumulation,[38] but it is not due to an inhibition of β-VLDL binding, or delivery of cholesterol to the cells.

On the basis of these studies, it is reasonable to speculate that atherosclerosis develops in the dog when the lipoprotein pattern favors the net influx of cholesterol into the cells. The capability of cholesterol-rich, d<1.006 lipoproteins to deliver cholesterol to macrophages is balanced by normal levels of typical HDL. This, coupled with the findings discussed previously, leads to the postulate that typical HDL pick up excess cellular cholesterol, become HDL$_c$, and then deliver cholesterol to the liver for excretion. In hyperresponders, the β-VLDL may deliver excessive quantities of cholesterol to the tissues. Because of the low levels of typical HDL, the reverse cholesterol transport system is impaired in the hyperresponders and atherosclerosis results. Such a scheme is speculative and open to criticism. However, it is offered as a reasonable interpretation of the data.

SUMMARY

The dog is an interesting animal model for the study of lipoprotein metabolism and diet-induced atherogenesis. Many of the characteristics of canine lipoproteins resemble those of human lipoproteins, including chylomicrons, VLDL, LDL, and HDL. Studies of the dog have highlighted the importance of two different subclasses of HDL: the HDL-with apo-E (HDL$_1$ and HDL$_c$) and HDL-without apo-E. Unique cholesterol-induced lipoproteins, referred to as apo E HDL$_c$, are cholesteryl ester-rich lipoproteins that resemble LDL in many respects, but differ from LDL in the absence of apo B. The apo E HDL$_c$ possess the E apoprotein as the exclusive protein constituent. The apo-E HDL$_c$ have proved to be invaluable in elucidating the role of the E apoprotein in lipoprotein metabolism. A study of these lipoproteins led to the discovery that the LDL (apo B,E) receptor interacts with lipoproteins containing either the B or E apoproteins. In addition, apo-E HDL$_c$ have been useful in establishing the importance of the E apoprotein in mediating lipoprotein uptake by the liver.

Alterations in plasma lipoproteins induced by diets high in saturated fat and cholesterol can be dramatically illustrated in the dog. These changes include the occurrence of β-VLDL, an increase in LDL, the appearance of cholesterol-rich HDL (HDL$_c$), and a decrease in typical HDL (HDL-without apo-E). These same changes occur in many other animals, including man, after the consumption of diets high in fat and cholesterol.

ACKNOWLEDGMENTS

We wish to acknowledge the important contributions of colleagues with whom it continues to be our pleasure to be associated: Drs. T. P. Bersot, M. Fainaru, T. L. Innerarity, R. E. Pitas, S. C. Rall, and Mrs. K. S. Holcombe. Also, thanks are expressed to Mrs. K. S. Holcombe, Mr. J. F. Andres, Mr. R. P. Levine, and Mr. R. Wolfe for editorial assistance and manuscript preparation.

APPENDIX

Geon-Pevikon Block Electrophoresis

The Geon-Pevikon block electrophoretic procedure, based on the method of Barth, et al.[39] has been used widely to separate plasma lipoproteins.[5,6,8,9,16,17] The apparatus is composed of two buffer chambers, each with three baffle plates and a platinum electrode (Figure 15). Each buffer chamber has a 4-ℓ capacity and is fabricated of 1/4-in. Plexiglass. The chambers measure 6 × 12 in. and are 6 1/2-in. deep. A Plexiglass tray, which supports the Geon-Pevikon medium, is placed between the buffer chambers (Figure 15). The tray measures 11 3/8-in. wide by 15-in. long by 5/8-in. deep. Plexiglass covers for the tray and the buffer chamber are made of 1/8-in. Plexiglass. A power supply to deliver a constant current of 50 mA is required. Other apparatus are a coarse sintered-glass filter (125 mm) mounted on a 4-ℓ vacuum flask, several smaller coarse sintered-glass filters (45 mm), sidearm test tubes (200 × 25 mm), Telfa surgical dressing pads (Kendall Hospital Products, Boston, 3 × 8-in.), and Parafilm (large rolls).

Approximately 8 ℓ of barbital buffer (pH 8.6, ionic strength 0.06) are needed. The lipoproteins used for markers are prestained with Sudan black B. A stock solution of Sudan black B, 1% in diethylene glycol, is prepared by heating to 110°C and filtering; this solution can be stored for several months. Prior to each experiment, the stock solution is diluted 1:20 with distilled H$_2$O, mixed with Tween 20® (1:100), and used for prestaining lipoproteins.

Several different types of Geon and Pevikon have been used over the years, and specific types have been found to vary from batch to batch. We now recommend that Pevikon be used without Geon. Consistently, we find that Pevikon C870 is suitable for all separations (Mercer Consolidated Corporation, 216 Lake Avenue, Yonkers, N.Y. 10701).

To wash the Pevikon prior to use, 600 g are placed in a 2-ℓ beaker with 900 mℓ of distilled H$_2$O and stirred to form a slurry. After the slurry stands 10 min, the fine particles are removed by aspiration, and then the slurry is transferred to a large sintered-glass filter on a vacuum flask and washed two more times (900 mℓ of distilled H$_2$O). After the last H$_2$O wash, the mixture is almost dried by suction and then resuspended in the filter with 800 mℓ of barbital buffer. The Pevikon is again almost dried by suction. The stem of the filter is covered with Parafilm to stop draining, 500 mℓ of barbital buffer are added, and the slurry is resuspended in the filter and poured into a 2-ℓ beaker (avoid bubbles). The Plexiglass tray is lined with Parafilm and four Telfa pad wicks are placed across each end of the tray. The wicks, soaked in barbital buffer, are applied to the bottom of the tray and extend inward about 1 in. from the end of the tray.

The Pevikon slurry is poured slowly into the tray (~1100 mℓ of slurry are needed). Care is taken to keep the

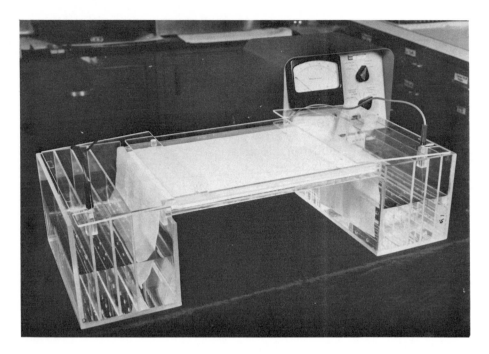

FIGURE 15.   Photograph of the apparatus used for Geon-Pevikon block electrophoresis. The actual plans for constructing the apparatus can be obtained from the authors.

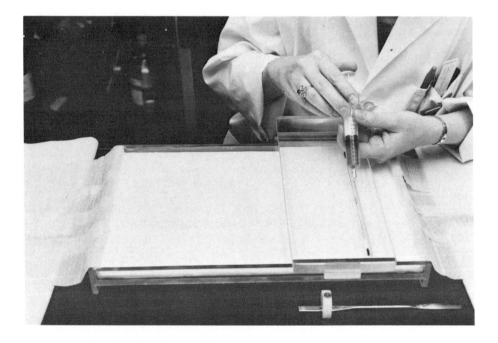

FIGURE 16.   Photograph of the tray containing the Geon-Pevikon support medium. The application of the sample to the Geon-Pevikon medium is illustrated.

wicks submerged in the medium. Excess buffer is allowed to drain from the medium through the wicks into towels (Figure 16) for about 15 to 20 min. Enough buffer is removed through the wicks so that when a slit is made in the medium with a spatula, the walls of the slit do not collapse. Care is taken to avoid overdrying the block. Once the medium is of the correct consistency, the wicks are rolled up and placed on top of the Pevikon block to prevent further drying of the block while applying the sample.

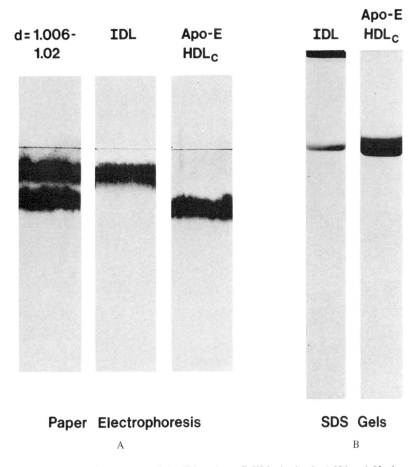

FIGURE 17.   Paper electrophoretograms of the IDL and apo E HDL$_c$ in the d = 1.006 to 1.02 ultracentrifugal fraction (A) and the Geon-Pevikon purified IDL and apo E HDL$_c$ are shown for comparison. The SDS-polyacrylamide gels (B) demonstrate that the apo E HDL$_c$ contain the E apoprotein as the only apoprotein constituent, whereas, apo B is a major protein moiety of the IDL.

With a flat spatula, a slit is made in the Pevikon, 7 cm from one end of the tray, extending to within about 2 mm of the bottom of the support medium. The sample is taken up in a syringe and applied slowly into the slit through a 19-gauge needle (Figure 16). If the entire width of the block is used, approximately 12 mℓ total volume can be applied. Good resolution can be obtained with a load of up to 200 mg of lipoprotein protein. The slit is closed by packing its sides together with a small spatula. Equal volumes of Sudan black B and lipoprotein (∼0.5 mℓ; 0.5 to 2 mg of lipoprotein protein) are mixed and applied to a small slit at either side of the block to serve as a marker for cutting and eluting the unstained bands of lipoproteins.

The tray is transferred to a cold room (5 to 10°C), placed between the buffer chambers, each containing approximately 3 ℓ of buffer, and the wicks are submerged in the buffer. The block is covered with a plexiglass cover to prevent excessive evaporation from the surface during the electrophoresis. After the block is allowed to equilibrate approximately 5 min in the cold, a constant current of 50 mA is applied for 18 hr.

At the end of the run the tray is removed and the excess buffer allowed to drain through the wicks into towels (∼5 min). The tray is placed on an X-ray viewing box in a dark room, and the lipoproteins can be observed as sharp bands. Canine lipoproteins are fluorescent when observed under a UV light source. The areas of the block that contain the desired lipoproteins are cut out with a spatula, transferred to small sintered-glass filters draining into sidearm test tubes, and eluted with 30 to 40 mℓ of saline. The washing process can be hastened by applying a small amount of suction (do not cause foaming). The eluates are transferred to conical test tubes and centrifuged at 7000 × g for 20 min to remove small amounts of Pevikon particles from solution. Fractions are concentrated to desired volumes.

The method has been applied successfully to the purification of lipoproteins with overlapping densities and particle sizes. The separation of LDL or IDL from apo E-containing HDL (HDL$_1$ or HDL$_c$) serves as an example. As shown in Figure 17, the d = 1.006 to 1.02 ultracentrifugal fraction contains two distinct lipoproteins, IDL and apo-E HDL$_c$. These lipoproteins are easily separated and almost quantitatively recovered by Pevikon block electrophoresis. The apo-E HDL$_c$ lack apo B, as shown on the SDS gels in Figure 17, whereas IDL predominantly contain B apoprotein. The method has also been used to separate LDL and HDL$_1$ or HDL$_c$ in several species,[6,8,15,18,22] monkey LDL and Lp(a),[13,35] and canine β-VLDL and VLDL.[16]

REFERENCES

1. **Mahley, R. W.,** Alterations in plasma lipoproteins induced by cholesterol feeding in animals including man, in *Disturbances in Lipid and Lipoprotein Metabolism,* Dietschy, J. M., Gotto, A. M., Jr., and Ontko, J. A., Eds., American Physiological Society, Bethesda, 1978, 181.
2. **Mahley, R. W.,** Dietary fat, cholesterol and accelerated atherosclerosis, in *Atherosclerosis Reviews,* Paoletti, R. and Gotto, A. M., Jr., Eds., Raven Press, New York, 1979, 1.
3. **Mahley, R. W.,** Cellular and molecular biology of lipoprotein metabolism in atherosclerosis, *Diabetes,* 30 (Suppl.), 60, 1981.
4. **Mahley, R. W.,** Atherogenic hyperlipoproteinemia: the cellular and molecular biology of plasma lipoproteins altered by dietary fat and cholesterol, in *Medical Clinics of North America: Lipid Disorders,* Vol. 66, Havel, R. J., Ed., W. B. Saunders, Philadelphia, 1982, 375.
5. **Mahley, R. W., Weisgraber, K. H., and Innerarity, T. L.,** Canine lipoproteins and atherosclerosis. II. Characterization of the plasma lipoproteins associated with atherogenic and nonatherogenic hyperlipidemia, *Circ. Res.,* 35, 722, 1974.
6. **Mahley, R. W., Innerarity, T. L., Weisgraber, K. H., and Fry, D. L.,** Canine hyperlipoproteinemia and atherosclerosis. Accumulation of lipid by aortic medial cells *in vivo* and *in vitro, Am. J. Pathol.,* 87, 205, 1977.
7. **Mahley, R. W., Nelson, A. W., Ferrans, V. J., and Fry, D. L.,** Thrombosis in association with atherosclerosis induced by dietary perturbations in dogs, *Science,* 192, 1139, 1976.
8. **Mahley, R. W. and Weisgraber, K. H.,** Canine lipoproteins and atherosclerosis. I. Isolation and characterization of plasma lipoproteins from control dogs, *Circ. Res.,* 35, 713, 1974.
9. **Mahley, R. W. and Weisgraber, K. H.,** An electrophoretic method for the quantitative isolation of human and swine plasma lipoproteins, *Biochemistry,* 13, 1964, 1974.
10. **Kane, J. P., Hardman, D. A., and Paulus, H. E.,** Heterogeneity of apolipoprotein B: isolation of a new species from human chylomicrons, *Proc. Natl. Acad. Sci. U.S.A.,* 77, 2465, 1980.
11. **Krishnaiah, K. V., Walker, L. F., Borensztajn, J., Schonfeld, G., and Getz, G. S.,** Apolipoprotein B variant derived from rat intestines, *Proc. Natl. Acad. Sci. U.S.A.,* 77, 3806, 1980.
12. **Weisgraber, K. H., Troxler, R. F., Rall, S. C., and Mahley, R. W.,** Comparison of the human, canine, and swine E apoproteins, *Biochem. Biophys. Res. Commun.,* 95, 374, 1980.
13. **Mahley, R. W., Weisgraber, K. H., Innerarity, T. L., and Brewer, H. B., Jr.,** Characterization of the plasma lipoproteins and apoproteins of the Erythrocebus Patas monkey, *Biochemistry,* 15, 1928, 1976.
14. **Mahley, R. W. and Holcombe, K. S.,** Alterations of the plasma lipoproteins and apoproteins following cholesterol feeding in the rat, *J. Lipid Res.,* 18, 317, 1977.
15. **Weisgraber, K. H., Mahley, R. W., and Assmann, G.,** The rat arginine-rich apoprotein and its redistribution following injection of iodinated lipoproteins into normal and hypercholesterolemic rats, *Atherosclerosis,* 28, 121, 1977.
16. **Fainaru, M., Mahley, R. W., Hamilton, R. L., and Innerarity, T. L.,** Structural and metabolic heterogeneity of β-very low density lipoproteins from cholesterol-fed dogs and from humans with Type III hyperlipoproteinemia, *J. Lipid Res.,* 23, 702, 1982.
17. **Innerarity, T. L. and Mahley, R. W.,** Enhanced binding by cultured human fibroblasts of apo-E-containing lipoproteins as compared with low density lipoproteins, *Biochemistry,* 17, 1440, 1978.
18. **Innerarity, T. L., Mahley, R. W., Weisgraber, K. H., and Bersot, T. P.,** Apoprotein (E-A-II) complex of human plasma lipoproteins. II. Receptor binding activity of a high density lipoprotein subfraction modulated by the apo (E-A-II) complex, *J. Biol. Chem.,* 253, 6289, 1978.
19. **Mahley, R. W. and Weisgraber, K. H.,** Subfractionation of high density lipoproteins into two metabolically distinct subclasses by heparin affinity chromatography and Geon-Pevikon electrophoresis, in *Report of the High Density Lipoprotein Methodology Workshop,* Lippel, K. L., Ed., Lipid Metabolism Branch, National Heart, Lung and Blood Institute, Bethesda, Md., 1979, 356.
20. **Weisgraber, K. H. and Mahley, R. W.,** Subfractionation of human high density lipoproteins by heparin-Sepharose affinity chromatography, *J. Lipid Res.,* 21, 316, 1980.

21. **Mahley, R. W., Innerarity, T. L., Bersot, T. P., Lipson, A., and Margolis, S.,** Alterations in human high-density lipoproteins, with or without increased plasma-cholesterol, induced by diets high in cholesterol, *Lancet,* II, 807, 1978.

22. **Mahley, R. W., Weisgraber, K. H., Innerarity, T. L., Brewer, H. B., Jr., and Assmann, G.,** Swine lipoproteins and atherosclerosis. Changes in the plasma lipoproteins and apoproteins induced by cholesterol feeding, *Biochemistry,* 14, 2817, 1975.

23. **Mahley, R. W. and Innerarity, T. L.,** Interaction of canine and swine lipoproteins with the low density lipoprotein receptors of fibroblasts as correlated with heparin/manganese precipitability, *J. Biol. Chem.,* 252, 3980, 1977.

24. **Mahley, R. W. and Innerarity, T. L.,** Properties of lipoproteins responsible for high affinity binding to cell surface receptors of fibroblasts and smooth muscle cells, in *Drugs, Lipid Metabolism, and Atherosclerosis,* Kritchevsky, D., Paoletti, R., and Holmes, W. L., Eds., Plenum Press, New York, 1978, 99.

25. **Hui, D. Y., Innerarity, T. L., and Mahley, R. W.,** Lipoprotein binding to canine hepatic membranes: metabolically distinct apo-E and apo-B,E receptors, *J. Biol. Chem.,* 256, 5646, 1981.

26. **Mahley, R. W., Hui, D. Y., Innerarity, T. L., and Weisgraber, K. H.,** Two independent lipoprotein receptors on hepatic membranes of the dog, swine, and man: the apo-B,E and apo-E receptors, *J. Clin. Invest.,* 68, 1197, 1981.

27. **Mahley, R. W., Innerarity, T. L., and Weisgraber, K. H.,** Alterations in metabolic activity of plasma lipoproteins following selective chemical modifications of the apoproteins, *Ann. N. Y. Acad. Sci.,* 348, 265, 1980.

28. **Mahley, R. W., Innerarity, T. L., Weisgraber, K. H., and Oh, S. Y.,** Altered metabolism (*in vivo* and *in vitro*) of plasma lipoproteins after selective chemical modification of lysine residues of the apoproteins, *J. Clin. Invest.,* 64, 743, 1979.

29. **Sherrill, B. C., Innerarity, T. L., and Mahley, R. W.,** Rapid hepatic clearance of the canine lipoproteins containing only the E apoprotein by a high affinity receptor: identity with the chylomicron remnant transport process, *J. Biol. Chem.,* 255, 1804, 1980.

30. **Havel, R. J., Chao, Y., Windler, E. E., Kotite, L., and Guo, L. S. S.,** Isoprotein specificity in the hepatic uptake of apolipoprotein E and the pathogenesis of familial dysbetalipoproteinemia, *Proc. Natl. Acad. Sci. U.S.A.,* 77, 4349, 1980.

31. **Kovanen, P. T., Bilheimer, D. W., Goldstein, J. L., Jaramillo, J. J., and Brown, M. S.,** Regulatory role for hepatic low density lipoprotein receptors *in vivo* in the dog, *Proc. Natl. Acad. Sci. U.S.A.,* 78, 1194, 1981.

32. **Shelburne, F., Hanks, J., Meyers, W., and Quarfordt, S.,** Effect of apoproteins on hepatic uptake of triglyceride emulsions in the rat, *J. Clin. Invest.,* 65, 652, 1980.

33. **Windler, E., Chao, Y., and Havel, R. J.,** Regulation of the hepatic uptake of triglyceride-rich lipoproteins in the rat: opposing effects of homologous apolipoprotein E and individual C apoproteins, *J. Biol. Chem.,* 255, 8303, 1980.

34. **Heiss, G., Johnson, N. J., Reiland, S., Davis, C. E., and Tyroler, H. A.,** The epidemiology of plasma high-density lipoprotein cholesterol levels, *Circulation,* 62, 116, 1980.

35. **Mahley, R. W., Weisgraber, K. H., and Innerarity, T. L.,** Atherogenic hyperlipoproteinemia induced by cholesterol feeding the Patas monkey, *Biochemistry,* 15, 2979, 1976.

36. **Goldstein, J. L., Ho, Y. K., Brown, M. S., Innerarity, T. L., and Mahley, R. W.,** Cholesteryl ester accumulation in macrophages resulting from receptor-mediated uptake and degradation of hypercholesterolemic canine β-very low density lipoproteins, *J. Biol. Chem.,* 255, 1839, 1980.

37. **Mahley, R. W., Innerarity, T. L., Brown, M. S., Ho, Y. K., and Goldstein, J. L.,** Cholesteryl ester synthesis in macrophages: stimulation by β-very low density lipoproteins from cholesterol-fed animals of several species, *J. Lipid Res.,* 21, 970, 1980.

38. **Innerarity, T. L., Pitas, R. E., and Mahley, R. W.,** Modulating effects of canine high density lipoproteins on cholesteryl ester synthesis induced by β-very low density lipoproteins in macrophages, *Atherosclerosis,* 2, 114, 1982.

39. **Barth, W. F., Wochner, R. D., Waldmann, T. A., and Fahey, J. L.,** Metabolism of human gamma macroglobulins, *J. Clin. Invest.,* 43, 1036, 1964.

PLASMA LIPOPROTEINS OF SWINE:
COMPOSITION, STRUCTURE, AND METABOLISM

Richard L. Jackson

INTRODUCTION

During the last two decades, the pig (*Sus domesticus*) has been used extensively for the study of experimental atherosclerosis. As is discussed by Ratcliffe and Luginbuhl,[1] there are several important reasons for this animal model in atherosclerosis research. Anatomically, the origin and distribution of the coronary arteries in the pig are similar to those in man. The pig also develops spontaneous atherosclerotic lesions. These lesions are located in the intima of the aorta and coronary arteries and have lipid compositions similar to those which develop in man. Another important reason for the pig as an experimental animal model is the close similarity between pig and human serum lipoproteins.

The purpose of the present review is to bring together current information concerning the composition, structure, and metabolism of swine lipoproteins. Because of the vast literature, it is not the intent of this review to discuss pig lipoproteins as they relate to human or other animal lipoproteins. For the comparative aspects of animal lipoproteins, the reader is referred to an excellent review by Chapman[2] or to other chapters in this volume. A number of review articles are also available describing human lipoproteins.[3-7]

Most of the experimental work with the pig has been performed with the Hormel mini pig. In general, the fasting lipid and lipoprotein compositions in the various species of swine when fed a standard chow diet are similar to the concentration in man. The values for pig plasma triglycerides and total cholesterol are approximately 60 and 150 mg/dℓ, respectively. However, as is discussed below, cholesterol feeding in the pig markedly increases plasma lipid levels and has major effects on lipoprotein structure and metabolism.

COMPOSITION

Very Low-Density (VLD) Lipoproteins

Havel et al.[8] were the first to quantitate pig serum lipoprotein VLDL, LDL, and HDL cholesterol. The lipoproteins were fractionated by successive ultracentrifugal flotation in salt solutions at density 1.019 and 1.063 g/mℓ. An interesting finding in this early study was the observation that the ratio of total cholesterol to phospholipid in the three classes of lipoproteins was more similar to that of human lipoproteins than lipoproteins from the rabbit, rat, hamster, dog, or monkey.

By electromicroscopy after negative staining, pig VLDL are heterogeneous with particle sizes varying from 260 to 780 Å with a mean diameter of 493 Å.[9] On a chow diet, pig plasma does not appear to contain chylomicra in the fasting state. Azuma and Komano[9] separated VLDL into three fractions by isoelectric focusing between pH 4 and 6; the isoelectric points of the three fractions were 4.81, 5.05, and 5.21. Although the lipid content was not determined on each fraction, polyacrylamide gel electrophoresis (PAGE) in 8 M urea of the apoproteins suggests some slight differences among the three subfractions. However, since the individual apoproteins were not isolated and characterized, it is difficult to explain the differences in the isoelectric points.

Pig VLDL heterogeneity has also been demonstrated by chromatography on concanavalin A-Sepharose.[10] Azuma et al.[10] separated VLDL into a fraction which did not bind to concanavalin A-Sepharose and a fraction which was eluted with 0.3 M methyl α-D-mannopyranoside; the ratio of the unbound to bound lipoproteins was approximately 1.5. The unretained fraction had a slower mobility on a 2 to 5% gradient gel, suggesting larger particles. The major difference between the two lipoprotein fractions was the presence of low-molecular-weight components in the unbound fraction. Based on the migration pattern, one of these proteins may correspond to apo E.

VLDL which are isolated at d<1.007 g/mℓ contain approximately 90% lipid and 10% protein.[11] Janado et al.[11] calculated a mean molecular weight from sedimentation flotation values of approximately 16 × 10⁶. The major apoprotein constituent of VLDL is apo B. VLDL apo B has chemical and immunochemical properties similar to the major protein of pig LDL.[12] Based on its insolubility in 1 *M* acetic acid, Knipping et al.[12] showed that apo B accounts for 52% of the total protein mass in VLDL. The amino acid composition of pig serum apo B isolated from VLDL is nearly identical to that from LDL.[12] In addition to apo B, VLDL also contain soluble apoproteins which have been fractionated by chromatography on DEAE-cellulose in the presence of urea.[12-14] In contrast to human apo VLDL which contain three major apoprotein constituents, termed apo C-I, apo C-II, and apo C-III, pig VLDL contain only one major soluble apoprotein, the molecular weight of which is approximately 8000 to 9000; its COOH-terminal amino acid is serine.[12-14] Based on its similarity in migration on PAGE and amino acid composition to human apo C-II,[15] the protein has also been called apo C-II. Knipping et al.[12] and Jackson et al.[14] have shown that pig apo C-II also activates bovine milk lipoprotein lipase. Further studies are required in order to determine the relationship of the other minor VLDL apoproteins to human apo C-I and apo C-III.

Low-Density Lipoproteins (LDL)

Janado et al.[11,16,17] were the first to recognize that pig LDL are more heterogeneous than human LDL. LDL of pig serum show a bimodal distribution of approximately equal concentration when analyzed by analytic ultracentrifugation. The LDL subfractions have been isolated preparatively between d 1.007 to 1.060 g/mℓ (LDL$_1$) and 1.060 to 1.090 g/mℓ (LDL$_2$). LDL$_1$ contain 79.3% protein and 20.4% lipid; the values for LDL$_2$ are 72.3% and 27.3%, respectively.[11] LDL$_1$ and LDL$_2$ both bind to concanavalin A-Sepharose and are quantitatively eluted with 0.3 *M* methyl α-D-mannopyranoside.[10] An interesting finding in this study, and consistent with the heterogeneity observed for human LDL, was that the amount of protein per LDL$_1$ or LDL$_2$ particle remained relatively constant between 6.1 to 7.3 × 10⁵ daltons. Pig LDL heterogeneity is also apparent by density gradient ultracentrifugation and electrophoresis in agarose gel.[18] However, chromatography on Bio-Gel® A-5m does not separate the two LDL subfractions.[19]

More recently, pig LDL have been fractionated into LDL$_1$ and LDL$_2$ by rate-zonal ultracentrifugation.[20-22] By this method, the amount of LDL$_1$ is approximately equal to that of LDL$_2$ in chow-fed Large White pigs[20] or Yorkshire swine.[22] The physical and chemical properties of LDL isolated by the zonal method appear to be similar to those isolated by fixed angle ultracentrifugation.

A number of investigators have isolated and characterized apo B, the major apoprotein constituent of LDL.[12,20,23-27] Although the apoprotein is insoluble in aqueous buffers, it can be solubilized in sodium decyl sulfate. Apo B isolated from either LDL$_1$ or LDL$_2$ is indistinguishable with respect to its amino acid composition and immunochemical properties. As determined by PAGE in sodium dodecyl sulfate (SDS) apo B shows two major protein constituents with apparent molecular weight of 350,000 and 180,000, suggesting a monomer-dimer relationship.[26] The confusion concerning the molecular weight of human apo B⁴ is also apparent with pig apo B. Walkley and Husbands[27] succinylated apo B and fractionated the modified protein on Sephadex® G-100 in the presence of urea; two major fractions were obtained. Based on the elution profile with proteins of known molecular weight, these investigators concluded that the two apo B components have molecular weights of 34,000 and 22,000. Although no detailed characterization studies were reported, it is interesting that the amino acid compositions of the proteins were uniquely different. No information is available on the amino acid sequence of pig apo B. However, Azuma et al.[28] digested the protein with pronase and isolated three unique glycopeptides; the carbohydrate moieties in each are linked to the polypeptide chain through N-acetylglucosaminylasparagine bonds.

High-Density Lipoproteins (HDL)

Pig HDL are polydisperse as determined by analytical ultracentrifugation[11] or gradient pore gel electrophoresis.[29] However, pig HDL are more homogeneous than human HDL;[30] the flotation pattern resembles that for human HDL_3. Pig plasma HDL have a flotation range of $F°_{1.20}$ 0.5 to 6.0 with a peak maximum at $F°_{1.20}$ 2.83. Unfractionated HDL have a molecular weight of approximately 250,000 and contain approximately 51% protein;[11] HDL_3 (d 1.12 to 1.21 g/mℓ) have a molecular weight of 210,000[31] and a molecular diameter ranging between 100 to 110 Å.[32] The major apoprotein constituent of pig HDL is apo A-I;[12,25,29-33] it accounts for 60% of the total HDL protein. Pig apo A-I has a molecular weight of approximately 26,000 and physical and chemical properties similar to human apo A-I. The protein has a COOH-terminal glutamine and an NH_2-terminal aspartic acid. Apo A-I contains a high percentage of α-helical structure and it activates human lecithinolesterol acyltransferase (LCAT).[33] In contrast to human HDL, pig HDL does not contain a significant amount, if any, of the human HDL equivalent of the dimeric form of apo A-II.[33] In addition to apo A-I, pig HDL also contain lower-molecular-weight protein components, including apo C-II. However, these minor proteins have not been rigorously characterized so as to compare their properties to human HDL apoproteins.

Cholesteryl Ester-Rich Lipoproteins

It is well known from a number of studies in the pig that an increase in the amount of dietary cholesterol results in the development of hypercholesterolemia and marked changes in the type and amount of plasma lipoproteins and in their lipid and protein constituents.[22,34-41] With increasing concentrations of dietary cholesterol, there is a dramatic change in the zonal ultracentrifugal pattern of LDL;[22] LDL_1 increases in amount whereas LDL_2 decreases. Furthermore, the size of LDL_1 increases and the density decreases. As determined by paper electrophoresis, plasma lipoproteins from cholesterol-fed swine also show a different electrophoretic pattern than that for chow-fed animals. In addition to the characteristic β-migrating (LDL) and α-migrating (HDL) lipoproteins, the cholesterol-fed pig contains a second lipoprotein with α-mobility. Mahley et al.[40] have termed this abnormal lipoprotein HDL_c. The cholesterol-fed pig also contains an abnormal β-migrating lipoprotein (termed β-VLDL) which is isolated at d<1.006 g/mℓ. Mahley and co-workers[40,41] have isolated and characterized each of these plasma lipoproteins from the cholesterol-fed Hormel mini pig. The hypercholesterolemic lipoproteins were isolated by a combination of ultracentrifugation and Geon-Pevikon block electrophoresis.[41] The β-VLDL have diameters ranging between 200 to 350 Å. The lipoprotein is cholesterol enriched and contains apo B, apo E, and apo C. HDL_c range in size from 100 to 240 Å and are isolated between d 1.02 to 1.09 g/mℓ. As the HDL_c become more enriched with cholesterol, the size becomes larger and the density lower. HDL_c do not contain apo B; the major apoprotein constituents are apo E, apo A-I, and apo C. As is discussed below, HDL_c interact with specific receptors on cells and compete with apo B-containing lipoproteins.

STRUCTURE

Most of the available information on the structure of pig lipoproteins pertains to LDL, HDL, and cholesterol-rich lipoproteins; essentially nothing is known concerning the structure of triglyceride-rich lipoproteins from pig plasma.

The optical rotatory dispersion spectra of pig LDL shows a trough at 233 nm.[42] Based on the mean residue optical rotation at 233 nm, the protein constituent of LDL contains approximately 25% α-helical structure.[42] This amount of α-helix is similar to that in human LDL. In the presence of 2 to 4 *M* guanidine-HCl, LDL undergoes a reversible denaturation. Denatured LDL can be treated with subtilisin causing a release of approximately 30% of LDL protein. Compared to native LDL, the fluorescence emission spectra of protease-treated LDL is unaltered suggesting that the tryptophan residues of LDL are located in a nonpolar environment and protected from proteolytic digestion.[42]

The structure of LDL from pigs on chow diets and that produced from cholesterol feeding has been examined by nuclear magnetic resonance (NMR),[43] small-angle X-ray scattering,[44] and differential scanning calorimetry.[22,45] Based on the restriction of motion of the choline groups of LDL phospholipids, Finer et al.[43] concluded from NMR studies that LDL contain both an inner core and an outer surface monolayer of protein, each associated with phospholipid. However, small-angle X-ray scattering measurements show that human LDL have a micellar structure with an outer surface of protein and phospholipids and a central core of only neutral lipids and no protein or phospholipid.[4]

Atkinson et al.[44] determined the structure of HDL_c by small-angle X-ray scattering. At 10°C, a temperature which is well below the transition temperature, HDL_c have a single peak in the electron density which arises from an overlap of steroid moieties of the cholesteryl esters. Calculations from the electron density profiles indicate that the diameter of the cholesteryl ester core region is 65 Å; the distance between the outer edge of the steroid peak and the aqueous phase is 15 to 20 Å, a distance sufficient to accommodate the phospholipid choline groups and the protein.

LDL from cholesterol-fed swine have a reversible thermal transition between 30 to 50°C with a peak transition temperature at 42 to 44° C.[22,45] This temperature range is approximately 10 to 12° higher than that reported for human LDL. The higher transition temperature in hypercholesterolemic LDL is caused by a decreased content of triglyceride and an increased content of saturated fatty acyl moieties in the cholesteryl esters. HDL_c also show a higher transition temperature than normal LDL.[45] Again, one possible reason is the increased cholesteryl ester/triglyceride ratio.

The molecular structure of pig HDL_3 (d 1.120 to 1.210 g/mℓ) has been determined by NMR[43] and low-angle X-ray scattering methods.[46] Based on these techniques, HDL has a micellar structure with a central core of neutral lipids and an outer surface monolayer of phospholipid, protein, and unesterified cholesterol. Furthermore, most of the phospholipid choline head groups are in contact with water as evidenced by their accessibility to ferricyamide-induced NMR shifts. The X-ray scattering technique shows that HDL_3 are spherical particles with an average diameter of approximately 108 Å. The lipoprotein particle has a central core of neutral lipid of 84 Å in diameter. The outer surface is 12 Å in thickness and accommodates the choline moieties of phospholipids and the proteins,[46] which are presumably highly helical and interact with phosphatidylcholine through amphipathic helical segments.

The mechanism by which the HDL apoproteins interact with phospholipids has been assessed by reconstitution studies with delipidated apo HDL and a synthetic phospholipid, dimyristoyl phosphatidylcholine (DMPC).[47-49] The complex formed by association of DMPC and apo HDL is a discoidal structure as determined by electron microscopy after negative staining. Based on X-ray scattering studies, the isolated lipid-protein complex has an oblate shape with a major axis of 110 Å and a minor axis of 55 Å. The interaction of apo HDL with DMPC is associated with an increase in α-helical structure, suggesting that apo A-I, the major apoprotein constituent of pig HDL, contains amphipathic helical segments similar to those described for human apo A-I.

METABOLISM

Turnover Studies

Wolfe and Belbeck[50] measured the splanchnic and hepatic secretion of triglycerides in glucose-infused mini swine; the net secretion of plasma triglyceride fatty acids was approximately 2 μmol/min·kg. Under steady-state conditions, the secretion of most plasma triglyceride fatty acids is attributable to the secretion of VLDL triglycerides; less than 7% of the released fatty acid is as free fatty acids (FFA). The $t_{1/2}$ of the pig plasma VLDL-triglyceride pool is approximately 20 min. An interesting finding in the study by Wolfe and Belbeck[50] was that the fraction of VLDL-triglyceride fatty acids derived from plasma FFA was 9% whereas that derived from glucose during a constant infusion of radiolabeled glucose was only 2%. Apparently in the 220 min of the experimental protocol, glucose mixes with a larger pool of other substrates, possibly glycogen, and these other substrates are preferentially incorporated into VLDL triglycerides.

In a more recent report, Hannan et al.[51] carried out triglyceride turnover studies in female Large White pigs using radiolabeled glucose and glycerol. Based on multicompartmental analyses for the incorporation of radiolabeled glycerol into glucose and into liver triglyceride, these investigators concluded that the intestine is responsible for the secretion of 20 to 30% of newly synthesized VLDL triglyceride. In contrast to man, liver triglycerides turn over more slowly compared to plasma VLDL triglycerides; pig hepatic triglycerides turn over with a fractional rate constant of 0.006 min^{-1}, whereas the rate of plasma VLDL turnover is 0.085 min^{-1}.

Only limited information is available on the synthesis of pig lipoprotein apoproteins. Nakaya et al.[52,53] used the isolated perfused pig liver and measured the incorporation of radiolabeled leucine into LDL apo B and immunoassayable apo B and apo A-I. The perfused liver obtained from young fasted pigs secreted higher levels of apo B than the livers obtained from the fed animals. An unexpected result in these studies[52,53] was the finding that a greater amount of apo B was present in the LDL fraction as compared to VLDL. Since ^{125}I-labeled VLDL were not degraded, these authors[52,53] concluded that LDL were synthesized directly by the liver and did not result as a product of VLDL catabolism, the major pathway in man.

The in vivo turnover of LDL has been assessed by measuring the disappearance of injected ^{125}I-labeled LDL$_1$ and LDL$_2$.[54-56] In the study by Sniderman et al.,[54] the disappearance of radiolabeled LDL in the mini pig was biexponential with half-lives of 0.83 hr for the rapid phase and 22.5 hr for the slow one. Using a three-compartment model, a mean fractional catabolic rate of 0.041 hr^{-1} was determined.[54] Marcel et al.[55] also used the mini pig and determined a synthetic rate for apo LDL synthesis of 5.6 mg/hr. A fractional catabolic rate of 0.015 hr^{-1} was obtained using a three-exponential equation to analyze the data. In the Large White pig, a fractional catabolic rate of 0.058 hr^{-1} was calculated.[56] In each of these studies,[54-56] the disappearance of plasma radioactivity was complex and not monoexponential. Furthermore, a fractional catabolic rate of 0.04 to 0.06 hr^{-1} corresponds to turnover values which are three to four times higher than values reported in man.

The tissue site of LDL degradation is somewhat confusing. In the studies of Sniderman et al.[54] and Calvert et al.,[56] the liver accounts for the highest concentration of [125]I. Two days after the injection of ^{125}I-labeled LDL, the liver contains about 33% of the total radioactivity present in the plasma compartment; after 4 days the liver contains 54%.[56] The conclusion from these studies that the liver is the primary site for the catabolism of LDL appears to be contrary to the results of Sniderman et al.[57] They performed LDL turnover studies before and after hepatectomy and found that the half-life of LDL clearance dropped significantly from a $t_{1/2}$ of approximately 20 hr to 9 hr after hepatectomy. As is discussed by Pittman et al.,[58] interpretation of the LDL turnover data using radiolabeled LDL is complicated by the fact that the liver contains a significant fraction of the total extravascular extracellular LDL pool which exchanges rapidly with the plasma compartment. Thus, the rapid phase[54-56] corresponds to LDL uptake by the liver but not degradation. To evaluate the rapid hepatic exchange, Pittman et al.[58] have used [^{14}C]sucrose-labeled LDL. These investigators have shown that the amount of [^{14}C]sucrose which accumulates in cultured cells correlates with the amount of LDL bound. Based on this [^{14}C]sucrose method, Pittman et al.[58] reported that the sum of apo LDL degradation occurring in all extrahepatic tissues in young Duroc or Hampshire swine was approximately equal to that occurring in the liver. This result suggests that the liver does play an important role in the degradation of LDL and is consistent with experiments carried out in hepatocyte cultures described below. Although the liver quantitatively accounts for the greatest amount of LDL degradation, on a per gram basis, the concentration of [^{14}C]sucrose in the adrenal is two to five times greater than the amount in the liver.[58]

Finally, with respect to metabolism, LDL turnover studies have been performed in swine after portacaval shunt.[59] The portacaval shunt in swine causes a significant decrease in plasma levels of cholesterol and triglycerides and in VLDL triglycerides and LDL cholesterol.[60-62] In the study of Nestruck et al.,[62] portacaval shunt caused a 47% reduction in VLDL triglycerides and a 17% reduction in LDL cholesterol. Nestruck et al.[62] and Carew et al.[59] also showed that HDL cholesterol was reduced after portacaval shunting. Part of the reduction in VLDL triglycerides may be due to a decrease in the lipid composition of the VLDL relative to the protein.[62] Portacaval shunting does not affect the fractional catabolic rate of LDL, the calculated fractional catabolic rate being approximately 0.06 hr^{-1} in both the control and shunted animals.[59] Since LDL cholesterol decreases after shunting, Carew et al.[59] have suggested that changes in the rate of LDL production account for the decrease in plasma LDL cholesterol.

Cell Culture Studies

It is now well established that hepatic and extrahepatic cells in culture are capable of degrading plasma lipoproteins. The interaction of pig lipoproteins with swine and human cells include studies with smooth muscle cells,[63-65] fibroblasts,[66,67] and hepatocytes.[68,69] The addition of homologous VLDL, LDL, HDL, and HDL$_c$ to swine smooth muscle cells causes a three- to sevenfold increase in the growth rate of the cells. VLDL, LDL, and HDL$_c$ also cause a concentration dependent decrease in the activity of 3-hydroxy-3-methylglutaryl coenzyme A (HMG-CoA) reductase, the rate-limiting enzyme in cholesterol biosynthesis from acetyl CoA; HDL (d = 1.10 to 1.21 g/mℓ) do not suppress HMG-CoA reductase activity. These results suggest that apo B and apo E containing lipoproteins are capable of regulating cholesterol biosynthesis in arterial smooth muscle cells in culture. Direct evidence for the binding and internalization of pig LDL has been provided by Weinstein et al.[65] The uptake of ^{125}I-labeled LDL by swine arterial smooth muscle cells is dependent on time and lipoprotein concentration. In these studies, when the cultured cells were exposed to 300 µg/ mℓ LDL they degraded 6 µg of LDL protein per milligram cell protein in 4 hr. Even with some uncertainties, Weinstein et al.[65] concluded that the extrahepatic degradation of pig LDL must account for a large percentage of total LDL catabolism.

Studies similar to those described by Weinstein et al.[65] in muscle cells have been carried out using pig lipoproteins and human fibroblasts.[66,67] HDL$_c$, which does not contain apo B, binds to the LDL receptor (an apo B/apo E receptor) causing suppression of HMG-CoA reductase activity and enhancement of the incorporation of [^{14}C]oleate into cholesteryl esters.[66] As with LDL, the binding and internalization of pig HDL$_c$ causes an accumulation of both free and esterified cholesterol in human fibroblasts. In a more recent study, Mahley and Innerarity[67] further fractionated HDL$_c$ by precipitation with heparin/manganese. The fractions which were the most effective in regulating cholesterol biosynthesis were those which were the most readily precipitated by heparin. The characteristic compositional feature of the precipitated fraction was that it contained apo E as its major protein constituent; those fractions which did not contain apo E did not compete with LDL for either lipoprotein binding and internalization or degradation, suggesting that apo E plays a fundamental role in the interaction with cell receptors.

Pig hepatocytes prepared from neonatal animals and maintained in nonproliferating monolayer culture also take up and degrade LDL by a high-affinity mechanism.[68] LDL degradation is saturable, specific for LDL, and inhibited by treatment of the cells with pronase; the number of LDL receptors is also subject to down regulators. Pig HDL (d 1.090 to 1.210 g/mℓ) do not compete for ^{125}I-labeled LDL binding and internalization.[68] Furthermore, desialytion of pig LDL does not appear to alter the rate of uptake by either cultured hepatocytes or the fractional catabolic rate[69] when injected into pigs.

The studies described above suggest that the pig liver has two types of lipoprotein receptors, one of which recognizes LDL- or apo B- and/or apo E-containing lipoproteins and a second receptor which recognizes only apo E-containing lipoproteins, HDL$_c$. In an attempt to characterize hepatic lipoprotein receptors, Bachorik et al.[70] described the isolation of a plasma membrane fraction from adult pig livers. The partially purified microsome fraction shows high-affinity binding for radiolabeled LDL; no attempt was made in this study[70] to measure HDL$_c$ binding. In a more recent report, Mahley and co-workers[71] attempted to identify the two hepatic receptors. In these studies using dog and swine, these investigators showed that the pig has two independent hepatic receptors. One receptor recognizes swine lipoproteins containing either apo B- and/or apo E- and the second only apo E-containing lipoproteins (HDL$_c$). The interesting finding in these studies was that the apo B/apo E receptor was present in fetal pigs but was not detectable in adult animals. Similar findings were also reported in the dog and man.[71]

ACKNOWLEDGMENTS

The author gratefully acknowledges the editorial assistance of Ms. Janet Simons in the preparation of this review article.

REFERENCES

1. **Ratcliffe, H. L. and Luginbuhl, H.,** The domestic pig: a model for experimental atherosclerosis, *Atherosclerosis,* 13, 133, 1971.
2. **Chapman, M. J.,** Animal lipoproteins: chemistry, structure and comparative aspects, *J. Lipid Res.,* 21, 789, 1980.
3. **Osborne, J. C., Jr. and Brewer, H. B., Jr.,** The plasma lipoproteins, in *Advances in Protein Chemistry,* Vol. 31, Academic Press, New York, 1977, 253.
4. **Jackson, R. L., Morrisett, J. D., and Gotto, A. M., Jr.,** Lipoprotein structure and metabolism, *Physiol. Rev.,* 56, 259, 1976.
5. **Scanu, A. M.,** Structural studies on serum lipoproteins, *Biochim. Biophys. Acta,* 265, 471, 1972.
6. **Schaefer, E. J., Eisenberg, S., and Levy, R. I.,** Lipoprotein apoprotein metabolism, *J. Lipid Res.,* 19, 667, 1978.
7. **Scanu, A. M., Edelstein, C., and Keim, P.,** Serum lipoproteins, in *The Plasma Proteins: Structure, Function and Genetic Control,* Vol. 1, 2nd ed., Academic Press, New York, 1975, 317.
8. **Havel, R. J., Eder, H. A., and Bragdon, J. H.,** The distribution and chemical composition of ultracentrifugally separated lipoproteins in human serum, *J. Clin. Invest.,* 34, 1345, 1955.
9. **Azuma, J. and Komano, T.,** Studies on pig serum lipoproteins. VI. Surface charge of very low density lipoprotein, *J. Biochem.,* 83, 1789, 1978.
10. **Azuma, J., Kashimura, N., and Komano, T.,** Studies on pig serum lipoproteins. III. Affinity chromatography of native lipoproteins on concanavalin A-Sepharose, *Biochim. Biophys. Acta,* 439, 380, 1976.

11. **Janado, M., Martin, W. G., and Cook, W. H.,** Separation and properties of pig-serum lipoproteins, *Can. J. Biochem.,* 44, 1201, 1966.
12. **Knipping, G. M. J., Kostner, G. M., and Holasek, A.,** Studies on the composition of pig serum lipoproteins: isolation and characterization of different apoproteins, *Biochim. Biophys. Acta,* 393, 88, 1975.
13. **Fidge, N. H.,** Characterisation of the small molecular weight apolipoproteins from pig plasma very low density lipoprotein, *Biochim. Biophys. Acta,* 424, 253, 1976.
14. **Jackson, R. L., Chung, B. H., Smith, L. C., and Taunton, O. D.,** Physical, chemical and immunochemical characterization of a lipoprotein lipase activator protein from pig plasma very low density lipoproteins, *Biochim. Biophys. Acta,* 490, 385, 1977.
15. **Jackson, R. L., Baker, H. N., Gilliam, E. B., and Gotto, A. M., Jr.,** Primary structure of very low density apolipoprotein C-II of an plasma, *Proc. Natl. Acad. Sci. U.S.A.,* 74, 1942, 1977.
16. **Janado, M. and Martin, W. G.,** Evidence for protein subunits in low density lipoprotein from pig serum, *Agric. Biol. Chem.,* 37, 2835, 1973.
17. **Janado, M. and Martin, W. G.,** Molecular heterogeneity of a pig serum low-density lipoprotein, *Can. J. Biochem.,* 46, 875, 1968.
18. **Kalab, M. and Martin, W. G.,** Electrophoresis of pig serum lipoproteins in agarose gel, *Anal. Biochem.,* 24, 218, 1968.
19. **Kalab, M. and Martin, W. G.,** Gel filtration of native and modified pig serum lipoproteins, *J. Chromatogr.,* 35, 230, 1968.
20. **Calvert, G. D. and Scott, P. J.,** Properties of two pig low density lipoproteins prepared by zonal ultracentrifugation, *Atherosclerosis,* 22, 583, 1975.
21. **Danielsson, B., Ekman, R., Johansson, B. G., and Petersson, B. G.,** Zonal ultracentrifugation of plasma lipoproteins from normal and cholestatic pigs, *Clin. Chim. Acta,* 65, 187, 1975.
22. **Pownall, H. J., Jackson, R. L., Roth, R. I., Gotto, A. M., Patsch, J. R., and Kummerow, F. A.,** Influence of an atherogenic diet on the structure of swine low density lipoproteins, *J. Lipid Res.,* 21, 1108, 1980.
23. **Fidge, N. H. and Smith, G. D.,** Isolation and characterisation of pig lipoprotein subfractions within the density range 1.006 to 1.090 g/mℓ, *Artery,* 1, 406, 1975.
24. **Jackson, R. L., Taunton, O. D., Segura, R., Gallagher, J. G., Hoff, H. F., and Gotto, A. M., Jr.,** Comparative studies on plasma low density lipoproteins from pig and man, *Comp. Biochem. Physiol.,* 53B, 245, 1976.
25. **Fidge, N.,** The isolation and properties of pig plasma lipoproteins and partial characterisation of their apoproteins, *Biochim. Biophys. Acta,* 295, 258, 1973.
26. **Chapman, M. J. and Goldstein, S.,** Comparison of the serum low density lipoprotein and of its apoprotein in the pig, rhesus monkey and baboon with that in man, *Atherosclerosis,* 25, 267, 1976.
27. **Walkley, C. S. and Husbands, D. R.,** Polypeptides from serum low-density lipoproteins of pigs *(Sus domesticus),* *Aust. J. Biol. Sci.,* 29, 301, 1976.
28. **Azuma, J., Kashimura, N., and Komano, T.,** Studies on pig serum lipoproteins. IV. Isolation and characterization of glycopeptides from pig serum low density lipoprotein, *J. Biochem.,* 81, 1613, 1977.
29. **Janado, M., Doi, Y., Azuma, J., Onodera, K., and Kashimura, N.,** Heterogeneity of pig serum high density lipoprotein, *Artery,* 1, 166, 1975.
30. **Forte, T. M., Nordhausen, R. W., Nichols, A. V., Endemann, G., Miljanich, P., and Bell-Quint, J. J.,** Dissociation of apolipoprotein A-I from porcine and bovine high density lipoproteins by guanidine hydrochloride, *Biochim. Biophys. Acta,* 573, 451, 1979.
31. **Cox, A. C. and Tanford, C.,** The molecular weights of porcine plasma high density lipoprotein and its subunits, *J. Biol. Chem.,* 243, 3083, 1968.
32. **Davis, M. A. F., Henry, R., and Leslie, R. B.,** Comparative studies on porcine and human high density lipoproteins, *Comp. Biochem. Physiol.,* 47B, 831, 1974.
33. **Jackson, R. L., Baker, H. N., Taunton, O. D., Smith, L. C., Garner, C. W., and Gotto, A. M., Jr.,** A comparison of the major apolipoprotein from pig and human high density lipoproteins, *J. Biol. Chem.,* 248, 2639, 1973.
34. **Reisen, R., Sorrels, M. F., and Williams, M. C.,** Influence of high levels of dietary fat and cholesterol on atherosclerosis and lipid distribution in swine, *Circ. Res.,* 7, 833, 1959.
35. **Jackson, R. L., Morrisett, J. D., Pownall, H. J., Gotto, A. M., Jr., Kamio, A., Imai, H., Tracy, R., and Kummerow, F. A.,** Influence of dietary *trans*-fatty acids on swine lipoprotein composition and structure, *J. Lipid Res.,* 18, 182, 1977.
36. **Hill, E. G., Lundberg, W. O., and Titus, J. L.,** Experimental atherosclerosis in swine. II. Effects of methionine and menhaden oil on an atherogenic diet containing tallow and cholesterol, *Mayo Clin. Proc.,* 46, 621, 1971.
37. **Calvert, G. D. and Scott, P. J.,** Serum lipoproteins in pigs on high-cholesterol — high-triglyceride diets, *Atherosclerosis,* 19, 485, 1974.

38. **Goldsmith, D. P. J. and Jacobi, H. P.,** Atherogenesis in swine fed several types of lipid-cholesterol diets, *Lipids,* 13, 174, 1978.

39. **Hill, E. G. and Silbernick, C. L.,** Development of hyperbetalipoproteinemia in pigs fed atherogenic diet, *Lipids,* 10, 41, 1975.

40. **Mahley, R. W., Weisgraber, K. H., Innerarity, T., Brewer, H. B., Jr., and Assmann, G.,** Swine lipoproteins and atherosclerosis: changes in the plasma lipoproteins and apoproteins induced by cholesterol feeding, *Biochemistry,* 14, 2817, 1975.

41. **Mahley, R. W. and Weisgraber, K. H.,** An electrophoretic method for the quantitative isolation of human and swine plasma lipoproteins, *Biochemistry,* 13, 1964, 1974.

42. **Ikai, A.,** Denaturation and proteolytic digestion of porcine low-density lipoprotein in aqueous guanidine hydrochloride solutions, *J. Biochem.,* 77, 321, 1975.

43. **Finer, E. G., Henry, R., Leslie, R. B., and Robertson, R. N.,** NMR studies of pig low- and high-density serum lipoproteins, *Biochim. Biophys. Acta,* 380, 320, 1975.

44. **Atkinson, D., Tall, A. R., Small, D. M., and Mahley, R. W.,** Structural organization of the lipoprotein HDL_c from atherosclerotic swine: structural features relating the particle surface and core, *Biochemistry,* 17, 3930, 1978.

45. **Tall, A. R., Atkinson, D., Small, D. M., and Mahley, R. W.,** Characterization of the lipoproteins of atherosclerotic swine, *J. Biol. Chem.,* 252, 7288, 1977.

46. **Atkinson, D., Davis, M. A. F., and Leslie, R. B.,** The structure of a high density lipoprotein (HDL_3) from porcine plasma, *Proc. R. Soc. Lond. B.,* 186, 165, 1974.

47. **Hauser, H., Henry, R., Leslie, R. B., and Stubbs, J. M.,** The interaction of apoprotein from porcine high-density lipoprotein with dimyristoyl phosphatidylcholine, *Eur. J. Biochem.,* 48, 583, 1974.

48. **Atkinson, D., Smith, H. M., Dickson, J., and Austin, J. P.,** Interaction of apoprotein from porcine high-density lipoprotein with dimyristoyl lecithin. I. The structure of the complexes, *Eur. J. Biochem.,* 64, 541, 1976.

49. **Andrews, A. L., Atkinson, D., Barratt, M. D., Finer, E. G., Hauser, H., Henry, R., Leslie, R. B., Owens, N. L., Phillips, M. C., and Robertson, R. N.,** Interaction of apoprotein from porcine high-density lipoprotein with dimyristoyl lecithin. II. Nature of lipid-protein interaction, *Eur. J. Biochem.,* 64, 549, 1976.

50. **Wolfe, B. M. and Belbeck, L. W.,** Splanchnic and hepatic triglyceride secretion during hypercaloric intravenous glucose infusion in conscious swine, *J. Lipid Res.,* 16, 19, 1975.

51. **Hannan, S. F., Baker, N., Rostami, H., Lewis, K., and Scott, P. J.,** Very low density lipoprotein metabolism in domestic pigs, *Atherosclerosis,* 37, 55, 1980.

52. **Nakaya, N., Chung, B. H., and Taunton, O. D.,** Synthesis of plasma lipoproteins by the isolated perfused liver from the fasted and fed pig, *J. Biol. Chem.,* 252, 5258, 1977.

53. **Nakaya, N., Chung, B. H., Patsch, J. R., and Taunton, O. D.,** Synthesis and release of low density lipoproteins by the isolated perfused pig liver, *J. Biol. Chem.,* 252, 7530, 1977.

54. **Sniderman, A. D., Carew, T. E., and Steinberg, D.,** Turnover and tissue distribution of ^{125}I-labeled low density lipoprotein in swine and dogs, *J. Lipid Res.,* 16, 293, 1975.

55. **Marcel, Y. L., Nestruck, A. C., Bergseth, M., Bidallier, Robinson, W. T., and Jeffries, D.,** Low density lipoprotein turnover in swine, *Can. J. Biochem.,* 56, 963, 1978.

56. **Calvert, G. D., Scott, P. J., and Sharpe, D. N.,** The plasma and tissue turnover and distribution of two radio-iodine-labelled pig plasma low density lipoproteins, *Atherosclerosis,* 22, 601, 1975.

57. **Sniderman, A. D., Carew, T. E., Chandler, J. G., and Steinberg, D.,** Paradoxical increase in rate of catabolism of low-density lipoproteins after hepatectomy, *Science,* 183, 526, 1974.

58. **Pittman, R. C., Attie, A. D., Carew, T. E., and Steinberg, D.,** Tissue sites of degradation of low density lipoprotein: application of a method for determining the fate of plasma proteins, *Proc. Natl. Acad. Sci. U.S.A.,* 76, 5345, 1979.

59. **Carew, T. E., Saik, R. P., Johansen, K. H., Dennis, C. A., and Steinberg, D.,** Low density and high density lipoprotein turnover following portacaval shunt in swine, *J. Lipid Res.,* 17, 441, 1976.

60. **Chase, H. P. and Morris, T.,** Cholesterol metabolism following portacaval shunt in the pig, *Atherosclerosis,* 24, 141, 1976.

61. **Nestruck, A. C., Bergseth, M., Bidallier, M., Davignon, J., and Marcel, Y. L.,** Lipid and lipoprotein secretion following portacaval shunt in swine, *Atherosclerosis,* 29, 355, 1978.

62. **Nestruck, A. C., Lussier-Cacan, S., Bergseth, M., Bidallier, M., Davignon, J., and Marcel, Y. L.,** The effect of portacaval shunt on plasma lipids and lipoproteins in swine, *Biochim. Biophys. Acta,* 488, 43, 1977.

63. **Assmann, G., Brown, B. G., and Mahley, R. W.,** Regulation of 3-hydroxy-3-methylglutaryl coenzyme A reductase in cultured swine aortic smooth muscle cells by plasma lipoproteins, *Biochemistry,* 14, 3996, 1975.

64. **Brown, B. G., Mahley, R., and Assmann, G.,** Swine aortic smooth muscle in tissue culture: some effects of purified swine lipoproteins on cell growth and morphology, *Circ. Res.,* 39, 415, 1976.

65. **Weinstein, D. B., Carew, T. E., and Steinberg, D.,** Uptake and degradation of low density lipoprotein by swine arterial smooth muscle cells with inhibition of cholesterol biosynthesis, *Biochim. Biophys. Acta,* 424, 404, 1976.

66. **Bersot, T. P., Mahley, R. W., Brown, M. S., and Goldstein, J. L.,** Interaction of swine lipoproteins with the low density lipoprotein receptor in human fibroblasts, *J. Biol. Chem.,* 251, 2395, 1976.

67. **Mahley, R. W. and Innerarity, T. L.,** Interaction of canine and swine lipoproteins with the low density lipoprotein receptor of fibroblasts as correlated with heparin/manganese precipitability, *J. Biol. Chem.,* 252, 3980, 1977.

68. **Pangburn, S. H., Newton, R. S., Chang, C-M., Weinstein, D. B., and Steinberg, D.,** Receptor-mediated catabolism of homologous low density lipoproteins in cultured pig hepatocytes, *J. Biol. Chem.,* 256, 3340, 1981.

69. **Attie, A. D., Weinstein, D. B., Freeze, H. H., Pittman, R. C., and Steinberg, D.,** Unaltered catabolism of desialylated low-density lipoprotein in the pig and in cultured rat hepatocytes, *Biochem. J.,* 180, 647, 1979.

70. **Bachorik, P. S., Kwiterovich, P. O., and Cooke, J. C.,** Isolation of a porcine liver plasma membrane fraction that binds low density lipoproteins, *Biochemistry,* 17, 5287, 1978.

71. **Mahley, R. W., Hui, D. Y., Innerarity, T. L., and Weisgraber, K. H.,** Two independent lipoprotein receptors on hepatic membranes of dog, swine and man: apo-B,E and apo-E receptors, *J. Clin. Invest.,* 68, 1197, 1981.

BOVINE SERUM LIPOPROTEINS

Donald L. Puppione

INTRODUCTION

Most comparative lipoprotein studies have been undertaken with the expectation that the characterization of lipoproteins in a given species would provide a suitable model for, or would answer questions about, the development of the disease processes associated with human hyperlipoproteinemia. On the other hand, research on bovine lipoproteins has been conducted primarily to develop ways to maintain the important resources of the meat and dairy industries. Although not always explicitly stated, the underlying goal of these bovine studies has been to understand how circulating lipoproteins can be related either to animal growth or to the amount of fat synthesized and secreted by the mammary gland. Rather than focusing on these questions, the emphasis of this chapter will be on the literature dealing with the physicochemical properties of bovine lipoproteins and how they are affected by certain metabolic processes.

The data will be presented in three sections: Lymph and Plasma Lipids, Physicochemical Properties of Lipoproteins, and Lipoprotein Compositional Data. In the final section, these data will be discussed in terms of metabolic and physiological processes peculiar to cattle.

BOVINE LIPOPROTEIN DATA

Lymph and Plasma Lipids

In a ruminant, the composition of circulating lipids is influenced by many factors, particularly the biohydrogenating action of the various ruminal microorganisms.[1] In addition, bacteria and protozoa entering the small intestine from the rumen also are digested[2] and the bulk of these fatty acids in the microorganisms are highly saturated.[2] As a result, the majority of fatty acids that reach the small intestine are saturated. The two major fatty acids which are absorbed in the small intestine are stearate and palmitate. Lipids of thoracic and intestinal lymph consist mostly of triglycerides (70 to 80%) and phospholipids (12 to 15%).[3,4] A high percentage (55 to 75%) of the lymph triglyceride fatty acids are stearate and palmitate.[3,4] Lymph phospholipids are less saturated, with stearate and palmitate comprising 45 to 50% of the total fatty acids; linoleate (24%) and oleate (13%) are the major unsaturated fatty acids.[3,4] The lipid concentration in the lymph varies between 500 and 2000 mg/dℓ.[3]

In the plasma, the lipid concentration ranges between 200 and 500 mg/dℓ.[3] Plasma triglycerides are as saturated as those in the lymph;[3-8] however, the plasma triglyceride concentration, 5 to 30 mg/dℓ,[2,3,9] is considerably less than in lymph. Plasma phospholipids have the same content of stearate and palmitate as those in the lymph, but they contain more oleate (25%) and less linoleate (11%).[3] More than 70% of the phospholipids in lymph and plasma is phosphatidylcholine.[2] Approximately 25% of the plasma lipids are phospholipids.[3-8] In adult bovine plasma, the major lipid class (50 to 60%) is cholesteryl esters.[3-8] Remarkably, in spite of ruminal microbial action, plasma cholesteryl esters are very unsaturated with a high content of linoleate, 50 to 75%.[3-8] A small percentage of dietary linoleic acid is able to escape ruminal degradation or biohydrogenation by being incorporated into microorganisms, primarily protozoa.[1] After the protozoa are digested in the intestinal lumen, the component fatty acids are absorbed. In a similar manner, when the animal is grazing, a small percentage of linolenate can be absorbed, and the plasma level of cholesteryl linolenate, under these circumstances, will vary.[10] The eventual incorporation of these unsaturated fatty acids into cholesteryl esters will be discussed at the end of this chapter. It should be noted that the distribution of the various fatty acids, esterified to each type of lipid, and the concentration of the lipid classes will change in response to dietary components or physiological status.[3]

The effect of diet on lipid metabolism in ruminants has been reviewed by Christie.[11] However, some brief general statements concerning dietary effects on lipid concentrations will be made. The following are known to increase the plasma concentrations of cholesterol and phospholipids: (1) incorporation of either saturated or unsaturated fats into the diet,[11-14] and (2) the feeding of fats encapsulated in formaldehyde treated casein pellets.[15-17] The fats contained in such pellets are not acted on by the bacteria and protozoa in the rumen, but are digested in the lumen of the small intestine in a manner similar to what occurs in a nonruminant animal. The feeding of these so-called "protected fats" will also result in plasma triglycerides attaining levels in excess of 100 mg/dℓ.[17] Total lipid levels, exceeding 500 mg/dℓ, are not uncommon in animals on diets containing either protected[15,16] or unprotected fats.[12-14] More specific details on diets and their effects are discussed in the review of Christie.[11]

In addition to diet, factors such as age, pregnancy, and stage of lactation will influence the concentration and composition of plasma lipids.[2,3] At birth, free fatty acid (FFA) levels are approximately 20 mg/dℓ in calf plasma.[11,18,19] Within 2 days, FFA levels decrease by more than a factor of two. The levels continue to fall and by the 6th month the FFA concentration is 5 mg/dℓ. At birth, phospholipids predominate among the different lipid classes, but by 10 days the plasma levels of cholesteryl esters and phospholipids are comparable.[19] The fatty acids of all lipid classes are highly saturated in the newborn calf.[18-20] In striking contrast to the plasma cholesteryl esters of the adult animal, calf plasma cholesteryl esters contain very little linoleate (5%). Similarly, the content of linoleate in plasma phospholipids is lower in the calf (2%)[3,19,20] than in the adult (12%).[3]

In the cow, fluctuations in plasma lipid levels occur during the lactation and gestation cycles.[6,9,21-25] At parturition, the levels are low and then rise sharply with the onset of lactation. By the second week of lactation, the cholesterol concentration can increase from less than 100 mg/dℓ to over 300 mg/dℓ.[9] Levels of FFA will also increase at parturition and reportedly are correlated with milk yield. Observed changes in both plasma and lymph lipids are associated with fluctuation in the distribution of the major lipoproteins.

Physicochemical Properties of Bovine Lipoproteins

As is the case for most mammals, the major class of lipoproteins in the adult animal is alpha lipoproteins (α-Lp).[6,26-30] A major band, with the mobility of human α-Lp can be seen in the electrophoretogram shown in Figure 1. In agreement with the electrophoretic data, a high percentage of plasma cholesterol (80 to 90%) remains in the supernatant following heparin manganese precipitation.[9,31] Unlike most terrestrial mammals, the α-Lp of the adult bovine animal comprise a highly polydisperse system. They vary over a wide range in size (9 to 16 nm),[32-34] in density (1.21 to 1.03 g/mℓ)[30] and in molecular weight (4 × 10^5 to 1 × 10^6).[30,32] The size heterogeneity of the α-Lp can be inferred from the schlieren data shown in Figure 2. The majority of bovine α-Lp can be isolated in the ultracentrifugal density fraction, 1.063 to 1.21 g/mℓ,[6,30] also referred to as the high-density lipoprotein (HDL) class. When floated in a solution having a density of 1.21 g/mℓ, bovine HDL will generate a schlieren pattern with a major, albeit broad, peak. As can be seen in Figure 2, the schlieren pattern extends on both sides of the major peak. In general, as the lipoprotein decreases in size, the ratio of lipid to protein also decreases and the particle density increases. The variation in size and density will cause the lipoproteins to float at different rates in a solution of density 1.20 g/mℓ. Flotation rates or F rates are measured in negative Svedberg units.[36] $F_{1.20}$ rate is defined as Svedbergs of flotation, measured in a solution with a density of 1.20 g/mℓ. In Figure 2, the HDL distribution is plotted in these units. For those unfamiliar

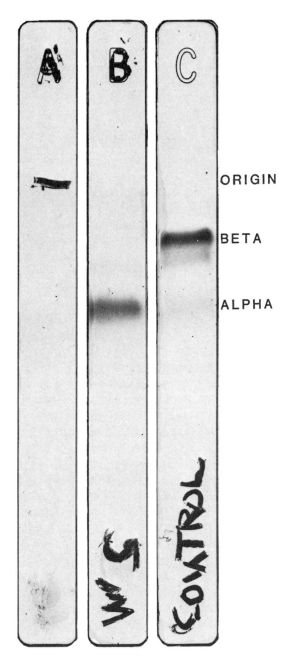

FIGURE 1.   Agarose electrophoretograms of lactating cow
serum (B) and human plasma (C). Minor band in cow sera
comigrated with human (pre-β) lipoproteins.

with schlieren analysis, the pattern can be considered as a plot of concentration (ordinate)
vs. density (abscissa). Lipoproteins having $F_{1.20}$ values of 0, 3.5, and 20 would have approximate densities of 1.20, 1.12, and 1.063 g/mℓ, respectively. The HDL, generating the
pattern to the right of the peak, are smaller and less dense than those generating the pattern
to the left of the peak. Lipoproteins generating the peak of the schlieren pattern are intermediate in size and density to the two groups of lipoproteins comprising the outer portions
of the distribution.

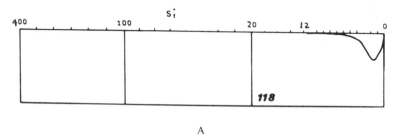

A

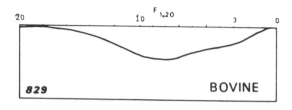

B

FIGURE 2.   Computer-derived plot of the schlieren patterns obtained during analytical ultracentrifugal analyses of lactating cow lipoproteins floated in salt solutions of d = 1.063 g/mℓ, in (A); and of d = 1.21 g/mℓ, in (B) (3). Values for serum concentrations (mg/dℓ) are given in boldface numerals.

Table 1
PHYSICOCHEMICAL PROPERTIES OF THE MAJOR HDL IN DIFFERENT BOVINE PLASMAS

Property	Cow	Steer	Calf	Fetus
Electrophoretic mobility	α	α	α	α
Diameter (nm)	13.6	12.6	9.0	8.2
Flotation rate ($F_{1.20}$)	6.7—8.6	6.9	3.2	2.0
Sedimentation rate ($S_{20,w}$)	5.6	ND[a]	ND	ND
Hydrated density (g/mℓ)	1.069—1.083	ND	ND	ND
Diffusion coefficient (Ficks)	3—4	ND	ND	ND
Stokes radius (nm)	5.3—6.5	ND	ND	ND
Molecular weight ($\times 10^{-5}$)	4.0—7.4	ND	ND	ND
Ref.	30,32,33,35	34	34	34

[a]   ND, no data available.

The physicochemical parameters of the major components of the HDL isolated from various bovine sera are given in Table 1. The flotation rate of the major peak in the cow distribution has an $F_{1.20}$ value of 7.8. Steer HDL have a similar distribution with a peak $F_{1.20}$ value of 6.9.[34] However, in neonatal and fetal sera, the HDL are smaller and have lower flotation rates than the HDL of older animals. The schlieren peaks have $F_{1.20}$ values of 3.2 for calf HDL and 2.0 for fetal HDL.[34]

Low-density lipoproteins (LDL) (d 1.020 to 1.063 g/mℓ) also are polydisperse and this fraction contains two distinct electrophoretic groups.[26-30] As noted above, the distribution of α-Lp extended beyond the lower density limit of the HDL class, i.e., 1.063 g/mℓ. These α-Lp are designated α-LDL. Another class of lipoproteins, β-LDL, are present in this ultracentrifugal density fraction. This second type of LDL comprise the minor band seen in the electrophoretogram in Figure 1. However, this band's mobility, which is greater than human beta lipoproteins (β-Lp), indicates that the bovine lipoproteins are more negatively charged than their human counterparts. The possibility also exists that these lipoproteins correspond to a bovine form of Lp (a).[39] For the purpose of this chapter, this second type of bovine LDL will be referred to as β-LDL.

A third type of lipoprotein also may be isolated in the ultracentrifugal LDL fraction. Once the blood has been drawn, a group of lipoproteins can form if any procedure is carried out below a critical temperature, somewhere between 25 and 38°C. Puppione et al.[40] have shown these lipoproteins to be large, polydisperse triglyceride-rich lipoproteins which can be isolated in the VLDL, LDL, and IDL (intermediate density lipoprotein) fractions. Using preparative ultracentrifugations, 40 to 50% of these lipoproteins can be isolated in the IDL fraction, 1.006 to 1.020 g/ml. These TG-rich IDL are roughly 10- to 15-fold larger than a typical IDL. This broad density range exhibited by these lipoproteins is thought to be due to the core lipids undergoing a phase transition from a supercooled liquid state to a crystalline state.[40,41] If TG-rich lipoproteins are isolated under standard conditions, e.g., 4 to 20°C, the highly saturated core lipids will solidify, causing dramatic changes in the shape and density of the lipoproteins.[40,41] Such a change will undoubtedly affect the composition of the lipoprotein as well. Following the phase transition, the surface components (phospholipids, unesterified cholesterol, and protein) will be inadequate to cover the core lipids in the transformed particles although sufficient to cover spherical oily droplets of triglycerides. Figure 3B illustrates the change in morphology which occurs when the TG-rich lipoproteins isolated at 37°C are allowed to cool to 16°C. To compensate for this increase in surface area, apoproteins, certain serum proteins with extensive hydrophobic regions and phospholipids would have to be transferred to the surface of these structures. The acquisition of these additional components could take place during the time the blood is being kept on ice or while the plasma is stored in the refrigerator. Further alterations, such as fusion of particles, may occur during ultracentrifugal isolation because the additional proteins are still insufficient to minimize the constraints on the surface components. In support of the possibility that fusion of particles might have occurred, the isolated IDL tend to be more irregular in appearance than the particles which form upon cooling (compare Figure 3B with Figure 4). After undergoing the phase transition the bovine TG-rich lipoproteins retain their altered shape even at 37°C.[40,43]

Schlieren analyses of lipoproteins floating in a salt solution with a density of 1.063 g/ml provide concentration data on three different classes: (1) the TG-rich VLDL, (2) the IDL, and (3) the LDL. The symbol, S_f^o, is used to designate the flotation rate in a solution with a density of 1.063 g/ml. The superscript "o" indicates that the flotation rates have been corrected for concentration dependence. The VLDL correspond to S_f^o 20 to 400, the IDL to S_f^o 12 to 20, and the LDL to S_f^o 0 to 12.[35,36]

Because the concentration of triglycerides in bovine plasma is low, it is not surprising that the levels of VLDL also are low. Thus, in Figure 2, there is no demonstrable VLDL in the serum of the lactating cow. This is true in neonatal and fetal sera.[34] However, the VLDL in steer plasma are sufficiently concentrated (20 mg/dl) to produce a discernible pattern.[34] Steer VLDL exhibit a flotation rate with S_f values between 20 and 250. However, because cooling will cause a phase transition among the core triglycerides, the flotation rates are undoubtedly less than what the native lipoproteins would have. Having higher concentrations than VLDL bovine LDL produce discernible schlieren patterns. In Figure 2, the schlieren peak of cow LDL floated with an S_f^o value of 1.3.[35] In fetal sera, the flotation rate was higher (S_f^o 8.9).[34] Flotation rates intermediate to the values for cow and fetal LDL are found in steer (S_f^o 5.3) and calf (S_f^o 4.9) sera.[34] Data, summarized in Table 2, show that for bovine LDL there is a direct relationship between size and flotation rate.

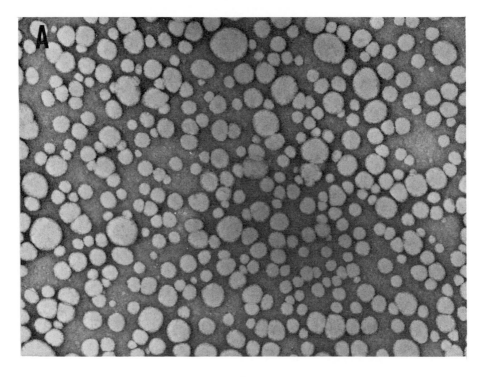

A

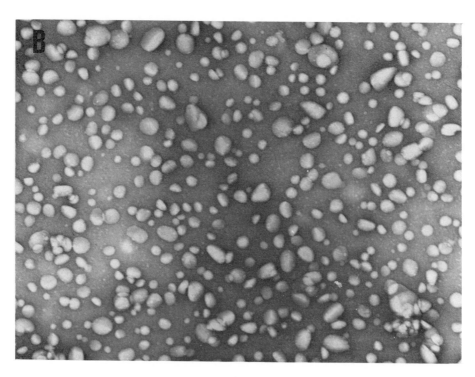

B

FIGURE 3.   Electron micrographs of negatively stained plasma lipoproteins in the d <1.006 g/mℓ density fraction. (A) Lipoproteins were isolated and negatively stained at 37°C. (B) Lipoproteins were isolated at 37°C and then cooled to 16°C, after which they were negatively stained at room temperature. (Magnification × 78,000.)

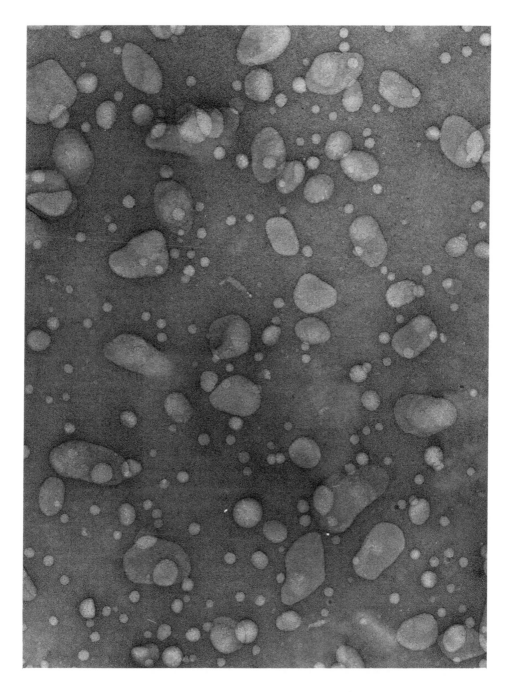

FIGURE 4.   Electron micrograph of negatively stained crystallized bovine IDL isolated from calf lymph. (Magnification × 120,000.)

Representative concentrations of LDL and HDL for animals at different ages are given in Table 3. Only the fetal calf serum had a higher concentration of LDL than HDL. The concentrations for calf and fetal HDL are consistent with the low lipid concentration in younger animals. Also differences between lactating and dry cows, seen in Table 3, agree with changes in the concentration of HDL known to occur during the gestation and lactation cycles. However, it should not be inferred that the HDL levels in the cows and steer

Table 2
SIZE AND FLOTATION RATE (S_f^o) OF
BOVINE LDL

Property	Cow	Steer	Calf	Fetus
S_f^o rate	1.3	5.3	4.9	8.9
Diameter (nm)	16.0	18.8	19.4	26.0
Ref.	32	34	34	34

Table 3
BOVINE LIPOPROTEIN
CONCENTRATIONS (mg/dℓ)
DERIVED FROM SCHLIEREN
ANALYSIS

	LDL	HDL	Ref.
Fetus	44	35	34
Calf	31	167	34
Steer	73	365	34
Lactating cow	118	829	35
Dry cow	N.D.	301	35

necessarily differ to this extent. A combination of factors, such as components in the diet of the steer and stage of lactation in the cow, can result in higher HDL levels in a steer than in a cow.

Composition of Bovine Lipoproteins

Obtaining representative data for TG-rich lipoproteins and LDL circulating in bovine blood is complicated by both the temperature constraints and electrophoretic heterogeneity just described. The majority of plasma triglycerides can be recovered in an ultracentrifugal density fraction, d <1.020 g/mℓ, even if the lipoproteins have undergone a phase transition.[8] However, if ultracentrifugal isolations are done at 16°C, all of the TG-rich lipoproteins will not float at d = 1.006 g/mℓ, with 50% of the plasma triglycerides being recovered in the d <1.006 g/mℓ fraction and 40% in the 1.006 to 1.020 g/mℓ fraction.[40] Data that appear in Table 4 were obtained on TG-rich lipoproteins isolated between 25 and 38°C.[42]

Most attempts in separating bovine LDL into β- and α-Lp have not been successful. The compositional data in Table 4 for beta LDL were obtained after the lipoproteins had been precipitated with dextran sulfate, solubilized in a sodium citrate solution, and then isolated ultracentrifugally in the density interval, 1.039 to 1.060 g/mℓ.[8] Gradient gel data indicated that the lipoproteins isolated in this way migrated as β-Lp. The data for α-LDL in Table 4 were obtained from compositional analyses of an ultracentrifugal density fraction containing mostly α and very little β-LDL.[30] There appears to be little difference in the gross composition of the two electrophoretic forms of bovine LDL.

Based on numerous analyses, bovine HDL in the adult animal have a high content of cholesteryl esters.[2,3] When analyses are done on HDL fractionated in an isopycnic gradient, such a composition is found to be peculiar to the major fraction. However, as the lipoprotein density becomes greater than 1.090 g/mℓ, the size and cholesteryl ester content decrease steadily, with a corresponding increase in protein and phospholipid content.[30] Assuming that all HDL particles have a spherical shape, the percent of surface components (protein, phospholipids, and unesterified cholesterol) would be expected to increase as the lipoprotein decreased in size. To enable a comparison to be made between composition and physico-

Table 4
COMPOSITION OF THE MAJOR BOVINE LIPOPROTEIN CLASSES

| Class | Percent by weight | | | | | Ref. |
	Pro	CE	PL	TG	UC	
Chylomicron	3	2	4	87	4	42
VLDL	8	5	7	74	7	42
β-LDL[a]	24.0	39.3	29.2	1.8	5.7	8
α-LDL[a]	21.8	38.6	32.2	—	7.5	30
HDL lactating cow	33.4	33.3	29.5	—	3.9	30
Steer	37.5	37.3	22.4	—	2.8	34
Calf	45.1	29.0	20.6	—	2.7	34
Fetus	50.3	22.9	23.8	—	3.0	34

Note: Protein (Pro), cholesteryl esters (CE), phospholipids (PL), triglyc-erides (TG), and unesterified cholesterol (UC).

[a] See text for description.

chemical properties, the HDL data appearing in Table 4 correspond to those given in Table 1. Consonant with the size difference, cholesteryl ester content is higher in the larger HDL of cows and steers in comparison with the smaller HDL found in newborn calf and fetal sera. However, from the flotation rate and size data, steer HDL would be expected to have a lower cholesteryl ester and a higher phospholipid content than cow HDL. In each of the four types of bovine HDL, the triglyceride content of HDL was less than 1%.

Apoproteins

The major apoprotein of bovine plasma has a molecular weight of approximately 28,000.[45] This apoprotein comprises more than 80% of the protein moiety of α-Lp with hydrated densities between 1.05 and 1.21 g/mℓ.[30] Amino acid analyses have revealed that the bovine apoprotein has an amino acid content similar to that for the major apoprotein of human HDL,[45-47] i.e., apo A-I. By analogy with the human apoprotein, the major bovine apoprotein will be referred to as apo A-I. The amino acid composition of bovine apo A-I is given in Table 5. Isoelectric focusing studies of bovine apo A-I have revealed the existence of isoforms which are more basic than the corresponding isoforms found in other mammals.[47] Based on the work of Swaney,[47] as many as six molecules of apo A-I have been estimated to be on the surface of bovine HDL.[30] Consistent with these results, the protein content of lipoproteins in major fractions in an isopycnic gradient was approximately 30% and the average molecular weight was 580,000.[30] In other words, $6 \times 28,000$ approximately equals $580,000 \times 0.3$. α-Lp with densities less than 1.05 g/mℓ also contain a high content of apo A-I,[41] however, the relative content of apo A-I in these large, less dense lipoproteins has not been determined.

Minor apoproteins also are associated with α-Lp. These include a group of from five to seven low-molecular-weight (7×10^3 to 1×10^4) apoproteins.[48,49] Amino acid analyses for five of these apoproteins are given in Table 5. Because DEAE ion-exchange chroma-tography was used in their purification, they are designated D_2 through D_5, corresponding to their pattern of elution. The amino acid data for D_3, D_4, and D_5 are quite distinct from any of the apo C found on human lipoproteins. However, D_4 and D_5 contain 1 and 2 mol, respectively, of sialic acid. As a result, D_3, D_4, and D_5 are thought to correspond in an analogous way to human apo C-III-0, apo C-III-1, and apo C-III-2.[49] A close correspondence was found when amino acid composition data for D_2 were compared with those for various mammalian forms of apo A-II.[49] However, Lim and Scanu[48] reported that this apoprotein

Table 5
AVERAGE VALUES (MOL/MOL)[a] OBTAINED
FROM PUBLISHED AMINO ACID[b] DATA FOR
BOVINE APOPROTEINS

Amino acid	A-I	D_2	D_3	D_4	D_5
Asp	21	6.2	4.7	5.1	4.8
Thr	8	5.9	5.1	5.7	4.8
Ser	13	5.8	7.9	10.7	9.9
Glu	48	16.2	11.9	13.0	12.3
Pro	14	2.9	2.5	2.9	2.6
Gly	10	3.2	3.0	3.6	4.6
Ala	24	7.0	5.4	6.0	5.5
Val	13	3.0	2.6	3.0	2.6
Met	1		0.9	1.6	1.4
Ileu	4	1.1	0.5		
Leu	32	7.8	4.7	4.1	4.0
Tyr	7	2.6	1.9	1.7	1.4
Phe	6	5.7	4.0	3.9	3.3
Lys	21	9.1	6.3	6.3	5.2
His	3				0.8
Arg	13		0.9	1.1	1.1
Try	3		0.3	1.7	1.6
Refs.	45—47	48,49	48,49	48,49	48,49

[a] Assumed molecular weights were apo A-I (28,000); apo D_2 (8800); apo D_3 (7100); apo D_4 (7800), and apo D_5 (7300).
[b] Cysteine was not detected in any of the above apoproteins.

had a valine residue at the amino terminus, rather than having a blocking pyroglutamic acid residue that has been found in other mammalian apo A-II.[49] These same authors[48] also reported that D_1 and D_3 activated bovine milk lipase. D_1 was three to five times more active than D_3. Again by analogy with human apoprotein, D_1 may correspond to apo C-II. Using both isopycnic gradient and gel filtration, these low-molecular-weight apoproteins have been found associated primarily with the large α-Lp.[30,50]

Low amounts of high-molecular-weight apoproteins, possibly apo A-IV and apo A-V, also have been found on bovine α-Lp.[30] In contrast to the low-molecular-weight apoproteins, these large apoproteins were found primarily on lipoproteins with densities greater than 1.10 g/mℓ. Apoproteins corresponding to human apo E appear to be present in trace amounts[33] on bovine α-Lp.

As has been noted in other mammals,[51,52] there also appears to be two forms of bovine apo B. The amino acid compositions and molecular weights of these two bovine forms of apo B as well as their sites of synthesis are not known.

DISCUSSION

Synthesis of Lipid Constituents of Bovine TG-Rich Lipoproteins

The physicochemical properties which have been described for TG-rich lipoproteins are a consequence of certain key metabolic steps associated with the synthesis of the three major lipid classes found in these lipoproteins. The core triglycerides are derived from fats entering the small intestine from the rumenreticulum. These consist primarily of unesterified saturated fatty acids complexed to particulate digesta.[3,53,54] As a result, the major mode of triglyceride synthesis is the α-glycerolphosphate pathway. Absorbing mostly stearate and palmitate, bovine enterocytes synthesize and secrete chylomicrons and VLDL with highly saturated core triglycerides. As already noted these two fatty acids comprise as much as 70% of the total fatty acids in plasma and lymph triglycerides. Monoglycerides are minor products of fat digestion in ruminants. Nevertheless, ruminants do have pancreatic enzymes capable of hydrolyzing esterified lipids even at the extremely acidic conditions prevailing in the duodenum.[55] A pancreatic lipase has been described for sheep.[56] The absorption of protected fats in cattle indicate that a similar enzyme can participate in the digestion of fats of large ruminants as well.

Because of their size differences, chylomicrons (80 to 500 nm) and VLDL (30 to 80 nm) require different relative amounts of surface and core lipids for their formation. For example, a VLDL with a diameter of 40 nm would have three times the surface to volume ratio of 120-nm chylomicron. The synthetic rates of the various component lipid classes very likely are important determinants of the range in size of the TG-rich lipoproteins.

The studies of Leat and Harrison[57] have provided insight into the synthesis of the major lipid components on the surface and in the core of the TG-rich lipoproteins. Studying the fate of different radioactive fatty acids, Leat and Harrison[57] found a high incorporation of saturated fatty acids into core triglycerides and of polyenoic acids, 18:2 and 18:3 into surface phospholipids. Phospholipids, on the surface of enteric TG-rich lipoproteins, are thought to be derived from phosphatidylcholine molecules present in bile.[54] Through the action of pancreatic phospholipase A, these molecules are converted into lyso forms that are absorbed from the intestinal lumen and then esterified to form phosphatidylcholine within the enterocytes. Based on the relatively low specific activity of lymph phospholipids, Leat and Harrison concluded that the alpha glycerolphosphate pathway was probably of minor importance in the synthesis of surface phospholipids. However, because lymph phospholipids are not associated exclusively with newly synthesized TG-rich lipoproteins, the same conclusions might not have been drawn if the specific activity had been determined for phospholipids on the surface of chylomicrons and VLDL isolated from lymph.

Based on studies of small ruminants, goats, and sheep, it would appear that *de novo* synthetic pathways are utilized for the biosynthesis of cholesterol, the other lipid on the surface of chylomicrons and VLDL.[58] Lacking a dietary source of cholesterol, it is not surprising that these animals must rely on *de novo* synthesis of this sterol. Moreover, rather than being derived from the enterohepatic circulation or from digested ruminal protozoa, Scott et al.[58] have reported that cholesterol molecules needed for the surface of TG-rich lipoproteins are synthesized within individual enterocytes and that sterolgenesis[58] can be stimulated within mucosal cells by increasing the amount of fat in the diet. In other words, the amount of triglycerides synthesized and secreted by the intestinal mucosal cells dictates the amount of cholesterol which these cells need to produce. At the same time, this increase in enteric sterolgenesis apparently suppresses cholesterol synthesis in the liver.[58]

The degree to which bovine enterocytes preferentially secrete VLDL or chylomicrons is unresolved.[3,35,39] The composition of TG-rich lipoproteins[2,3,8,34,44] has been the basis for concluding that chylomicrons were not major carriers. In particular, because of the high phospholipid content, bovine TG-rich lipoproteins have been thought of as being smaller particles needing more surface components than larger chylomicrons.[3,60] The composition of different human VLDL fractions[61] has revealed that the smallest class still has a lower phospholipid content than reported values for the total TG-rich lipoprotein fraction isolated from bovine plasma. Carrying out procedures between 25 and 38°C, Ferreri and Elbein[42] found the phospholipid content to be one third to one half previously published values for bovine TG-rich lipoproteins. The previously reported values indicating a higher phospholipid content, as suggested earlier, could be due to acquisition of additional surface components when the lipoproteins undergo a temperature-induced change in morphology.

Gage and Fish, using dark field microscopy, reported the first observation of chylomicrons in bovine blood in 1924.[62] However, supportive physicochemical data only recently were obtained. Electron microscopic studies have shown that bovine plasma does contain large lipoproteins with the dimensions of chylomicrons.[34,41,42] As the electron micrograph in Figure 3A demonstrates, the particles have the expected morphology if the lipoproteins are isolated at 37°C. Moreover, even upon cooling, not all of the lipoproteins undergo a shape change (see Figure 3B). Consistent with these observations, a small percentage of TG-rich lipoproteins that were isolated at 16°C, exhibited flotation rates with S_f values greater than 400 in the analytical ultracentrifuge.[43] This small group of lipoproteins with high flotation rates probably contained a sufficient amount of unsaturated fatty acids among their core triglycerides to prevent them from undergoing a phase change. Flotation rate studies still need to be done on bovine TG-rich lipoproteins isolated at 37°C. The inability to observe a chylomicron band in an electrophoretogram may be due to the poor stainability of saturated triglycerides[63] as well as to the low triglyceride concentration in bovine plasma. It therefore would appear that bovine enterocytes secrete more chylomicron-sized particles than previously thought. The question still needs to be answered concerning the relative amounts of chylomicrons and VLDL that enter the thoracic duct.

The physicochemical properties of lipoproteins secreted by the bovine liver are unknown at this time. Although *de novo* synthesis of fatty acids in bovine hepatocytes is thought to be a minor pathway,[64] uptake of circulating fatty acids for hepatic VLDL production is apparently an active mechanism in the cow during the first 3 to 4 weeks of lactation.[65] Being in negative energy balance during the initial phase of lactation, stored fats,[68,69] rather than dietary fat, are the main source of circulating triglycerides which the cow utilizes for milk fat synthesis. The net uptake of FFA bound to albumen by the mammary gland is almost negligible.[66,67] Rather, hepatic VLDL are presumably the main vehicles whereby mobilized fatty acids in early lactation are able to reach the mammary gland for milk fat production.[65] Nevertheless, to be taken up by the mammary gland the fatty acids present on the triglycerides first must be hydrolyzed through the action of the enzyme, lipoprotein lipase. In later lactation, the diet becomes the almost exclusive source of circulating triglycerides. By midlactation, the daily influx of lymph triglycerides is approximately 400 g. It has been estimated that as much as 76% of the circulating triglycerides in the cow are utilized for milk fat synthesis.[70] The relative utilization of hepatic and enteric TG-rich lipoproteins in other animals such as bulls, steers, and calves, is unknown.

Metabolic Relationship of TG-Rich Lipoproteins to α-Lp

The fats delivered by chylomicrons and VLDL to different tissues in the body can be utilized for energy, stored as depot fat, or converted into milk fat. Their core triglycerides are not taken up directly by the tissue but are hydrolyzed through the action of lipoprotein lipase, an enzyme located on the outer surface of the endothelial cells. Studies indicate that, at a given site, as much as 70% of the core triglycerides undergo hydrolysis.[71] The bulk of the released fatty acids are taken up rapidly by the tissue at the site of hydrolysis. The residual triglycerides form the core for what have been termed remnant particles.[72] The fate of the remnants in cattle has not been studied.

Remnant particles from enteric TG-rich lipoproteins are taken up by the liver in several mammalian species,[73-76] but hepatic VLDL remnants are converted into β-LDL. Both chylomicron and VLDL remnants in the rat are removed by the liver to a very high degree.[76] It has been suggested that this preferential hepatic removal of both types of remnant particles in the rat might explain why the levels of β-LDL in these animals are much lower than in humans.[77] Because the concentration of β-LDL is low in the bovine plasma, hepatic uptake could be the major metabolic fate of the remnant particles in cattle as well. This also could be due to rapid turnover of β-LDL.

The remnant particle is always smaller in size than the precursor lipoprotein.[71] The lipolytic action of lipoprotein lipase results in rapid removal of core lipids; however, little degradation of surface lipids occurs. To adjust for the decrease in size, the excess surface components dissociate from the remnant particles.[78,79] The physicochemical nature of these shed components has not been determined. Because the surface components are mixtures of cholesterol, phospholipid, and apoprotein, they are thought to form bilayer sheets, discoidal lamellar aggregates, and vesicular structures.[79] The smaller structures of these complexes can be incorporated into existing α-Lp, or along with the larger structures, can be converted into α-Lp through the action of the enzyme, lecithin cholesterol acyl transferase (LCAT).[78,79]

Hyperalphalipoproteinemia in Lactation

The LCAT enzyme, which is activated by apo A-I, removes the β fatty acid from lecithin and transesterifies it to cholesterol.[80] During lactation when the circulating triglycerides in the cow are being taken up rapidly by the mammary gland, an LCAT-mediated process has been proposed to explain the marked increase in the concentrations of HDL_1 (lipoproteins with hydrated densities between 1.063 to 1.09 g/mℓ) and α-LDL.[35] As noted earlier, studies on small ruminants have indicated that linoleate absorbed in the small intestine is incorporated into phospholipids, principally phosphatidylcholine. Similarly, during periods of grazing, linolenate molecules would be absorbed and incorporated into the surface phospholipids of enteric TG-rich lipoproteins. Approximately 95% of linoleate and linolenate in bovine phosphatidylcholine is located in the β position, and linoleate is the major fatty acid in this position.[2,3,81] What percentage of the polyenoic acids, 18:2 and 18:3, used in the enteric synthesis of the surface phospholipids of TG-rich lipoproteins comes from the diet as opposed to being drawn from existing body pools still remains to be determined. The surface lecithin of hepatic TG-rich lipoproteins presumably has a very similar distribution of polyenoic acids. At the present time, no definitive data are available on the relative contribution of hepatic and enteric TG-rich lipoproteins to the plasma cholesteryl ester pool. The high content of cholesteryl linoleate in the core of bovine α-Lp is consistent with these cholesteryl esters being derived from surface components which have been acted on by LCAT.

It has already been noted that hepatic VLDL are thought to be the major contributor to milk fat synthesis in early lactation. In midlactation when feed intake has peaked, the majority of triglycerides in the circulation would be presumed to be intestinal in origin. The clearance of enteric TG-rich lipoproteins, as described above, becomes the major source of the plasma cholesteryl esters. Furthermore, this means of generating cholesteryl esters explains the findings of Maynard et al.,[22] who were able to produce fluctuations in the serum cholesterol concentrations of the lactating cow by varying the fat content in the feed.

Developmental Changes in the HDL Distribution

This same series of reactions, involving LCAT in the processing of surface components for conversion or incorporation into HDL, also might be the reason for the discernible difference in HDL size between young and adult cattle. In the bovine fetus, two factors may be contributing to the small size of the HDL. First, synthesis of TG-rich lipoproteins could be very low in the fetal state. If so, there would be little in the way of LCAT reactants present in components shed from the surface of TG-rich lipoproteins for the production of HDL. Secondly, in fetal plasma all lipid classes, including lecithin, are enriched in saturated fatty acids.[2,3,19] In the absence of the appropriate polyenoic acids in the β position of lecithin, LCAT reactivity might be markedly reduced in bovine fetal plasma. In the neonate, dietary milk fats would stimulate the production of chylomicrons and VLDL in calf enterocytes. Moreover, although linoleate is not a major fatty acid in milk fat, Nobel et al.[19] reported that the content of linoleate in the cholesteryl esters and phospholipids in plasma increase dramatically with age, with a sharp rise occurring within the first week. As long as the animal continues to grow and to increase its feed intake, the plasma levels of unsaturated cholesteryl esters also will rise steadily. Concomitantly, the amount of feed intake will influence directly the rate of synthesis of both core and surface components of TG-rich lipoproteins. Apparently, the enteric production of TG-rich lipoproteins, the formation of cholesteryl linoleate, and the concentration of HDL are all interrelated.

Possible Mechanisms Involved in the Formation of Large α-Lp

While providing a means for increasing the plasma level of cholesteryl linoleate, the interaction of the surface components from TG-rich lipoproteins with LCAT does not explain the unusually large mean size found for bovine α-Lp. More than 50% of the bovine α-Lp in the lactating cow can be isolated within the density interval of 1.063 to 1.090 g/mℓ.[30] Moreover, Forte et al.[34,46] have found that in the adult animal the mean diameter of bovine HDL is larger than that found in other mammals. There are several mechanisms which might be involved in the formation of these large α-Lp. Glomset and Norum[82] have proposed that LCAT mediates the conversion of small α-Lp into larger structures. Such a process might be stimulated in the lactating cow in which the rapid clearing of triglycerides is providing substrate for the LCAT reaction. Alternatively, as a result of this rapid clearance, the number of lipoproteins in bovine plasma needed to transport the resulting cholesteryl esters would be reduced by the formation of large α-Lp. This might be important if the amount of the apo A-I, the major apoprotein on the surface of these lipoproteins, was limited. A third possibility for the size of bovine α-Lp might be a metabolic response to increase the available surface for the low-molecular-weight apoproteins. As has been proposed for other mammalian systems,[83] such a reservoir would provide the C apoproteins to chylomicron and VLDL, enabling them to interact with lipoprotein lipase. Following triglyceride hydrolysis, the C apoproteins would return to the surface of the large α-Lp.

Each of the above possibilities involves an adaptation to the action of the two enzymes, lipoprotein lipase and LCAT. However, because rapid clearance of triglycerides can be demonstrated in a variety of mammals, the question might be raised why the adult bovine is unique in terms of the size distribution of its α-Lp. Although bovine forms of lipoprotein lipase[84-86] and LCAT[82,87] have been demonstrated, studies of Etienne et al.[86] indicate that hepatic lipase, a third enzyme associated with lipoprotein metabolism, may not be present in bovine liver. Being able to hydrolyze the phospholipids and triglycerides of HDL$_2$, the larger of the two HDL subclasses in human plasma, hepatic lipase is thought to participate in converting HDL$_2$ into HDL$_3$, the smaller subclass of human HDL.[88,89] Lacking this enzyme, bovine liver would be unable to participate in such a conversion. The accumulation of the predominantly large α-Lp in bovine plasma may be due to an absence of a major catabolic pathway.

CONCLUSION

To recapitulate, the absorption of fat and the formation of triglycerides appear to be major contributors to the hypercholesterolemia which develop in cattle. Hepatocytes are also sites of synthesis and secretion of TG-rich lipoproteins. The contribution of the liver to the blood cholesterol pool is variable. During periods of high feed intake, hepatic sterolgenesis is probably limited and the intestinal mucosal cells become the major source of both cholesterol and triglycerides in the circulation.

The hypercholesterolemia is associated primarily with an increase in the concentration of polyenoic cholesteryl esters, that are products of the LCAT reaction. The substrates for this reaction are surface components shed from catabolized TG-rich lipoproteins. The action of lipoprotein lipase generates substrate for the LCAT reaction. The action of LCAT is probably a major factor in the formation of large α-Lp, which in the bovine can have densities less than 1.063 g/mℓ. Lacking hepatic lipase, the bovine may be unable to clear rapidly these lipoproteins from the blood.

The major apoprotein on the surface of the highly polydisperse bovine α-Lp is apo A-I. Other apoproteins, corresponding to human apoproteins, A-II, A-IV, A-V, B, C-II, C-III, and E, also appear to be present in trace amounts in bovine plasma.

ACKNOWLEDGMENTS

The author wishes to express his thanks to the various colleagues who took time to read and comment on the preliminary versions of this chapter. I am most indebted to Dr. Donald Palmquist for his time and suggestions. I also want to express my gratitude to Professor Stuart Patton for his support, encouragement, and inspiration during the past 12 years and for being solely responsible for extending my research interest into the area of bovine lipoproteins.

Preparation of this chapter was made possible through the support of NIH Atherosclerosis Training Grant (HL 07386). During my stay at the Molecular Biology Institute, I have had the active collaboration of three individuals whom I wish to thank. They are Professor Verne N. Schumaker, Dr. Steven T. Kunitake, and Martin L. Phillips. The author expresses his gratitude to Ms. Elaine Stieglitz for her time and effort in the preparation of this manuscript.

REFERENCES

1. **Harfoot, C. G.,** Lipid metabolism in the rumen, in *Progress in Lipid Research,* Vol. 17, Holman, R. T., Ed., Pergamon Press, New York, 1978, 21.
2. **Nobel, R. C.,** Digestion, absorption and transport of lipids in ruminant animals, in *Progress in Lipid Research,* Vol. 17, Holman, R. T., Ed., Pergamon Press, New York, 1978, 55.
3. **Christie, W. W.,** The composition, structure and function of lipids in the tissues of ruminant animals, in *Progress in Lipid Research,* Vol. 17, Holman, R. T., Ed., Pergamon Press, New York, 1978, 111.
4. **Leat, W. M. F. and Hall, J. G.,** Lipid composition of lymph and blood plasma of the cow, *J. Agric. Sci. Camb.,* 71, 189, 1968.
5. **Evans, L., Patton, S., and McCarthy, R. D.,** Fatty acid composition of the lipid fractions from bovine serum lipoproteins, *J. Dairy Sci.,* 44, 475, 1961.
6. **Duncan, W. R. H. and Garton, G. A.,** Blood lipids. III. Plasma lipids of the cow during pregnancy and lactation, *Biochem. J.,* 89, 414, 1963.
7. **Hanahan, D. J., Watts, R. M., and Pappajohn, D.,** Some chemical characterization of the lipids of human and bovine erthrocytes and plasma, *J. Lipid Res.,* 1, 421, 1960.
8. **Stead, D. and Welch, V. A.,** Lipid composition of bovine serum lipoproteins, *J. Dairy Sci.,* 58, 122, 1975.
9. **Puppione, D. L., Smith, N. E., Clifford, C. K., and Clifford, A. J.,** Relationships among serum lipids, milk production and physiological status in dairy cows, *Comp. Biochem. Physiol.,* 65A, 319, 1980.
10. **Leat, W. M. F. and Baker, J.,** Distribution of fatty acids in the plasma lipids of herbivores grazing pasture: a species comparison, *Comp. Biochem. Physiol.,* 36, 153, 1970.
11. **Christie, W. W.,** The effect of diet and other factors on the lipid composition of ruminant tissues and milk, in *Progress in Lipid Research,* Vol. 17, Holman, R. T., Ed., Pergamon Press, New York, 1978, 245.
12. **Marchello, J. A., Dryden, F. D., and Hale, W. H.,** Bovine serum lipids. I. The influence of added animal fat to the ration, *J. Anim. Sci.,* 32, 1008, 1971.
13. **MacLeod, G. K., Wood, A. S., and Yao, Y. T.,** Influence of dietary fat on rumen fatty acids, plasma lipids and milk fat composition in the cow, *J. Dairy Sci.,* 55, 446, 1972.
14. **Brumby, P. E., Storry, J. E., and Sutton, J. D.,** Metabolism of cod-liver oil in relation to milk fat secretion, *J. Dairy Res.,* 39, 167, 1972.
15. **Scott, T. W., Cook, L. J., and Mills, S. C.,** Protection of dietary polyunsaturated fatty acids against microbial hydrogenation in ruminants, *J. Am. Oil Chem. Soc.,* 43, 358, 1971.
16. **Bitman, J., Dryden, L. P., Goering, H. D., Wrenn, T. R., Yoncoskie, R. A., and Edmonson, L. F.,** Efficiency of transfer of polyunsaturated fats into milk, *J. Am. Oil Chem. Soc.,* 50, 93, 1973.

17. **Yang, Y. T., Rohde, J. M., and Baldwin, R. L.,** Dietary lipid metabolism in lactating dairy cows, *J. Dairy Sci.,* 61, 1400, 1978.
18. **Shannon, A. D. and Lascelles, A. K.,** Changes in the concentration of lipids and some other constituents in the blood plasma of calves from birth to 6 months of age, *Aust. J. Biol. Sci.,* 19, 831, 1966.
19. **Noble, R. C., Crouchman, M. L., McEwan Jenkinson, D., and Moore, J. H.,** Relationships between lipids in plasma and skin secretions of neonatal calf with particular reference to linoleic acid, *Lipids,* 10, 128, 1975.
20. **Leat, W. M. F.,** Fatty acid composition of the plasma lipids of newborn and maternal ruminants, *Biochem. J.,* 98, 598, 1966.
21. **Maynard, L. A. and McCay, C. M.,** The influences of a low fat diet upon fat metabolism during lactation, *J. Nutr.,* 2, 67, 1929.
22. **Maynard, L. A., Harrison, E. S., and McCay, C. M.,** The changes in the total fatty acids, phospholipid fatty acids and cholesterol of the blood during the lactation cycle, *J. Biol. Chem.,* 92, 263, 1931.
23. **Schaible, P. J.,** Plasma lipids in lactating and non-lactating animals, *J. Biol. Chem.,* 95, 79, 1932.
24. **Hartmann, P. E. and Lascelles, A. K.,** Variation in the concentration of lipids and some other constituents in the blood plasma of cows at various stages of lactation, *Aust. J. Biol. Sci.,* 18, 114, 1965.
25. **Arave, C. W., Miller, R. H., and Lamb, R. C.,** Genetic and environmental effects on serum cholesterol of dairy cattle of various ages, *J. Dairy Sci.,* 58, 423, 1975.
26. **Fleischer, B. and Fleischer, S.,** Interaction of the protein moiety of bovine serum alpha lipoproteins with phospholipid micelles. II. Isolation and characterization of the complex, *Biochim. Biophys. Acta,* 147, 552, 1967.
27. **Brumby, P. E. and Welch, V. A.,** Fractionation of bovine serum lipoproteins and their characterization by gradient gel electrophoresis, *J. Dairy Res.,* 37, 121, 1970.
28. **Raphael, B. C., Dimick, P. S., and Puppione, D. L.,** Electrophoretic characterization of bovine serum lipoproteins throughout gestation and lactation, *J. Dairy Sci.,* 56, 1411, 1973.
29. **Ferreri, L. F. and Gleockler, D. H.,** Electrophoretic characterization of bovine lipoprotein subfractions isolated by gel chromatography, *J. Dairy Sci.,* 62, 1577, 1979.
30. **Puppione, D. L., Kunitake, S. T., Toomey, M. L., Loh, E., and Schumaker, V. N.,** Physicochemical characterization of ten fractions of bovine alpha lipoproteins, *J. Lipid Res.,* 23, 371, 1982.
31. **Burstein, M. and Scholnick, H. R.,** Lipoprotein-polyanion-metal interaction, in *Advances in Lipid Research,* Vol. 11, Paoletti, R. and Kritchevsky, D., Eds., Academic Press, New York, 1973, 103.
32. **Puppione, D. L., Forte, G. M., Nichols, A. V., and Strisower, E. H.,** Partial characterization of serum lipoproteins in the density interval 1.04 — 1.06 g/ml, *Biochim. Biophys. Acta,* 202, 392, 1970.
33. **Tall, A. R., Puppione, D. L., Kunitake, S. T., Atkinson, D., Small, D. M., and Waugh, D.,** Organization of the core lipids of high density lipoproteins in the lactating bovine, *J. Biol. Chem.,* 256, 170, 1981.
34. **Forte, T. M., Bell-Quint, J. J., and Cheng, F.,** Lipoproteins of fetal and newborn calves and adult steer: A study of developmental changes, *Lipids,* 16, 240, 1981.
35. **Puppione, D. L.,** Implications of unique features of blood lipid transport in the lactating cow, *J. Dairy Sci.,* 61, 651, 1978.
36. **Jensen, L. C., Rich, T. H., and Lindgren, F. T.,** Graphic presentation of computer-derived schlieren lipoprotein data, *Lipids,* 5, 491, 1970.
37. **Dryden, F. D., Marchello, J. A., Adams, G. H., and Hale, W. H.,** Bovine serum lipids. II. Lipoprotein quantitative and qualitative composition as influenced by added animal fat diets, *J. Anim. Sci.,* 32, 1016, 1971.
38. **Stead, D. and Welch, V. A.,** Determination of physical properties of bovine serum lipoproteins by analytical ultracentrifugation, *J. Dairy Sci.,* 59, 9, 1976.
39. **Enholm, C., Garoff, H., Simon, K., and Aro, H.,** Purification and quantification of the human plasma lipoprotein carrying the Lp(a) antigen, *Biochim. Biophys. Acta,* 236, 431, 1971.
40. **Puppione, D. L., Kunitake, S. T., Hamilton, R. L., Phillips, M. L., Schumaker, V. N., and Davis, L. D.,** Characterization of unusual intermediate density lipoproteins, *J. Lipid Res.,* 23, 283, 1982.
41. **Small, D. M., Puppione, D. L., Phillips, M. L., Atkinson, D., Hamilton, J. A., and Schumaker, V. N.,** Crystallization of a metastable lipoprotein. Massive change of lipoprotein properties during routine preparation, *Arteriosclerosis,* 1, 96, 1981.
42. **Ferreri, L. F. and Elbein, R. C.,** Fractionation of plasma triglyceride-rich lipoproteins of the dairy cow: evidence of chylomicron sized particles, *J. Dairy Sci.,* 65, 912, 1982.
43. **Puppione, D. L.,** unpublished observations, 1980.
44. **Raphael, B. C., Dimick, P. S., and Puppione, D. L.,** Lipid characterization of bovine serum lipoproteins throughout gestation and lactation, *J. Dairy Sci.,* 56, 1025, 1973.
45. **Jonas, A.,** Physicochemical properties of bovine serum high density lipoproteins, *J. Biol. Chem.,* 247, 7767, 1972.

46. **Forte, T. M., Nordhausen, R. D., Nichols, A. V., Endermann, G., Miljanich, P., and Bell-Quint, J. J.,** Dissociation of apolipoprotein A-I from porcine and bovine high density lipoproteins by guanidine hydrochloride, *Biochim. Biophys. Acta,* 573, 451, 1979.
47. **Swaney, J. B.,** Characterization of the high density lipoprotein and its major apoprotein from human, canine, bovine and chicken plasma, *Biochim. Biophys. Acta,* 617, 489, 1980.
48. **Lim, C. T. and Scanu, A. M.,** Apoproteins of bovine serum high density lipoproteins: isolation and characterization of the small molecular weight components, *Artery,* 2, 483, 1976.
49. **Patterson, B. W. and Jones, A.,** Bovine apolipoprotein C. I. Isolation and spectroscopic investigations of the phospholipid binding properties, *Biochim. Biophys. Acta,* 619, 572, 1980.
50. **Ferreri, L. F.,** unpublished observations, 1982.
51. **Kane, J. P., Hardman, D. A., and Paulus, H. E.,** Heterogeneity of apolipoprotein B; isolation of a new species from human chylomicrons, *Proc. Natl. Acad. Sci. U.S.A.,* 77, 2465, 1980.
52. **Elovson, J., Huang, Y. O., Baker, N., and Kannan, R.,** Apolipoprotein B is structurally and metabolically heterogeneous, *Proc. Natl. Acad. Sci. U.S.A.,* 78, 157, 1981.
53. **Leat, W. M. F. and Harrison, F. A.,** Lipid digestion in the sheep: Effect of bile and pancreatic juices on the lipids of intestinal contents, *Quant. J. Exp. Physiol.,* 54, 187, 1969.
54. **Harrison, F. A. and Leat, W. M. F.,** Digestion and absorption of lipids in non-ruminant and ruminant animals: a comparison, *Proc. Nutr. Soc.,* 34, 203, 1974.
55. **Harrison, F. A. and Hill, K. J.,** Digestive secretions and the flow of digesta along the duodenum of the sheep, *J. Physiol. London,* 162, 225, 1962.
56. **Arienti, G., Harrison, F. A., and Leat, W. M. T.,** The lipase activity of sheep pancreatic juices, *Quant. J. Exp. Physiol.,* 59, 351, 1974.
57. **Leat, W. M. F. and Harrison, F. A.,** Origin and formation of lymph lipids in the sheep, *Quant. J. Exp. Physiol.,* 59, 131, 1974.
58. **Nestel, P. J., Poyser, A., Hood, R. L., Mills, S. C., Willis, M. R., Cook, L. J., and Scott, T. W.,** The effect of dietary fat supplements on cholesterol metabolism in ruminants, *J. Lipid Res.,* 19, 899, 1978.
59. **Palmquist, D. L.,** A kinetic concept of lipid transport in ruminants, a review, *J. Dairy Sci.,* 59, 355, 1976.
60. **Hartmann, P. E., Harris, J. G., and Lascelles, A. K.,** The effect of oil feeding and starvation on the composition and output of lipid in thoracic duct lymph in the lactating cow, *Aust. J. Biol. Sci.,* 19, 635, 1966.
61. **Sata, T., Havel, R. J., and Jones, A. L.,** Characterization of subfractions of triglyceride-rich lipoproteins separated by gel chromatography from blood plasma of normolipemic and hyperlipemic humans, *J. Lipid Res.,* 13, 757, 1972.
62. **Gage, S. H. and Fish, P. A.,** Fat digestion, absorption and assimilation in man and animals as determined by the dark-field microscope and a fat soluble dye, *Am. J. Anat.,* 34, 1, 1924.
63. **Hatch, F. T. and Lees, R. S.,** Practical methods for plasma lipoprotein analysis, in *Advances in Lipid Research,* Vol. 6, Paoletti, R. and Kritchevsky, D., Eds., Academic Press, New York, 1968, 1.
64. **Baldwin, R. L. and Smith, N. E.,** Intermediary aspects and tissue interactions of ruminant fat metabolism, *J. Dairy Sci.,* 54, 583, 1971.
65. **Puppione, D. L., Raphael, B., McCarthy, R. D., and Dimick, P. S.,** Variations in the electrophoretic distributions of low density lipoproteins of Holstein cows, *J. Dairy Sci.,* 55, 265, 1973.
66. **Glascock, R. F., Welch, V. A., Bishop, C., Davies, T., Wright, E. W., and Noble, R. C.,** An investigation of serum lipoproteins and of their contribution to milk fat in the dairy cow, *Biochem. J.,* 98, 149, 1966.
67. **Bishop, C., Davies, T., Glascock, R. F., and Welch, V. A.,** Studies on the origin of milk fat, *Biochem. J.,* 113, 629, 1969.
68. **Coppock, C. E., Noller, C. H., and Wolf, S. A.,** Effect of forage concentrate ratio in complete feeds fed *ad libitum* on energy intake in relation to requirements by dairy cows, *J. Dairy Sci.,* 57, 1371, 1974.
69. **Smith, N. E., Ufford, G. R., and Merrill, W. G.,** Complete ration-group feeding systems for dry and lactating cows, *J. Dairy Sci.,* 61, 584, 1978.
70. **Palmquist, D. L. and Mattos, W.,** Turnover of lipoproteins and transfer of milk fat of dietary (1-Carbon-14) linoleic acid in lactating cows, *J. Dairy Sci.,* 61, 561, 1978.
71. **Mjøs, O. D., Faergman, O., Hamilton, R. L., and Havel, R. J.,** Characterization of remnants produced during the metabolism of triglyceride-rich lioproteins of blood plasma and intestinal lymph in the rat, *J. Clin. Invest.,* 56, 603, 1975.
72. **Redgrave, T. G.,** Formation of cholesteryl ester-rich particulate lipid during metabolism of chylomicrons, *J. Clin. Invest.,* 49, 465, 1970.
73. **Goodman, De W. S.,** Metabolism of chylomicron cholesteryl esters in the rat, *J. Clin. Invest.,* 41, 1886, 1963.
74. **Nestel, P. J., Havel, R. J., and Bezman, A.,** Metabolism of constituent lipids of dog chylomicrons, *J. Clin. Invest.,* 42, 1313, 1963.

75. **Bergman, E. N., Havel, R. J., Wolfe, B. M., and Bohmer, T.,** Quantitative studies of the metabolism of chylomicron triglycerides and cholesterol by liver and extrahepatic tissues of sheep and dogs, *J. Clin. Invest.,* 50, 1831, 1971.

76. **Windler, E., Chao, Y., and Havel, R. J.,** Determinants of hepatic uptake of triglyceride-rich lipoproteins and their remnants in the rat, *J. Biol. Chem.,* 255, 5475, 1980.

77. **Eisenberg, S. and Levy, R. I.,** Lipoprotein metabolism in *Advances in Lipid Research,* Vol. 13, Paoletti, R. and Kitchevsky, D., Eds., Academic Press, New York, 1975, 1.

78. **Schumaker, V. N. and Adams, G. H.,** Circulating lipoproteins, in *Annual Revue of Biochemistry,* Vol. 38, Snell, E. E., Ed., Annual Revues Inc., Palo Alto, Calif., 1969, 113.

79. **Tall, A. R. and Small, D. M.,** Plasma high density lipoproteins, *New Engl. J. Med.,* 229, 1232, 1978.

80. **Fielding, C. J., Shore, V. G., and Fielding, P. E.,** A protein cofactor of lecithin; cholesterol acyltransferase, *Biochem. Biophys. Res. Commun.,* 46, 1493, 1972.

81. **Noble, R. C., O'Kelly, J. C., and Moore, J. H.,** Observations on the lecithin: cholesterol acyltransferase system in bovine plasma, *Biochim. Biophys. Acta,* 270, 519, 1972.

82. **Glomset, J. A. and Norum, K. R.,** The metabolic role of lecithin: cholesterol acyltransferase: perspectives from pathology, in *Advances in Lipid Research,* Vol. 11, Paoletti, R. and Kritchevsky, D., Eds., Academic Press, New York, 1973, 1.

83. **Havel, R. J., Kane, J. P., and Kashyap, M. D.,** Interchange of apoproteins between chylomicrons and high density lipoproteins during alimentary lipemia in man, *J. Clin. Invest.,* 52, 32, 1973.

84. **Askew, E. W., Emery, R. S., and Thomas, J. W.,** Lipoprotein lipase of the bovine mammary gland, *J. Dairy Sci.,* 53, 1415, 1970.

85. **Emery, R. S.,** Biosynthsis of milk fat, *J. Dairy Sci.,* 56, 1187, 1973.

86. **Etienne, J., Noé, L., Rossignol, M., Dosne, A. M., and Debray, J.,** Post-heparin lipolytic activity with no hepatic triacylglycerol lipase involved in mammalian species, *Biochim. Biophys. Acta,* 663, 516, 1981.

87. **Stokke, K. T.,** Cholesteryl ester metabolism in liver and blood plasma of various species, *Atherosclerosis,* 19, 393, 1974.

88. **Groat, P. H. E., Jansen, H., and VanTol, A.,** Selective degradation of the high density lipoprotein-2 subfraction by heparin releasible liver lipase, *FEBS Lett.,* 129, 269, 1981.

89. **Shirai, K., Barnhart, R. L., and Jackson, R. L.,** Hydrolysis of human plasma high density lipoproteins$_2$-phospholipid and triglycerides by hepatic lipase, *Biochem. Biophys. Res. Commun.,* 100, 591, 1981.

Lipoproteins of Some Avian Forms

LIPOPROTEINS OF CHICKENS

Kang-Jey Ho and C. Bruce Taylor

INTRODUCTION

Although our current knowledge about the lipoproteins of chickens is still far from complete, the following features are quite unique to chicken's lipoproteins: (1) In the avian species, exogenous fat after absorption is transported primarily as very low-density lipoprotein (VLDL) via the portal vein, designated as portomicrons, while in mammals exogenous fat is transported mainly as chylomicrons via the lymphatic system.[1,2] (2) High-density lipoprotein (HDL) is the major lipoprotein in cockerels, roosters, and nonlaying hens.[3-6] (3) Laying hens develop hyperlipidemia as the result of enhanced hepatic synthesis and secretion of VLDL.[7-9] (4) Certain specific lipoproteins rich in phosphopeptides also appear in the plasma of laying hens.[10] (5) Egg yolk contains selective lipoprotein species of plasma origin.[11,12] (6) Embryogenesis of the egg provides an excellent model for observation of the utilization of yolk lipoproteins by the developing embryo. (7) Extreme hyperlipidemia is associated with a special strain of hereditary nonlaying hens when they reach sexual maturity.[9] (8) Chickens are susceptible to atherosclerosis secondary to either dietary induced exogenous hyperlipidemia or egg-laying related endogenous hyperlipidemia.[13]

In view of the dynamic changes of the lipoprotein patterns in chickens under various conditions, the present review will be divided into the following six sections: Lipoproteins in Cockerels, Roosters, and Nonlaying Hens; Lipoproteins in Laying Hens; Lipoproteins in Egg Yolk; Lipoproteins in Chick Embryos; Lipoproteins in Hereditary Nonlaying Hens; and Lipoproteins in Dietary Induced Hyperlipidemia. Most of the data cited in this review are obtained from the studies on white leghorn chickens (*Gallus domesticus*). Some data based upon studies on turkeys and pigeons are also included for the purposes of comparison.

LIPOPROTEINS IN COCKERELS, ROOSTERS, AND NONLAYING HENS

Plasma Lipid and Lipoprotein Patterns

The plasma lipoprotein and lipid patterns of cockerels, roosters, and immature hens can be considered as essentially identical.[3-5,7,9,14-17] Dramatic change occurs with the onset of sexual maturity in hens.[7,8] When mature hens are not actively laying eggs or pass the reproductive age the lipoprotein and lipid profiles return to the prereproductive pattern.

The average plasma apolipoprotein and lipid concentrations are, respectively, 222 (ranging from 124 to 374) and 320 (134 to 554) mg/100 mℓ plasma while the average concentrations of cholesterol, triglyceride, and phospholipid are, respectively, 100 (70 to 150), 105 (29 to 229), and 146 (82 to 217) mg/dℓ (Table 1). The HDL, the major lipoprotein in the plasma, has a mean concentration of 400 (268 to 703) mg/dℓ, accounting for 76% (59 to 83) of the total plasma lipoproteins (Tables 2 and 3). The low-density lipoprotein (LDL), 110 (41 to 225) mg/dℓ, accounts for 20% (13 to 30) of the total plasma lipoproteins. The remaining 4% (0 to 11) or 20 (0 to 71) mg/dℓ is the VLDL (Tables 2 and 3). The VLDL is therefore quite low or even undetectable in nonlaying chickens. Plasma VLDL is also very low in mature male turkeys[18] and is reported to be undetectable in mature pigeons.[19] The relative and absolute concentrations of LDL and HDL in both turkeys and pigeons are comparable to those of chickens (Tables 2 and 3). Chylomicrons are absent from all these avian species.

Composition of Lipoproteins

Except for the discoid nascent HDL which is yet to be demonstrated in chickens, all

Table 1
ABSOLUTE CONCENTRATION OF THE TOTAL APOLIPOPROTEINS AND INDIVIDUAL LIPIDS IN THE PLASMA OF CHICKENS, TURKEYS, AND PIGEONS (mg/dℓ PLASMA)

Animals	Number	Protein	Total lipid	Cholesterol[a]	Triglyceride	Phospholipid	Ref.
Cockerels (4-week-old)	4	192	134	70		82	3
Cockerels (6—8-week-old)	6	238	203	94	29	117	4
Cockerels (8-week-old)	4	230	179	94		113	3
Cockerels (8-week-old)	3	374	554	150 (33/117)	187	217	5
Chicks (23-day-old)	18		315	145	27		14
Roosters	3	197		82	103	122	7
Roosters	14		437 ± 70[b]	86 ± 15 (27/59)	229 ± 88	189 ± 21	9
Roosters	2			118	99		15
Roosters	9				108 ± 70		16
Hens (9-week-old)	8	124	416	69	37		17
Nonlaying hens	3	200	1450	84	111	183	7
Laying hens	3	272		107	864	456	7
Laying hens	12		2760 ± 957	133 ± 35 (114 ± 39/19 ± 3)	2063 ± 635	1021 ± 362	9
Estrogen treatment							
Stilbesterol	3	2329 ± 580	16046	1001 ± 378	11494 ± 3,357	3551 ± 1051	5
Estradiol	2			185	625		15
Diethylstilbesterol							
0.1 mg/day	18			405	1839		14
1.0 mg/day	18			1145	10241		14
5.0 mg/day	18			1381	12371		14
Cholesterol feeding							
1% for 5 days	2			231	63		15
1% for 4 weeks	4	197	403	275		97	3
1% for 6—8 weeks	6	307	1175	881	83	225	4
1% for 8 weeks	4	314	527	481		129	3
1% for 8 weeks	3	626	2829	997	125	586	5

Turkeys						
Mature rooster	4	149	147	55 (10/45)	14	77	18
Pigeons (9-month-old)						
White Carneau	6	308	440	190	29	221	19
Show Racer	6	239	395	163	37	196	19

[a] Free and esterified cholesterol in parenthesis.
[b] Mean ± SD.

Table 2
ABSOLUTE CONCENTRATION OF THE TOTAL AND VARIOUS LIPOPROTEIN FRACTIONS IN THE PLASMA OF CHICKENS, TURKEYS, AND PIGEONS (mg/dℓ PLASMA)

Animals	Number	Total lipoprotein	VLDL	LDL	HDL	Ref.
Cockerels (4-week-old)	4	326	17	41	268	3
Cockerels (6—8-week-old)	6	441	7 ± 3[a]	81 ± 10	353 ± 37	4
Cockerels (8-week-old)	4	409	7	64	338	3
Cockerels (8-wek-old)	3	928	0	225	703	5
Roosters	3	510	54 ± 9	86 ± 31	370 ± 48	17
Roosters	3	489	Trace	104	385	6
Nonlaying hens	3	615	71 ± 12	183 ± 23	361 ± 49	7
Laying hens	3	1,704	1,401 ± 241	152 ± 6	151 ± 3	7
Laying hens	Pooled	627	424	124	78	8
Laying hens	Pooled	1,019	794	75	151	8
Stilbesterol treatment	3	18,687	17,149 ± 511	1,327 ± 388	211 ± 10	5
Cholesterol feeding						
1% for 4 weeks	4	600	287	147	166	3
1% for 6—8 weeks	6	1,483	1,086 ± 104	77 ± 16	320 ± 36	4
1% for 8 weeks	4	841	345	218	279	3
1% for 8 weeks	3	3,938	2,514	815	609	5
Mature male turkeys	4	295.6	8.7	61.7	225.2	18
Pigeons (9-month-old)						
White Carneau	6	748	0	191	557	19
Show Racer	6	634	0	135	499	19

[a]   Mean ± SD.

Table 3
RELATIVE CONCENTRATION OF VARIOUS LIPOPROTEIN FRACTIONS IN THE PLASMA OF CHICKENS, TURKEYS, AND PIGEONS

Animals	Number	VLDL	LDL	HDL	Ref.
Cockerels (4-week-old)	4	5.2[a]	12.6	82.2	3
Cockerels (6—8-week-old)	6	1.5	18.3	80.2	4
Cockerels (8-week-old)	3	0	24.2	75.8	5
Cockerels (8-week-old)	4	1.7	15.6	82.7	3
Roosters	3	10.6	16.9	72.5	17
Roosters	3	0	21.3	78.7	6
Nonlaying hens	3	11.5	29.8	58.7	7
Laying hens	3	82.2	8.9	8.9	7
Laying hens	Pooled	67.7	19.8	12.5	8
Laying hens	Pooled	77.9	7.3	14.8	8
Stilbesterol treatment	3	91.8	7.1	1.1	5
Cholesterol feeding					
1% for 4 weeks	4	47.8	24.5	27.2	3
1% for 6—8 weeks	6	73.2	5.2	21.6	4
1% for 8 weeks	4	41.0	25.9	33.1	3
1% for 8 weeks	3	63.8	20.7	15.5	5
Mature male turkeys (pooled plasma)	4	2.9	20.9	76.2	18
Pigeons (9-month-old)					
White Carneau	6	0	25.6	74.4	19
Show Racer	6	0	21.3	78.7	19

[a]   Mean (% of total).

plasma lipoproteins are spherical in shape with a lipid core composed of nonpolar triglyceride and cholesterol ester covered completely by a monolayer of phospholipid, unesterified cholesterol, and apolipoprotein. The proteins interdigitate between the interfacial lipids.[20]

VLDL

VLDL (density <1.006 g/mℓ) has a mean diameter of 34 nm, ranging from 27 to 54 nm and a flotation rate of S_f 20 to 400 with 75% in the range S_f 20 to 100.[8,21] Its absolute protein and lipid concentrations in the plasma vary greatly (Table 4) whereas the relative concentrations show much less variation (Table 5). The average apoprotein and lipid compositions are, respectively, 12.4 and 87.6%. Cholesterol, triglyceride, and phospholipid account for 20.1, 49.3, and 17.2% of the VLDL by weight (Table 5). The composition of turkey VLDL is not different from that of chickens (Tables 4 and 5).

LDL

LDL (density = 1.006 to 1.063 g/mℓ) has a mean diameter of 24 nm with individual particles ranging from 16 to 38 nm and a flotation rate of S_f 11 to 13, ranging S_f 12 to 20, which is higher than that of mammalian LDL (S_f 4 to 8).[8,21] When LDL of a narrower density range between 1.024 and 1.045 is collected, the mean diameter remains the same but the variation in the particle size is less (18 to 28 nm). Since VLDL-LDL form a more or less continuous spectrum, mutual contamination seems to be unavoidable. The higher density of LDL than VLDL is due to its relatively higher protein (20.5 vs. 12.4%) and lower lipid content (79.5 vs. 87.6%) (Tables 5 to 7). The decrease in lipid content is primarily in the triglyceride fraction. LDL appears to retain most, if not all, of the cholesterol ester carried in its parent VLDL for the percent esterification of total cholesterol increases from 46% in VLDL to 77% in LDL.

The LDL of turkeys contains lower triglyceride and higher cholesterol ester than the chicken and has a lower flotation rate (S_f = 7.4).[18] The pigeon's LDL is characterized by high protein and cholesterol with low triglyceride and phospholipid contents.[19]

HDL

HDL (density = 1.063 to 1.210 g/mℓ) has a mean diameter of 12 nm, with individual particles ranging from 6 to 18 nm.[8] The protein and lipid concentrations are, respectively, 196 (114 to 339) and 189 (97 to 364) mg/100 mℓ plasma, much higher than that of VLDL and LDL (Table 8). The cholesterol, triglyceride, and phospholipid contents are 63 (38 to 94), 41 (9 to 96), and 104 (73 to 174) mg/dℓ, respectively (Table 8). Its high density is therefore due to the high protein (48% with a range of 43 to 58%) and low triglyceride (8.8% with a range of 2.5 to 13.7%) contents. The percent esterification of cholesterol is 83% (76 to 90) and is comparable to that in LDL but higher than that in VLDL (Table 9).[4,5,23,24]

The protein and lipid composition of chicken HDL resembles more closely that of human HDL_2 than that of human HDL_3. On the other hand, the following physical characteristics of chicken HDL, such as sedimentation coefficient (3.99S), diffusion coefficient (4.36 × 10^{-6} cm^2 S^{-2}), molecular weight (1.73 × 10^5), partial specific volume (0.868 mℓ/g), and the anhydrous frictional ratio (1.24), are quite similar to those of human HDL_3.[23] The circular dichroic spectra of chicken HDL, however, indicate an α-helical content 20 to 30% greater than that in human HDL.[23,25] As a matter of fact, the α-helix content of chicken HDL appears to be the largest yet reported whether it is compared to that of man or to other animal species studied.

Turkey HDL contains more protein and less triglyceride and less cholesterol than chicken HDL, whereas pigeon HDL contains less protein and triglyceride but more cholesterol and phospholipid (Tables 8 and 9).

Table 4
ABSOLUTE CONCENTRATION OF APOLIPOPROTEIN AND INDIVIDUAL LIPIDS IN THE PLASMA VLDL OF CHICKENS AND TURKEYS (mg/dℓ PLASMA)

Animals	Number	Protein	Total lipid	Cholesterol[a]	Triglyceride	Phospholipid	Ref.
Cockerels (4-week-old)	4	0.6 ± 0.2[b]	16.4 ± 14.4	1.6 ± 0.5	14.1	0.7 ± 0.3	3
Cockerels (6—8-week-old)	6	1.0 ± 0.2	5.6	2.6 ± 0.8	1.8 ± 1.8	1.2 ± 0.5	4
Cockerels (8-week-old)	4	1.0 ± 0.2	5.8 ± 3.1	2.6 ± 0.8	2.0	1.2 ± 0.5	3
Roosters	3	8 ± 3	48 ± 4	5 ± 1	31 ± 5	10 ± 0.5	7
Nonlaying hens (9-week-old)	8	2		4	15		17
Nonlaying hens (12-week-old)	3	10 ± 2	61 ± 10	8 ± 3	39 ± 5	13 ± 2	7
Laying hens	3	177 ± 32	1,225 ± 209	71 ± 19	789 ± 146	360 ± 47	7
Stilbesterol treatment	3	1,952 ± 513	15,197	1,203 (485 ± 159/ 718 ± 426)	10,797 ± 3156	3,197 ± 952	5
Diethylstilbesterol treatment	5	780		490	4784		17
Cholesterol feeding							
1% for 4 weeks	4	49 ± 24	238 ± 80	156 ± 59		+/ ± 17	3
1% for 6—8 weeks	6	109 ± 16	976	767 ± 179 (193/574)	60 ± 10	176 ± 28	4
1% for 8 weeks	4	87 ± 31	258 ± 137	252 ± 121		49 ± 32	3
1% for 8 weeks	3	208	2,306	1,015 (157/858)	82	309	5
Mature male turkeys	4	1.2 ± 0.0	7.5	2.6 (0.3/2.3)	3.0	1.9 ± 0.2	18

a   Free and esterified cholesterol in parenthesis.
b   Mean ± SD.

Table 5
RELATIVE CONCENTRATION OF APOLIPOPROTEIN AND INDIVIDUAL LIPIDS IN THE PLASMA VLDL OF CHICKENS, AND TURKEYS (% BY WEIGHT)

Animals	Number	Protein	Total lipid	Cholesterol[a]	Triglyceride	Phospholipid	Ref.
Cockerels (4-week-old)	4	3.5	96.5	9.4	82.9	4.1	3
Cockerels (6—8-week-old)	6	15.2	84.8	39.4	27.3	18.2	4
Cockerels (8-week-old)	4	14.7	85.3	38.2	29.4	17.6	3
Roosters	3	14.3	85.7	8.9	55.4	17.9	7
Roosters	4	12.9	87.1	13.6 (7.4/6.2)	46.1	27.3	16
Nonlaying hens	3	14.1	85.9	11.3	54.9	18.3	7
Laying hens	3	12.6	87.4	5.1	56.3	25.7	7
Mature hens (d <1.01)	9	13.6	86.4	7.3 (4.7/2.5)	52.0	22.7	22
Stilbesterol treatment	3	11.4	88.6	7.0 (2.8/4.2)	63.0	18.6	5
Cholesterol feeding							
1% for 6—8 weeks	6	10.1	89.9	70.6 (17.8/52.8)	5.5	16.2	4
1% for 8 weeks	3	8.3	91.7	40.4 (6.3/34.1)	39.1	12.3	5
Mature male turkeys	4	13.8	86.2	30.3 (3.7/26.6)	34.2	21.7	18

[a] Free and esterified cholesterol in parenthesis.

Table 6
ABSOLUTE CONCENTRATION OF APOLIPOPROTEIN AND INDIVIDUAL LIPIDS IN THE PLASMA LDL OF CHICKENS, TURKEYS, AND PIGEONS (mg/dℓ PLASMA)

Animals	Number	Protein	Total lipid	Cholesterol[a]	Triglyceride	Phospholipid	Ref.
Cockerels (4-week-old)	4	20.5 ± 0.6[b]	20.6 ± 6.5	11.8 ± 4.1	—	8.5 ± 2.5	3
Cockerels (6—8-week-old)	6	21.5 ± 2.9	59.1	18.9 ± 2.9	18.1 ± 5.3	17.9 ± 4.6	4
Cockerels (8-week-old)	4	27.1 ± 1.5	36.8 ± 8.6	25.6 ± 6.7	—	13.7 ± 4.2	3
Cockerels (8-week-old)	9	35	190	56 (15/41)	91	43	5
Roosters	3	15 ± 4	71 ± 14	17 ± 4	25 ± 4	26 ± 6	7
Nonlaying hens (9-week-old)	9	8	—	11	5	—	17
Nonlaying hens	3	34 ± 3	150 ± 20	38 ± 5	36 ± 9	73 ± 12	7
Laying hens	3	27 ± 1	125 ± 6	14 ± 2	62 ± 8	47 ± 4	7
Laying hens	Pooled	22.5	77.5	36.9 (8.3/28.6)	19.4	21.2	8
Stilbesterol treatment	3	260 ± 68	1036	99 (44±21/55 ± 20)	649 ± 195	288 ± 90	5
Diethylstilbesterol treatment	5	574		195	1913		17
Cholesterol feeding							
1% for 4 weeks	4	44.9 ± 7.6	102.3 ± 32.1	78.1 ± 20.3	12.2 ± 3.4	27.7 ± 4.6	3
1% for 6—8 weeks	6	17.1 ± 4.3	59.6	36.8 ± 9.7		13.3 ± 4.3	4
1% for 8 weeks	4	59.9 ± 5.2	157.7 ± 47.1	152.3 ± 41.3		38.3 ± 8.8	3
1% for 8 weeks	3	112	703	343 (91/252)	211	149	5
Mature male turkeys	4	16.1 ± 4.1	46.1	26.8 (3.2 ± 0.6/23.6 ± 2.3)	3.2	16.2 ± 5.1	18
Pigeons (9-month-old)							
White Carneau	6	88.1 ± 19.2	103.4	59.7 ± 7.6 (24.7/35.0)	13.2 ± 3.3	30.5 ± 6.5	19
Show Racer	6	47.3 ± 13.3	88.0	49.5 ± 7.7 (13.8/35.7)	18.7 ± 4.7	19.8 ± 4.0	19

a   Free and esterified cholesterol in parenthesis.
b   Mean ± SD.

Table 7
RELATIVE CONCENTRATION OF APOLIPOPROTEIN AND INDIVIDUAL LIPIDS IN THE PLASMA LDL OF CHICKENS, TURKEYS, AND PIGEONS (% BY WEIGHT)

Animals	No.	Protein	Total lipid	Cholesterol[a]	Triglyceride	Phospholipid	Ref.
Cockerels (6—8 week)	6	27.4	72.6	23.4	22.4	22.2	4
Cockerels (8-week-old)	3	15.6	84.4	24.9 (6.7/18.2)	40.4	19.1	5
Roosters	3	17.4	82.6	19.8	29.1	30.2	7
Nonlaying hens	3	18.5	81.5	20.7	19.6	39.7	7
Chicken (unspecif.)	3	23.5	76.5	34.2 (6.4/27.8)	22.1	20.2	22
Laying hens	3	17.8	82.2	9.2	40.8	30.9	7
Laying hens (d = 1.024—1.045)	3	21.6 ± 1.3[b]	78.4	48.3 (13.4 ± 0.6/34.9 ± 0.3)	11.3 ± 2.0	18.8 ± 0.5	8
Laying hens (d = 1.024—1.045)	3	22.5 ± 4.8	77.5	36.9 (8.3 ± 1.2/28.6 ± 6.1)	19.4 ± 4.3	21.2 ± 2.9	8
Mature hens (d = 1.01—1.06)	9	24.6	75.4	17.1 (10.2/6.9)	33.3	18.2	12
Stilbesterol treatment	3	20.0	80.0	7.6 (3.4/4.2)	50.1	22.2	5
Cholesterol feeding							
1% for 6—8 weeks	6	22.3	77.7	48.0	15.9	17.3	4
1% for 8 weeks	3	13.7	86.3	42.1 (11.2/30.9)	25.9	18.3	5
Mature male turkeys	4	26.1	74	43.4 (5.2/38.2)	5.1	26.2	18
Pigeons (9-months-old)							
White Carneau	6	43.2 ± 1.5	56.8	34.0 ± 2.1 (14.1/19.9)	7.5 ± 0.7	15.3 ± 0.8	19
Show Racer	6	32.8 ± 2.9	67.2	37.6 ± 2.8 (10.5/27.1)	15.2 ± 2.2	14.5 ± 0.8	19

a Free and esterified cholesterol in parenthesis.
b Mean ± SD.

Table 8
ABSOLUTE CONCENTRATION OF APOLIPOPROTEIN AND INDIVIDUAL LIPIDS IN THE PLASMA HDL OF CHICKENS, TURKEYS, AND PIGEONS
(mg/dℓ PLASMA)

Animals	Number	Protein	Total lipid	Cholesterol[a]	Triglyceride	Phospholipid	Ref.
Cockerels (4-week-old)	4	171 ± 27[b]	97 ± 32	57 ± 20		73 ± 30	3
Cockerels (6—8-week-old)	6	215 ± 31	138	73 ± 12 (12/61)	9 ± 2	98 ± 35	4
Cockerels (8-week-old)	4	202 ± 21	136 ± 31	65 ± 16		98 ± 24	3
Cockerels (8-week-old)	3	339	364	94 (18/76)	96	174	5
Roosters	3	174 ± 21	196 ± 14	60 ± 8	47 ± 8	86 ± 2	7
Nonlaying hens	4	114		54	17		17
Nonlaying hens	3	156 ± 30	205 ± 20	38 ± 5	36 ± 9	97 ± 7	7
Laying hens	3	68 ± 10	100 ± 16	22 ± 1	13 ± 3	49 ± 3	7
Stilbesterol treatment	3	117 ± 5	94	12 (4 ± 2/ 8 ± 2)	48 ± 10	34 ± 6	5
Diethylstilbesterol treatment	5	15		16	16	2	17
Cholesterol feeding							
1% for 4 weeks	4	103.5 ± 24.1	62.6 ± 19.3	40.2 ± 14.2		21.6 ± 10.6	3
1% for 6—8 weeks	6	180.9 ± 18.6	139.1	77.9 ± 13.5 (11.7/66.2)	10.7 ± 2.0	35.7 ± 7.7	4
1% for 8 weeks	4	167.2 ± 21.5	111.5 ± 12.8	77.1 ± 13.9		41.4 ± 5.4	3
1% for 8 weeks	3	306	293	112 (19/103)	53	128	5
Mature male turkeys	4	132 ± 35	94	26 (6 ± 2/20 ± 3)	8	59 ± 6	18
Pigeons (9-month-old)							
White Carneau	6	220 ± 8	337	130 ± 12 (37/93)	16 ± 2	190 ± 26	19
Show Racer	6	192 ± 9	307	113 ± 16 (31/82)	18 ± 2	176 ± 17	19

a   Free and esterified cholesterol in parenthesis.
b   Mean ± SD.

Table 9

RELATIVE CONCENTRATION OF APOLIPOPROTEIN AND INDIVIDUAL LIPIDS IN THE PLASMA
HDL OF CHICKENS, TURKEYS, AND PIGEONS (% BY WEIGHT)

Animals	No.	Protein	Total lipid	Cholesterol[a]	Triglyceride	Phospholipid	Ref.
Cockerels (6—8-week-old)	6	58.1	41.9	20.6 (3.3/17.3)	2.5	27.7	4
Cockerels (8-week-old)	3	48.2	51.8	13.4 (2.6/10.8)	13.7	24.7	5
Roosters	5	43.9 ± 2.4[b]	56.1	21.0 (5.0 ± 0.3/16.0 ± 1.6)	6.4 ± 1.4	28.7 ± 2.27	23
Roosters	3	47.0	53.0	16.2	12.7	23.2	7
Nonlaying hens	3	43.2	56.8	10.5	10.0	26.9	7
Chickens (unspecified)		47.8	52.2	22.6 (22/20.4)	7.4	22.3	24
Laying hens	3	40.5	59.5	13.1	7.7	29.2	7
Mature hens (d = 1.06—1.20)	9	63.6	36.4	11.3 (1.4/9.1)	8.0	15.7	22
Stilbesterol treatment	3	55.4	44.6	5.7 (1.9/3.8)	22.7	16.1	5
Cholesterol feeding							
1% for 6—8 weeks	6	56.5	43.5	24.3 (3.7/20.7)	3.3	11.2	4
1% for 8 weeks	3	51.1	48.9	18.7 (3.2/17.2)	8.0	21.4	5
Mature male turkeys	4	58.5	41.5	11.6 (29/8.7)	3.5	26.4	18
Pigeons (9-month-old)							
White Carneau	6	40.3 ± 3.2	59.7	22.6 ± 2.8 (6.5/16.1)	2.9 ± 0.5	34.2 ± 3.7	19
Show Racer	6	39.0 ± 2.6	61.0	22.3 ± 2.1 (6.2/16.1)	3.8 ± 0.5	35.1 ± 1.8	19

[a] Free and esterified cholesterol in parenthesis.
[b] Mean ± SD.

Table 10
AMINO ACID COMPOSITION OF APOLIPOPROTEINS IN
THE PLASMA AND EGG YOLK OF CHICKENS (MOLES OF
AMINO ACID PER 100 MOL OF APOPROTEIN)

| | Plasma apolipoprotein | | | Egg yolk apolipoprotein | |
Amino acid	Apo-VLDL-II[a]	Apo-VLDL-B[b]	Apo-HDL[c]	Low molecular weight[d]	High molecular weight[e]
Lysine	7.3	8.0	8.5	7.9	8.8
Histidine	0	1.1	0.4	0	2.0
Arginine	8.5	4.1	6.9	8.0	4.5
Tryptophan	1.2	0.4	0.8	0.4	0
Aspartic acid	9.8	11.7	8.7	10.0	11.6
Threonine	6.1	6.5	5.1	5.8	7.0
Serine	4.9	8.5	4.7	4.7	8.7
Glutamic acid	9.8	12.5	19.0	10.3	11.4
Proline	2.4	3.5	3.9	2.5	3.2
Glycine	3.7	5.3	2.9	3.6	5.3
Alanine	9.8	7.0	9.7	9.6	6.9
Cysteine	1.2	0.2	0	1.4	0.5
Valine	8.5	5.6	6.3	8.6	5.4
Methionine	1.2	1.9	1.7	1.1	1.8
Isoleucine	8.5	5.6	3.4	7.9	4.8
Leucine	11.0	10.6	13.2	11.6	10.4
Tyrosine	3.7	2.9	2.6	4.8	3.9
Phenylalanine	2.4	4.5	2.0	2.3	4.5

[a]   Data obtained from Reference 30.
[b]   Mean of the data from References 8, 17, 20, 29.
[c]   Mean of the data from References 17, 21, 23, 25.
[d]   Mean of the data of apovitellenine 1 from References 33 and "apo C" of Reference 34.
[e]   Mean of the data of high-molecular-weight fraction (S1) of Reference 33 and "apo A" of Reference 34.

Structure of Apolipoproteins

The variety of apolipoproteins in each lipoprotein class as to their number and specificity has not yet been totally clarified. One of the reasons for the confusion is the heterogeneity of apolipoprotein molecule species in various classes of lipoproteins reported in the literature (8, 17, 20, 23, 26, 27). Such heterogeneity of apolipoproteins has been shown to arise through proteolytic cleavage of the parent apolipoprotein by the contaminated endogenous protease (28). The number of apolipoprotein species is drastically reduced when the endogenous proteolytic activity is adequately controlled during the process of separation.

Apolipoprotein of VLDL

There are two major apolipoproteins in chicken VLDL. One is of high molecular weight (mol wt 350,000), designated as apo VLDL-B,[28] another is of low molecular weight (mol wt 21,500), designated as apo VLDL-II.[21,29,32] While apo VLDL-II is the predominant apolipoprotein in VLDL of immature hens,[21] both apo VLDL-B and apo VLDL-II increase in laying hens with the increase of apo VLDL-B to such a great extent that it accounts for 46% of total VLDL apolipoprotein.[28]

The apo VLDL-B is characterized by stability to reduction and S-carboxymethylation. It is susceptible to cleavage by endogenous protease. The amino acid composition of apo VLDL-B is distinctly different from that of apo VLDL-II (Table 10) but is nearly identical to that of human apolipoprotein B.[28,34,35] While the amino terminal of human apo B is glutamic acid and carboxyl-terminal blocked, the N- and C-terminals of chicken apo VLDL-B are, respectively, lysine and tyrosine.[20]

Chicken apo VLDL-II has been isolated, characterized, and sequenced.[29,30] This protein contains two identical polypeptide chains of 82 amino acid residues each of which are linked by a single disulfide bond at residue 76.

Chicken VLDL also contains small-sized proteins with co-factor activity for lipoprotein lipase. Although these lipoprotein lipase activators are minor components of VLDL, two distinct proteins with molecular weights of 9000 and 5000, respectively, have been isolated. They are relatively soluble both at high and at low pH. They retain their co-factor activity even after denaturation in guanidium HCl and after reduction.[33]

Apolipoprotein of LDL

LDL is the remnant of VLDL after its intravascular lipolysis by lipoprotein lipase. LDL retains most, if not all, of the apo VLDL-B in its precursor VLDL. The lower-molecular-weight proteins such as apo VLDL-II and lipoprotein lipase activators are conspicuously absent from LDL.[17] Apo VLDL-B therefore comprises the great majority of the total protein in LDL. Whether or not apo HDL is present in chicken VLDL and LDL remains controversial.[20,26,36] Similar to human apo B, the chicken apo VLDL-B also possesses significant quantities of carbohydrate such as galactose, mannose and sialic acid.[18,37-39] The quantitative cross-reaction between chicken LDL and human LDL by immunological means is in the range of 1 to 10%.[40] Comparative aspects of animal lipoproteins have been recently reviewed.[41]

Apolipoprotein of HDL

The main apolipoprotein of chicken HDL, representing almost 90% of the total protein, has a molecular weight of approximately 28,000 as determined by polyacrylamide gel electrophoresis (PAGE) in sodium dodecyl sulfate (SDS). Its calculated molecular weight from its 234 constituent amino acids is 26,674.[23,25] Its amino acid composition and electrophoretic mobility are similar to those of human apo A-I except for the presence of isoleucine in chicken apolipoprotein.[17,21,23,25,42] This protein is therefore designated as chicken apo A-I. Chicken apo A-I is distinguished from apo VLDL-B and apo VLDL-II by its amino acid composition, particularly its high contents of glutamine and glutamic acid, and absence of half-s-cysteine (Table 10). Aspartic acid is the N-terminal amino acid residue, whereas there are two C-terminal amino acid residues, i.e., alanine and leucine[21] or alanine and glutamine.[25] Chicken apo A-I has a circular dichroic spectrum typical of an α-helical structure. The calculated helicity is 90% which is almost 30% greater than human apo A-I or apo A-II.[23,25] Delipidation reduces its high α-helicity by 20 to 30% otherwise the general conformation of apo A-I does not appear to depend on its lipid content.[43] Goat antisera against chicken apo A-I partially cross-react with human apo A-I.[25]

Other apolipoproteins in chicken HDL are less well defined. A protein of molecular weight (15,000) has been observed which, unlike the human apo A-II, is unaffected by reducing agents.[23] Small amounts of polypeptides having electrophoretic migratory rates similar to those of human apo C polypeptide, together with apo VLDL-II have also been detected in chicken HDL.[23]

Turkey HDL apolipoprotein consists of two polypeptides designated as apo A-I and apo A-II. Turkey apo A-I is the major HDL apolipoprotein with a molecular weight of 27,000 and amino acid composition and terminals that are similar to those of chicken apo A-I. Turkey apo A-II has a molecular weight of 10,000 and contains relatively less lysine, leucine, and arginine but more aspartic acid than apo A-I.[18,36]

Metabolic Interrelationship Among Various Lipoprotein Classes

The lymphatic system in chickens is not as well developed as that of mammals.[2,44] The

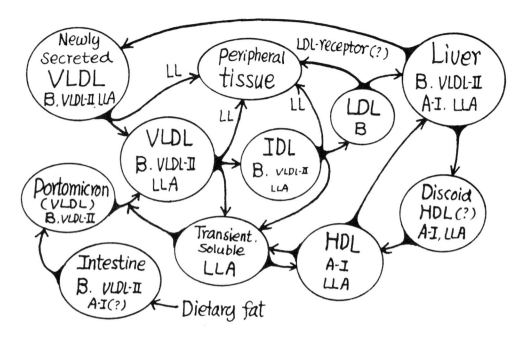

FIGURE 1.   Metabolic interrelationship among various lipoprotein classes in chickens. VLDL, IDL, LDL, and HDL are, respectively, very low, intermediate, low, and high-density lipoproteins. B. VLDL-II and A-1 are apolipoproteins. LL and LLA are, respectively, lipoprotein lipase and its activators. LCAT is lecithinolesteol acyltransferase.

great majority of dietary fat after absorption appears in the capillaries, instead of the lacteals of the intestinal villi and is transported via the portal vein in the form of VLDL or "portomicron".[1,2] Apparently the intestine is capable of synthesizing apo VLDL-B and probably also apo VLDL-II. Whether or not apo A-I can also be synthesized in the intestine as in human[45] remains to be investigated. Another unique feature of lipid metabolism in the avian species is that liver is the major organ for de novo fatty acid synthesis and adipose tissue is primarily for lipid storage rather than synthesis.[46-49] The liver also has an enormous reserve capacity for synthesis of all apolipoproteins.[50] Therefore, the liver is certainly another source of circulating VLDL (Figure 1).

There are at least two possibilities for the low plasma VLDL levels characteristic for chickens not actively laying eggs. One is the low fat content, usually 5% or less, of commercial chicken chow. Addition of fat to the diet increases the plasma lipid and VLDL levels slightly.[51-54] Another possibility is the rapid clearance of VLDL by lipoprotein lipase on the surface of the capillary endothelium.[6;16] The wide range of VLDL particle size and the great variation in its composition and in its plasma concentration reflect the dynamic character of VLDL metabolism. By in vivo blocking of lipoprotein lipase with a specific antibody, newly synthesized and secreted VLDL has been isolated from the plasma in roosters and characterized.[16] The average diameter of the newly secreted VLDL (60 to 65 nm) is almost twice the size of the "usual" VLDL, with a broad spectrum of particle size ranging from 30 to over 100 nm. As compared to the "usual" VLDL, the newly secreted VLDL contains more triglyceride (76.2 vs. 46.1%), less phospholipid (8.3 vs. 27.3%), and less cholesterol (4.8 vs. 13.6%). Only 29% of the cholesterol in the newly secreted VLDL is esterified, instead of 46% in the "usual" VLDL. The inability to detect the large size VLDL in the circulation, in almost all studies, suggests its rapid catabolism to the smaller size VLDL. The catabolic process is primarily the removal of triglyceride by lipoprotein lipase. A positive correlation exists between the triglyceride content of VLDL and its catabolic rate.[55]

If the portomicron synthesized in the intestine does not contain lipoprotein lipase activator, it must acquire these small apolipoproteins from HDL in the circulation. With the progressive degradation of triglycerides carried in VLDL by lipoprotein lipase and its activators, VLDL reduces in size and becomes an intermediate density lipoprotein (IDL). The removal of triglyceride from its lipid core results in a redundant surface membrane which is probably removed by HDL (Figure 1). As a consequence, VLDL progressively loses its surface unesterified cholesterol, phospholipid, apo VLDL-II, and lipoprotein lipase activator. Further lipolysis of IDL produces LDL which contains no apo VLDL-II and lipoprotein lipase activators. In man, LDL is primarily removed from the circulation through a specific LDL receptor on the cell surface.[56] Such a receptor has yet to be demonstrated in chickens.

Unlike humans, the liver is the major source of circulating HDL in chickens since chylomicrons, another source of HDL in man, are undetectable in chickens.[1,2,47] The newly secreted HDL by the liver in chickens could also be discoid in shape as in man.[57] By the action of lecithin: cholesterol acyltransferase (LCAT) which has been demonstrated in the plasma of chickens[58] and requires apo A-I for its activity, the discoid HDL is converted into spherical HDL.[55,57] Small apolipoproteins contained in HDL such as lipoprotein lipase activators are soluble and can be transported back and forth between HDL and VLDL (Figure 1). The major function of VLDL is therefore the transportation of either exogenous or endogenous triglycerides to peripheral tissues, while that of HDL might play the role of a regulator of the system which provides enzyme activators and removes redundant membrane, particularly unesterified cholesterol.

LIPOPROTEINS IN LAYING HENS

Lipoprotein Changes During Egg-Laying

When hens reach sexual maturity at an age of 5 to 6 months and are exposed to a photoperiod of 12 to 18 hr/day, they start to lay eggs. Egg-laying is accompanied by a dramatic change in the plasma lipid levels.[7,9] The plasma triglyceride increases more than tenfold, phospholipid fivefold, but cholesterol and apolipoprotein by only about 20% (Table 1). Among various classes of lipoproteins, VLDL increases 45-fold, LDL is unchanged, whereas HDL is consistently and remarkably decreased (Table 2).[7,8] VLDL thus becomes the predominant lipoprotein in laying hens, accounting for 76% of total plasma lipoprotein, instead of 4% in immature hens (Table 3). In the laying hen, HDL accounts for only 12% of total lipoprotein, instead of 76% in the nonlaying hen. The plasma concentration of VLDL apolipoprotein increases 56-fold, phospholipid 69-fold, triglyceride 46-fold, but cholesterol only 18-fold (Table 4). While the relative concentration of apolipoprotein in VLDL is unchanged, there is a relative increase in phospholipid content (140%), and a remarkable reduction in cholesterol content (30%) (Table 5). The composition of LDL remains essentially the same (Tables 6 and 7). The HDL contains relatively more triglyceride (150%), less phospholipid (80%) and less cholesterol (60%) (Tables 8 and 9).

Estrogen-Induced Hyperlipidemia

The above change of plasma lipoproteins is greatly exaggerated by administration of estrogen to nonlaying chickens including immature hens, cockerels, and roosters.[5,11,14,15,17,28,29,31,32,58,60] The response is proportional to the dose of estrogen administered.[14] Six hours after a single s.c. injection of diethylstilbesterol, plasma VLDL starts to increase, reaches a maximum in 24 to 48 hr and returns to the base-line by 72 hr.[29] A ''memory effect'' exists because the response is remarkably accelerated and enhanced by subsequent estrogen administration.

The liver is the primary, if not the only, organ which responds to estrogen stimulation for lipoprotein synthesis because without a liver the chicken fails to develop hyperlipidemia.[46]

At the peak of induction, newly synthesized VLDL represents as much as 8 to 18% of the total soluble protein synthesized in the liver.[15,29,49] The mechanism of estrogen-induction of hepatic synthesis of apo VLDL-II has been extensively studied. The hormone appears to bind first to specific receptors in the liver cell nucleus, resulting in enhanced RNA polymerase activity, an increased number of initiations of RNA synthesis, and an accumulation of apo VLDL mRNA, followed by an increased rate of VLDL synthesis and finally by hyperlipoproteinemia.[31] The in vitro translation product of apo VLDL-II mRNA is about 23 amino acids larger than apo VLDL-II isolated from the plasma.[32] It is therefore proposed that apo VLDL-II is synthesized initially as a larger protein and the N-terminal sequence is cleaved prior to completion of synthesis and secretion of the VLDL particle. The same mechanism is apparently also involved in the synthesis of apo VLDL-B. In estrogen-treated roosters apo VLDL-B and apo VLDL-II, both greatly increased, constitute 54 and 46% of the total VLDL protein, respectively.[29] The molar ratio of apo VLDL-B to apo VLDL-II monomer is estimated to be 1 to 32.[28]

In contrast to the lipoproteins from untreated chickens, lipoproteins of density less than 1.06 g/mℓ from estrogen-treated chicken cannot be clearly separated into discrete VLDL and LDL fractions; these two lipoproteins appear to be a single ultracentrifugal class.[17] The apolipoprotein patterns of VLDL and LDL from estrogen-treated chickens become indistinguishable from each other. The apolipoprotein composition of HDL is also markedly altered by estrogen administration: the major apolipoprotein, apo A-I, decreases to less than 5% of the total HDL protein while apo VLDL-B and apo VLDL-II become predominant.[17]

Vitellogenin

Besides VLDL, a new lipoprotein, designated as vitellogenin, appears in the plasma of laying hens.[59,62] Vitellogenin is absent from the plasma of cockerels, roosters, and nonlaying hens; it can be induced by administration of estrogen.[59,60,62] Plasma vitellogenin is composed of two polypeptides, each of which has a molecular weight of approximately 240,000 and contains within it, two phosvitins and one lipovitellin.[62] Phosvitin is a glycophosphoprotein of highly unusual amino acid composition; approximately 50% of its amino acid residues are serine, almost all of which are phosphorylated.[63,64] There are two phosvitins which differ in molecular weights, 28,000 and 34,000, and in amino acid composition. Lipovitellin is a lipophosphoprotein composed of 20% of lipid and 80% of protein by weight.[11,12] Most of the lipid is phosphorylated while 10% of the protein is made up of serine, about one half of the serine is phosphorylated. There are two lipovitellins, α and β forms. The molecular weights of three polypeptides obtained form α-lipovitellin are about 135,000, 105,000, and 40,000.[59] The polypeptide with molecular weight of 40,000 contains almost all the protein-bound phosphate in α-lipovitellin. β-Lipovitellin is composed of two polypeptide chains with molecular weights of 135,000 and 30,000, both of which are phosphopeptides.[59]

Liver is the site of vitellogenin synthesis which can be induced in roosters by estrogen administration.[59,60,62] The response is dose related.[65] Ten hours after a single injection of estrogen, vitellogenin starts to appear in the circulation but not until 24 hr after the injection does the rate of vitellogenin synthesis becomes maximal; this synthesis lasts for several days. As compared to VLDL, the initial rate of induction of vitellogenin synthesis is slower but it appears to stay in the circulation for a much longer time (6 to 7 vs. 3 days). After the second dose of estrogen, vitellogenin synthesis is detected sooner and its initial increase is more rapid than after the first dose.[60] This "memory effect" is undiminished even 50 days after the first hormone dose.

Vitellogenin is the precursor of lipovitellin and phosvitin in the egg yolk. Vitellogenin appears in the circulation as an intact complex.[62] The plasma level of laying hens is between 1.0 and 2.0 g/100 mℓ.[10] The induction of vitellogenin synthesis is associated with a compensatory reduction of the plasma albumin level.[59] Vitellogenin is a calcium-binding lipoprotein.[66,67] In fact calcium is essential for lipovitellin-phosvitin complex formation. The serum calcium in actively laying hens is thus markedly elevated, 24 mg/100 mℓ, as compared to 11 mg/100 mℓ in roosters.[9]

Lipoprotein Lipase Inhibitor

The elevated VLDL and vitellogenin in laying hens are primarily for the transport of lipid and protein from plasma to the egg yolk and not for peripheral tissue utilization. Of particular interest is that one of the constituents of vitellogenin, phosvitin, is a potent inhibitor of both extrahepatic and hepatic lipoprotein lipases; the former can be inhibited by protamine sulfate and requires a plasma activator for activity while the latter cannot be inhibited and does not require plasma co-factor.[68] The lipolytic activity of both extrahepatic and hepatic lipoprotein lipases is indeed greatly decreased in the post-heparin plasma of laying birds or estrogen-treated male birds.[68] The plasma VLDL in laying hens, destined solely for egg yolk, is therefore protected at least partially by phosvitin from lipolysis by lipoprotein lipase. The steady and tremendous accumulation of VLDL in estrogen-treated roosters or immature hens is not only due to the lack of a depository for VLDL, the developing egg follicle, but also due to the low lipolytic activity caused by lipoprotein lipase inhibitors.

LIPOPROTEINS IN EGG YOLK

Characterization of Yolk Lipoproteins

The egg yolk is composed entirely of proteins and lipoproteins, some of which form giant complexes termed granules. The granules comprise one fifth to one fourth of the total volume of egg yolk.[69] Centrifugation of egg yolk in diluted NaCl led to the formation of a yellow supernatant and a firm pellet, the granule. Upon recentrifugation in 10% NaCl, the supernatant fraction is dissolved into an upper layer of firm yellow gel composed of VLDL and a colorless lower portion composed of water soluble protein, livetin.[70] Upon dispersion in 0.5 M NaCl and recentrifugation, the pellet or granule yields a supernatant of pale yellow gel also composed of VLDL and a clear infranatant composed of lipovitellin and phosvitin. The amounts of the total yolk solid recovered in the above four fractions are, respectively, 65.2, 10.4, 5.5, and 18.7%.[69]

The VLDL of egg yolk has a highly apolar lipid core surrounded by phospholipid and proteins,[71] a molecular weight of 5.2×10^6, and a hydrated density of 0.962 g/mℓ. Its flotation rate ranges from S_f 16 to 126 with a major peak at S_f 37 and the diameter dimensions range from 17 to 37 nm with a maximum in the region of 25 to 27 nm.[72] Although up to ten polypeptides have been identified in egg yolk VLDL,[33,34,73,74] two of them seem to be the major apolipoproteins; one is of low molecular weight and another of high molecular weight.[33,34] The amino acid sequence of the low-molecular-weight apolipoprotein is identical to that of apo VLDL-II of the hen's plasma VLDL[75,76] (Table 10). The amino acid composition of high-molecular-weight apolipoproteins from the yolk resembles that of plasma apo VLDL-B (Table 10). The relative lipid composition of cholesterol, triglyceride, and phospholipid in yolk VLDL are, respectively, 6 to 8, 66 to 75, and 19 to 28%, similar to that of plasma VLDL of laying hens (Table 5).

Chicken apo A-I has been detected in the egg yolk by immunochemical methods.[77,78] The major HDL in the yolk, however, is α- and β-lipovitellin derived from plasma vitellogenin and present primarily in the yolk granules.[59,63,69]

Transport of Lipoproteins to Yolk

The developing ovarian follicle reaches its maturity within 9 to 10 days in chickens.[79] Over a period of 7 days preceding maturity and ovulation, the follicle increases an average from 8 to 37 mm in diameter and from 0.08 to 14 g in weight.[79,80] The similarity between the plasma and the yolk VLDL indicates that plasma VLDL is directly and specifically transported to the yolk in a more or less intact form.[27,37,72] Ultrastructural studies reveal the following mechanism of such transportation: VLDL passes through the fenestrated capillary in theca externa and interna which lacks a basal lamina, then penetrates the basal lamina of the granulosa layer and appears in the widened intercellular space, and finally enters the oocyte by pinocytosis.[80,81] Such transport of plasma VLDL to yolk does not appear to involve lipoprotein lipase which is inactivated by plasma inhibitors such as phosvitin.[67,82]

The transport of plasma vitellogenin, on the other hand, requires its cleavage or dissociation to lipovitellin and phosvitin prior to entering into the egg yolk for intact vitellogenin is not present in the yolk granules. Vitellogenin is highly susceptible to proteolysis.[62] Furthermore, chelating of bivalent cations, mainly Ca^{++}, may also facilitate the dissociation of phosvitin from vitellogenin.[83]

The electrophoretic mobility of various classes of lipoproteins from sera and egg yolk of chickens and sera of pigeons and turkeys is shown in Table 11.

LIPOPROTEINS IN CHICK EMBRYO

The developing chick embryo obtains all of its nutrients from the egg. During the first 2 weeks of embryogenesis, carbohydrate and protein are the main materials utilized and only a total of 0.5 to 0.6 g of lipid is transferred from the yolk to the embryo.[84] On the 15th to 17th day of embryogenesis there is a drastic rise in the transport of yolk lipid to the embryo. The granular fraction and its lipovitellin-phosvitin subfraction decrease remarkably. The yolk VLDL also decreases but at a slower rate while the water-soluble fraction which contains livetin increases in amount from 5 to 40% of the residual solid after 18 days of incubation.[85] The lipovitellin-phosvitin fraction contains most of the iron, calcium, phosphorus, and protein in the yolk; its rapid transport from yolk to embryo indicates its importance in embryogenesis.[85]

VLDL similar to yolk VLDL appears in the embryonic circulation at the time when utilization of yolk lipid becomes accelerated.[77,86] The mechanism of transport of yolk VLDL to embryo does not seem to involve enzymes because no lipolytic activity has been observed in yolk, yolk sac, or in the vitelline veins.[58,82] The lipoprotein lipase activity of the embryonic tissue, particularly adipose tissue, increases several fold on day 16 and 17 of incubation, coinciding with the time when the transportation of yolk lipid to the embryo becomes rapid.[84] Moreover, a lipoprotein lipase activator has been identified in the egg VLDL which when transported to the embryo could facilitate the uptake of lipid from VLDL by embryonic tissue.[33]

LDL and HDL are also present in the plasma of the chick embryo.[77] The liver of the chick embryo is capable of synthesizing lipoproteins.[87,88] A sudden doubling of the HDL concentration, at hatching, without a concomitant rise in VLDL and LDL suggests that a considerable amount of HDL is synthesized de novo by the embryonic liver.[77]

It is interesting to note that most of the cholesterol in egg yolk is unesterified, making it readily available for incorporation into various membranes in the cell of the developing embryo. In the later stage of embryogenesis enormous amounts of fat, more than 1 g per day, is transferred from the yolk to embryo.[84] The cholesterol not required for the cell growth of the embryo is stored in the liver as esterified cholesterol.[58,84] The accumulation occurs to such an extent that at hatching, esterified cholesterol accounts for about 70% of the total lipid and for about 30% of the total solid in the liver.[84] The esterification of cholesterol must involve Acyl Co A: cholesterol acyltransferase (ACAT) instead of LCAT since 80% of cholesterol ester in the liver is cholesterol oleate.[84] Furthermore, esterified cholesterol also appears in the circulation; it is related to the presence of HDL and an enzyme activity resembling LCAT in the plasma of the chick embryo.[58]

Table 11
ELECTROPHORETIC MOBILITY OF VARIOUS CLASSES OF LIPOPROTEINS FROM SERA AND EGG YOLK OF CHICKENS AND SERA OF PIGEONS AND TURKEYS

Species and treatment	Electrophoresis conditions	Mobility	Ref.
Serum lipoproteins isolated by preparative ultracentrifugation from adult white leghorn roosters and laying hens	0.5% Agarose gel on microscopic slide in 0.05 M barbital buffer pH 8.6 for 2 hr at 4 mA/slide, fixed in 5% acetic acid in 60% methanol and stained with 1% aniline blue black in 2% acetic acid	HDL migrates at the tailing edge of albumin, with a minor HDL component moving more slowly; VLDL migrates only a short distance from the origin; LDL migrates slightly further than VLDL	21
VLDL (d $<$1.006) and LDL (d = 1.006-1.063) isolated by ultracentrifugation from laying Rhode Island Red hens	Agarose-agar (8:2) gel mixture on a Cronar® polyester film in barbital buffer, 0.05 M, pH 8.6 for 2 hr at 10 mA/strip, stained with Sudan black B	VLDL migrates slower than human VLDL with a ratio of mobilities (human VLDL as 1.0) of 0.81:1.0; LDL migrates also slower than human LDL with a ratio of mobilities (human LDL as 1.0) of 0.41:1.0	8
VLDL and LDL separated from HDL by precipitation with heparin and manganese chloride and from each other by ultracentrifugation from white leghorn roosters and laying hens	Same as above	HDL migrates with albumin. VLDL migrates only slightly from the origin; LDL migrates between the origin and β-globulin position	26
Same as above	A stacking gel of 2.75% acrylamide and a separating gel of 3.5% acrylamide, with bromophenol blue as marker, stained overnight in 0.01% coomassic blue, 12.5% trichloroacetic acid	VLDL retains in the stacking gel with slight penetration into the separating gel; LDL migrates into the separating gel; HDL migrates with the tracking dye in a broad band at high concentration but two bands at low concentration	26
Sera from white leghorn roosters, laying hens, and hereditable nonlayers	Paper electrophoresis in albuminated barbital buffer, pH 8.6, ionic strength 0.075, for 14 hr at 8 mA/cell, stained with oil red O	Roosters have a thick band in albumin position and a thin band at origin; layers have a thick band migrating slightly ahead of origin, a thin band at origin and a faint band in α-globulin position; nonlayers have a thick band at origin and a faint band in α-globulin position	9
Sera from estrogenized, 12-week-old pullets	Paper electrophoresis in veronal buffer pH 8.6, ionic strength 0.05 for 24 hr at 5°C and 0.57 mA/in. width of paper, stained with Amidoschwarz-10B or oil red O	First band at origin (LDL); second band slightly ahead of origin (VLDL); third band in α_2 and α_3-globulin position (HDL); phosvitin migrates slightly ahead of albumin in veronal-citrate buffer	90
VLDL and livetin isolated from egg yolk by ammonium sulfate precipitation followed by DEAE-cellulose column chromatography	Acrylamide gel electrophoresis with spacer gel 4% cross-linked and running gel 7.5% cross-linked in Tris-glycine buffer, pH 8.2, stained with coomassic brillant blue	VLDL stays in the sample gel; livetin appears in several bands in the running gel	91

Table 11 (continued)
ELECTROPHORETIC MOBILITY OF VARIOUS CLASSES OF LIPOPROTEINS FROM SERA AND EGG YOLK OF CHICKENS AND SERA OF PIGEONS AND TURKEYS

Species and treatment	Electrophoresis conditions	Mobility	Ref.
Sera from pigeons at age from 1 day to 6 months	PAGE-disc with stacking and running gel concentrations of 3% and 4.25%, respectively, stained with Sudan black B	Two lipoprotein bands: one at β-globulin position (LDL), another at α-globulin position (HDL); α-migrating lipoprotein is predominant in 6-month-old pigeon whereas β-migrating lipoprotein is predominant in 1-day-old pigeon	92
Sera and lipoproteins prepared by ultracentrifugation from broad breasted white male turkeys	1% Agarose gel on glass slide in veronal buffer, pH 8.6, ionic strength 0.05, for 1 hr at 6.5 V/cm, fixed in 10% trichloroacetic acid and stained with amido black 10B or oil red O	HDL migrates in the α-globulin position; LDL migrates in the β-globulin position; VLDL has a mobility slightly faster than LDL	18

LIPOPROTEINS IN HEREDITARY NONLAYING HENS

The frequency of naturally occurring nonlaying hens has been reported as 1.38% of 13,065 white leghorn chickens.[89] The major causes of failure of egg-laying are a defective ovary and a defective oviduct. The most common ovarian abnormality is ovarian retrogression with the ovary resembling a bunch of grapes composed of follicles of about equal size. Most cases of ovarian retrogression are caused by exposure of the immature hens to excessively prolonged photoperiods. The oviduct abnormality can be either functional or organic. The organic oviduct abnormalities include a discontinuous oviduct, a closed infundibulum, and an impacted oviduct. Hyperlipidemia may, but does not invariably, occur in various types of nonlayers, particularly in ovarian retrogression. The nature of such hyperlipidemia has not yet been well characterized.

We have previously reported a special strain of white leghorn chickens wih a hereditable hyperlipidemia associated with failure of egg-laying.[9] The defective sex-linked recessive gene is transmitted from carrier roosters to one half of the female offspring. When the affected hens reach sexual maturity and are exposed to a photoperiod of 18 hr/day they fail to lay eggs and develop an extreme hyperlipidemia. Their total plasma lipid increases fivefold as compared to that of their laying siblings. While the relative composition of plasma cholesterol, triglyceride, and phospholipid of these nonlayers is quite similar to that of the layers, the electrophoretic patterns of the plasma lipoproteins are distinctly different. The kinetic study of cholesterol metabolism reveals that the rate of cholesterol synthesis in layers increases threefold, while that of nonlayers increases only slightly. The mean transit time of cholesterol is, however, greatly prolonged in nonlayers. The only anatomic abnormality found in the nonlayer is the failure of maturation of ovarian follicles which is different from ovarian retrogression. Gathering all the above information we have postulated that the hyperlipidemia of the nonlayers is due to the impaired mechanism for clearance of lipoprotein from the circulation by the liver and ovarian follicles.[9] Whether the basic defect is in the lipoprotein itself or in its cofactors remains to be elucidated.

LIPOPROTEINS IN CHOLESTEROL-FED CHICKENS

Effect of Cholesterol Feeding on Plasma Lipoproteins

Commercial chicken mash contains 5% or less of fat. The addition of neutral fat to chicken mash elevates the plasma VLDL only slightly.[51-54] The most dramatic change of the plasma lipid and lipoprotein occurs in chickens fed a diet supplemented with 1% of cholesterol with or without an additional 10% of neutral fat.[3-5] While the absolute plasma concentration of triglyceride remains unchanged and that of phospholipid increases slightly, the plasma cholesterol level is elevated sixfold 4 to 6 weeks after feeding a 1% cholesterol diet (Table 1). The primary increase is in the VLDL fraction of plasma lipoprotein. Plasma LDL increases only slightly whereas HDL remains unchanged or slightly decreases (Table 2). VLDL thus becomes the predominant plasma lipoprotein, instead of HDL as in the control chickens (Table 3). Although the protein/lipid ratios in various classes of lipoproteins remain unaltered, the relative composition of individual lipids is changed by cholesterol feeding. The relative cholesterol content of VLDL increases 2.5-fold whereas that of triglyceride decreases to one half, and that of phospholipid is unchanged. The relative cholesterol content in LDL increases 1.8-fold and that of triglyceride and phospholipid decrease about 20 and 30%, respectively. Similar changes are also observed in HDL. The percent esterification of cholesterol remains unchanged in LDL and HDL but increases from 45 to 70% in VLDL. Therefore, cholesterol feeding alters the plasma lipoprotein composition, primarily due to an increase in cholesterol content at the expense of triglyceride and phospholipid (Tables 4 to 9).

Dietary cholesterol is absorbed and reesterified in the intestine and then transported by VLDL to the portal circulation.[1,2] During its lipolysis by lipoprotein lipase a portion of the unesterified cholesterol in the redundant surface membrane of VLDL is probably detached and incorporated into HDL; most, if not all, of the esterified cholesterol carried in the lipid core of VLDL is then passed to its degradation remnant, namely LDL. The cholesterol ester-rich LDL is then taken up by the cells in various tissues, probably through an LDL receptor as described in human fibroblasts (Figure 1).

Atherosclerosis Secondary to Exogenous and Endogenous Hyperlipidemia

Exogenous hyperlipidemia caused by feeding 1% cholesterol diet for 2 months results in a significant increase of the cholesterol content in all tissues except brain, muscle, and adipose tissue.[13] The organ showing the most dramatic increase in the cholesterol content is the liver (3.1-fold), followed by colon (1.9-fold), small intestine (1.6-fold), aorta (1.6-fold), and skin (1.5-fold). Gross atheromas can be observed in the aorta.

The mild endogenous hyperlipidemia in laying hens is associated with a considerable reduction of cholesterol content in many tissues.[10,13] The paradox of hyperlipidemia associated with reduction of tissue cholesterol is probably the combined result of the low lipoprotein lipase activity and the urgent demand for lipid by egg follicles during egg-laying. The only organ in 2-year-old laying hens having an elevated cholesterol content is the aorta (3.4-fold increase), consonant with the occasional atheromata observed in their aortas.[10,13]

The prolonged extreme hyperlipidemia in the hereditable nonlaying hens, on the contrary, is not associated with reduction in tissue cholesterol content as in the laying hens. The cholesterol contents of many organs increase. The aorta is the organ showing the greatest increase in its cholesterol content, almost sevenfold that of the roosters.[13]

In short, a remarkable difference exists in various tissues in response to dietary induced exogenous hypercholesteremia and egg-laying related endogenous hyperlipidemia. In the presence of either endogenous or exogenous hyperlipidemia, the aorta of chickens exhibits a propensity for accumulation of excessive cholesterol derived from plasma lipoproteins and is thus susceptible to the development of atherosclerosis.

ACKNOWLEDGMENT

The original work of the authors cited in this chapter was supported by the Veterans Administration and Public Health Service Grant HL-13612 from the National Heart and Lung Institute.

REFERENCES

1. **Noyan, A., Lossow, W. J., Brot, N., and Chaikoff, I. L.,** Pathway and form of absorption of palmitic acid in the chicken, *J. Lipid Res.,* 5, 538, 1964.
2. **Bensadoun, A. and Rothfeld, A.,** The form of absorption of lipids in the chicken, *Gallus domesticus, Proc. Soc. Exp. Biol. Med.,* 141, 814, 1972.
3. **Kruski, A. W. and Narayan, K. A.,** Effect of orotic acid and cholesterol on the synthesis and composition of chicken (*Gallus domesticus*) serum lipoproteins, *Int. J. Biochem.,* 7, 635, 1976.
4. **Kruski, A. W. and Narayan, K. A.,** The effect of dietary supplementation of cholesterol and its subsequent withdrawal on the liver lipids and serum lipoproteins of chickens, *Lipids,* 7, 742, 1972.
5. **Hillyard, L. A., Entenman, C., and Chaikoff, I. L.,** Concentration and composition of serum lipoproteins of cholesterol-fed and stilbestrol-injected birds, *J. Biol. Chem.,* 223, 359, 1956.
6. **Fried, M., Wilcox, H. G., Faloona, G. R., Eoff, S. P., Hoffman, M. S., and Zimmerman, D.,** The biosynthesis of plasma lipoproteins in higher animals, *Comp. Biochem. Physiol.,* 25, 651, 1968.
7. **Yu, J. Y-L., Campbell, L. D., and Marquardt, R. R.,** Immunological and compositional patterns of lipoproteins in chicken (*Gallus domesticus*) plasma, *Poult. Sci.,* 55, 1626, 1976.
8. **Chapman, M. J., Goldstein, S., and Laudat, M. H.,** Characterization and comparative aspects of the serum very low and low density lipoproteins and their apoproteins in the chicken (*Gallus domesticus*), *Biochemistry,* 16, 3006, 1977.
9. **Ho, K. J., Lawrence, W. D., Lewis, L. A., Liu, L. B., and Taylor, C. B.,** Hereditary hyperlipidemia in nonlaying chickens, *Arch. Pathol.,* 98, 161, 1974.
10. **Redshaw, M. R. and Follett, B. K.,** Physiology of egg yolk production by the fowl: the measurement of circulating levels of vitellogenin employing a specific radioimmunoassay, *Comp. Biochem. Physiol.,* 55A, 399, 1976.
11. **Schjeide, O. A. and Urist, M. R.,** Proteins induced in plasma by estrogens, *Nature (London),* 188, 291, 1960.
12. **Cook, W. H.,** Proteins of hen's egg yolk, *Nature (London),* 190, 1173, 1961.
13. **Ho, K. J.,** Cholesterol contents of various tissues of chicken with exogenous or endogenous hypercholesteremia, *Am. J. Clin. Nutr.,* 29, 187, 1976.
14. **Kudzma, D. J., Hegstad, P. M., and Stoll, R. E.,** The chick as a laboratory model for the study of estrogen-induced hyperlipidemia, *Metabolism,* 22, 423, 1973.
15. **Luskey, K. L., Brown, M. S., and Goldstein, J. L.,** Stimulation of the synthesis of very low density lipoproteins in rooster liver by estradiol, *J. Biol. Chem.,* 249, 5939, 1974.
16. **Kompiang, I. P., Bensadoun, A., and Yang, M. W. W.,** Effect of an anti-lipoprotein lipase serum on plasma triglyceride removal, *J. Lipid Res.,* 17, 498, 1976.
17. **Kudzma, D. J., Swaney, J. B., and Ellis, E. N.,** Effects of estrogen administration on the lipoproteins and apoproteins of the chicken, *Biochim. Biophys. Acta,* 572, 257, 1979.
18. **Kelley, J. L. and Alaupovic, P.,** Lipid transport in the avian species. I. Isolation and characterization of apolipoproteins and major lipoprotein density classes of male turkey serum, *Atherosclerosis,* 24, 155, 1976.
19. **Langelier, M., Connelly, P., and Subbiah, T. R.,** Plasma lipoprotein profile and composition in White Carneau and Show Racer breeds of pigeons, *Can. J. Biochem.,* 54, 27, 1975.
20. **Schneider, H., Morrod, R. S., Colvin, J. R., and Tattrie, N. H.,** The lipid core model of lipoproteins, *Chem. Phys. Lipids,* 10, 328, 1973.
21. **Hillyard, L. A., White, H. M., and Rangburn, S. A.,** Characterization of apolipoproteins in chicken serum and egg yolk, *Biochemistry,* 11, 511, 1972.
22. **Evans, R. J., Fiegal, C. J., Foerder, C. A., Bauer, D. H., and LaVigne, M.,** The influence of crude cottonseed oil in the feed on the blood and egg yolk lipoproteins of laying hens, *Poult. Sci.,* 56, 468, 1977.
23. **Kruski, A. W. and Scanu, A. M.,** Properties of rooster serum high density lipoproteins, *Biochim. Biophys. Acta,* 409, 26, 1975.
24. **Mills, G. L. and Taylaur, C. E.,** The distribution and composition of serum lipoproteins in eighteen animals, *Comp. Biochem. Physiol.,* 40B, 489, 1971.
25. **Jackson, R. L., Lin, H-Y. U., Chan, L., and Means, A. R.,** Isolation and characterization of the major apolipoprotein from chicken high density lipoproteins, *Biochim. Biophys. Acta,* 420, 342, 1976.
26. **Hearn, V. and Bensadoun, A.,** Plasma lipoproteins of the chicken, *Gallus domesticus, Int. J. Biochem.,* 6, 295, 1975.
27. **Holdsworth, G., Michell, R. H., and Finean, J. B.,** Transfer of very low density lipoprotein from hen plasma into egg yolk, *FEBS Lett.,* 39, 275, 1974.
28. **Williams, D. L.,** Apoproteins of avian very low density lipoprotein: demonstration of a single high molecular weight apoprotein, *Am. Chem. Soc.,* 18, 1056, 1979.
29. **Chan, L., Jackson, R. L., O'Malley, B. W., and Means, A. R.,** Synthesis of very low density lipoproteins in the cockerel: effects of estrogen, *J. Clin. Invest.,* 58, 368, 1976.

30. **Jackson, R. L., Lin, H. Y., Chan, L., and Means, A. R.,** Amino acid sequence of a major apoprotein from hen plasma very low density lipoproteins, *J. Biol. Chem.,* 252, 250, 1977.
31. **Chan, L., Jackson, R. L., and Means, A. R.,** Regulation of lipoproteins synthesis: studies on the molecular mechanisms of lipoprotein synthesis and their regulation by estrogen in the cockerel, *Circ. Res.,* 8, 43, 1978.
32. **Chan, L., Bradley, W. A., Jackson, R. L., and Means, A. R.,** Lipoprotein synthesis in the cockerel liver: effects of estrogen on hepatic polysomal messenger ribonucleic acid activities for the major apoproteins in very low and high density lipoproteins and for albumin and evidence for precursors to these secretory proteins, *Endocrinology,* 106, 275, 1980.
33. **Bengtsson, G., Marklund, S. E., and Olivecrona, T.,** Protein components of very low density lipoproteins from hen's egg yolk, *Eur. J. Biochem.,* 79, 211, 1977.
34. **Raju, K. S. and Mahadevan, S.,** Protein components in the very low density lipoproteins of hen's egg yolks: identification of highly aggregating (gelling) and less aggregating (non-gelling) proteins, *Biochim. Biophys. Acta,* 446, 387, 1976.
35. **Chapman, M. J. and Goldstein, S.,** Comparison of the serum low density lipoprotein and of its apoprotein in36the pig, rhesus monkey, and baboon with that in man, *Atherosclerosis,* 25, 267, 1976.
36. **Kelley, J. L. and Alaupovic, P.,** Lipid transport in the avian species. II. Isolation and characterization of lipoprotein A and lipoprotein B, two major lipoprotein families of the male turkey serum lipoprotein system, *Atherosclerosis,* 24, 177, 1976.
37. **Abraham, S., Hillyard, L. A., and Chaikoff, I. L.,** Components of serum and egg yolk lipoproteins: galactose, mannose, glucosamine, and sialic acid, *Arch. Biochem. Biophys.,* 89, 74, 1960.
38. **Jackson, R. L., Morrisett, J. D., and Gotto, A. M. J.,** Lipoprotein structure and metabolism, *Physiol. Rev.,* 56, 259, 1976.
39. **Kane, J. P.,** Plasma lipoproteins: structure and metabolism, in *Lipid Metabolism in Mammals,* Vol. 1, Snyder, F., Ed., Plenum Press, New York, 1978, 209—257.
40. **Goldstein, S. and Chapman, J.,** Comparative immunochemical studies of the serum low-density lipoprotein in several animal species, *Biochem. Genet.,* 14, 883, 1976.
41. **Chapman, M. J.,** Animal lipoproteins: chemistry, structure, and comparative aspects, *J. Lipid Res.,* 21, 789, 1980.
42. **Jackson, R. L., Baker, H. N., Taunton, O. D., Smith, L. C., Garner, C. W., and Gotto, A. M.,** A comparison of the major apolipoprotein from pig and human high density lipoproteins, *J. Biol. Chem.,* 248, 2639, 1973.
43. **Hillyard, L. A., White, H. M., and Abraham, S.,** Microcomplement fixation and particle size of chicken lipoproteins, *J. Lipid Res.,* 21, 913, 1980.
44. **Long, J. F.,** Gastric secretion in unanesthetized chickens, *Am. J. Physiol.,* 212, 1303, 1967.
45. **Glickman, R. M., Green, P. H. R., Lees, R. S., and Tall, A.,** Apoprotein A-1 synthesis in normal intestinal mucosa and in Tangier disease, *New Engl. J. Med.,* 299, 1424, 1978.
46. **Ranney, R. E. and Chaikoff, I. L.,** Effect of functional hepatectomy upon estrogen-induced lipemia in the fowl, *Am. J. Physiol.,* 165, 600, 1951.
47. **O'Hea, E. K. and Leveille, G. A.,** Lipid biosynthesis and transport in the domestic chick (*Gallus domesticus*), *Comp. Biochem. Physiol.,* 30, 149, 1969.
48. **Pearce, J.,** Some differences between avian and mammalian biochemistry, *Int. J. Biochem.,* 8, 269, 1977.
49. **Tarlow, D. M., Watkins, P. A., Reed, R. E., Miller, R. S., Zwergel, E. E., and Lane, M. D.,** Lipogenesis and the synthesis and secretion of very low density lipoprotein by avian liver cells in nonproliferating monolayer culture: hormonal effects, *J. Cell Biol.,* 73, 332, 1977.
50. **Schaefer, E. J., Eisenberg, S., and Levy, R. I.,** Lipoprotein apoprotein metabolism, *J. Lipid Res.,* 19, 667, 1978.
51. **Leveille, G. A. and Sauberlich, H. E.,** Lipid changes in plasma, alpha lipoproteins, liver and aorta of chicks fed different fats, *Proc. Soc. Exp. Biol. Med.,* 112, 300, 1963.
52. **Epley, R. R. and Balloun, S. L.,** Dietary lipid effect on atherogenesis and plasma lipids in cockerels, *Poult. Sci.,* 49, 1705, 1970.
53. **Leveille, G. A., Feigenbaum, A. S., and Fisher, H.,** The effect of dietary protein, fat and cholesterol on plasma cholesterol and serum protein components of the growing chick, *Arch. Biochem. Biophys.,* 86, 67, 1960.
54. **Jones, R. J. and Dobrilovic, L.,** Aortic cholesterol and the plasma lipoproteins of the cholesterol-fed cholesterol, *Proc. Soc. Exp. Biol. Med.,* 130, 163, 1969.
55. **Higgins, J. M. and Fielding, C. J.,** Lipoprotein lipase: mechanisms of formation of triglyceride-rich remnant particles from very low density lipoproteins and chylomicrons, *Biochemistry,* 14, 2288, 1975.
56. **Goldstein, J. L. and Brown, M. S.,** The low-density lipoprotein pathway and its relation to atherosclerosis, *Ann. Rev. Biochem.,* 46, 897, 1977.
57. **Felker, T. E., Fainaru, M., Hamilton, R. L., and Havel, R. J.,** Secretion of the arginine-rich and A-1 apolipoproteins by the isolated perfused rat liver, *J. Lipid Res.,* 18, 465, 1977.

58. **Bengtsson, G., Hernell, O., and Olivecrona, T.,** Are egg yolk lipoproteins metabolized by the chick embryo in the same manner as plasma lipoproteins are in the adult?, *Int. J. Biochem.,* 8, 587, 1977.

59. **Bergink, E. W. and Wallace, R. A.,** Estrogen-induced synthesis of yolk proteins in roosters, *Am. Zool.,* 14, 1177, 1974.

60. **Bergink, E. W., Kloosterboer, H. J., Gruber, M., and AB, G.,** Estrogen-induced phosphoprotein synthesis in roosters: kinetics of induction, *Biochim. Biophys. Acta,* 294, 497, 1973.

61. **McIndoe, W. M.,** A lipophosphoprotein complex in hen plasma associated with yolk production, *Biochem. J.,* 72, 153, 1959.

62. **Deeley, R. G., Mullinix, K. P., Wetekam, W., Kronenberg, H. M., Meyers, M., Eldridge, J. D., and Goldberger, R. F.,** Vitellogenin synthesis in the avian liver: vitellogenin is the precursor of the egg yolk phosphoproteins, *J. Biol. Chem.,* 250, 9060, 1975.

63. **Clark, R. C.,** The isolation and composition of two phosphoproteins from hens, *Biochem. J.,* 118, 537, 1970.

64. **Clark, R. C.,** Amino acid sequence of cyanogen bromide cleavage peptide from hen's egg phosvitin, *Biochim. Biophys. Acta,* 310, 174, 1973.

65. **Beuving, G. and Gruber, M.,** Induction of phosvitin synthesis in roosters by estradiol injection, *Biochim. Biophys. Acta,* 232, 529, 1971.

66. **Grunder, A. A., Guyer, R. B., Buss, E. G., and Clagett, C. O.,** Calcium-binding proteins in serum: quantitative differences between thick and thin shell lines of chickens, *Poult. Sci.,* 59, 880, 1980.

67. **Guyer, R. B., Grunder, A. A., Buss, E. G., and Clagett, C. O.,** Calcium-binding proteins in serum of chickens: vitellogenin and albumin, *Poult. Sci.,* 59, 874, 1980.

68. **Kelley, J. L., Ganesan, D., Bass, H. B., Thayer, R. H., and Alaupovic, P.,** Effect of estrogen on triacylglycerol metabolism: inhibition of post-heparin plasma lipoprotein lipase by phosvitin, an estrogen-induced protein, *FEBS Lett.,* 67, 28, 1976.

69. **Gornall, D. A. and Kuksis, A.,** Resolution of egg yolk lipoproteins by chromatography on thin layers of hydroxylapatite, *Can. J. Biochem.,* 49, 44, 1971.

70. **McIndoe, W. M. and Culbert, J.,** The plasma albumins and other livetin proteins in egg yolk of the domestic fowl *(Gallus domesticus), Int. J. Biochem.,* 10, 659, 1979.

71. **Kamat, V. B., Lawrence, G. A., Barratt, M. D., Darke, A., Leslie, R. B., Shipley, G. G., and Stubbs, J. M.,** Physical studies of egg yolk low density lipoprotein, *Chem. Phys. Lipids,* 9, 1, 1972.

72. **Nichols, A. V., Forte, G. M., and Coggiola, E. L.,** Electron microscopy of very low density lipoproteins from egg yolk using negative staining, *Biochim. Biophys. Acta,* 175, 451, 1969.

73. **Burley, R. W.,** Isolation and properties of a low molecular weight protein (apovitellenin I) from the high-lipid lipoprotein of emu egg yolk, *Biochemistry,* 12, 1464, 1973.

74. **Culbert, J. and McIndoe, W. M.,** A comparison of the heterogeneity of phosvitin of egg-yolk and blood-plasma of the domestic fowl *(Gallus domesticus), Int. J. Biochem.,* 2, 617, 1971.

75. **Dopheide, T. A. and Inglis, A. S.,** Primary structure of apovitellenin I from hen egg yolk and its comparison with emu apovitellenin I, *Aust. J. Biol. Sci.,* 29, 175, 1976.

76. **Inglis, A. S. and Burley, R. W.,** Determination of the amino acid sequence of apovitellenin from duck's egg yolk using an improved sequenator procedure. A comparison with other avian species, *FEBS Lett.,* 73, 33, 1977.

77. **Schjeide, O. A., Rieffer, G., Kelley, J. L., and Alaupovic, P.,** Apolipoproteins and lipoprotein families in chicken embryos and egg yolk, *Comp. Biochem. Physiol.,* 58B, 349, 1977.

78. **Gornall, D. A. and Kuksis, A.,** Alterations in lipid composition of plasma lipoproteins during deposition of egg yolk, *J. Lipid Res.,* 14, 197, 1973.

79. **Gilbert, A. B.,** The egg: its physical and chemical aspects, in *Physiology and Biochemistry of the Domestic Fowl,* Vol. 3, Bell, D. J. and Freeman, B. M., Eds., Academic Press, New York, 1971, 1379.

80. **Perry, M. M., Gilbert, A. B., and Evans, A. J.,** Electron microscope observations on the ovarian follicle of the domestic fowl during the rapid growth phase, *J. Anat.,* 125, 481, 1978.

81. **Evans, A. J., Perry, M. M., and Gilbert, A. B.,** The demonstration of very low density lipoprotein in the basal lamina of the granulosa layer in the hen's ovarian follicle, *Biochim. Biophys. Acta,* 573, 184, 1979.

82. **Gornall, D. A., Kuksis, A., and Morley, N.,** Lipid-metabolizing enzymes in the ovary of the laying hen, *Biochim. Biophys. Acta,* 280, 225, 1972.

83. **Harduf, Z. and Alumot, E.,** The chelating effect of citrate buffer on plasma and yolk lipophosphoproteins, *Int. J. Biochem.,* 10, 815, 1979.

84. **Noble, R. C. and Moore, J. H.,** Studies on the lipid metabolism of the chick embryo, *Can. J. Biochem.,* 42, 1729, 1964.

85. **Saito, Z., Martin, W. G., and Cook, W. H.,** Changes in the major macromolecular fractions of egg yolk during embryogenesis, *Can. J. Biochem.,* 43, 1755, 1965.

86. **Schjeide, O. A.,** Lipoproteins of the fowl-serum, egg and intracellular, in *Progress in the Chemistry of Fats and Other Lipids,* Vol. 6, Pergamon Press, Oxford, 1963, 253.

87. **Schjeide, O. A. and Wilkens, M.,** Parameters of estrogen-stimulated protein synthesis, *Nature (London),* 201, 42, 1964.
88. **Schjeide, O. A., Nicholls, T., and Prince, R.,** Correlation of changes occurring in livers and serums of developing turkey embryos, *Cytobiologie,* 9, 407, 1974.
89. **Hutt, F. B., Goodwin, K., and Uban, W. D.,** Investigations of non-laying hens, *Cornell Vet.,* 46, 257, 1956.
90. **McCully, K. A. and Common, R. H.,** Zone electrophoresis of protein-bound phosphorus of fowl's serum, *Can. J. Biochem. Physiol.,* 39, 1451, 1961.
91. **Raju, K. S. and Mahadevan, S.,** Isolation of hen's egg yolk very low density lipoproteins by DEAE-cellulose chromatography, *Anal. Biochem.,* 61, 538, 1974.
92. **Jensen, P. F., Jensen, G. L., and Smith, S. C.,** Serum lipoprotein profiles of young atherosclerosis-susceptible White Corneau and atherosclerosis-resistant Show Race pigeons, *Comp. Biochem. Physiol.,* 60B, 67, 1978.

LIPIDS AND LIPOPROTEINS OF PIGEONS

Lena A. Lewis

It has been recognized for many years that some animal species in addition to the human being may develop atherosclerosis spontaneously or that its development may be greatly accelerated by use of certain atherogenic agents.[1-4] Pigeons are such a species. Atherosclerosis may occur spontaneously and feeding a diet high in cholesterol content may greatly accelerate its development. Pigeons are suited for study of experimental atherosclerosis for it has been recognized for at least the last 15 years that one strain of pigeon, the White Carneau (WC), is especially susceptible to development of atherosclerosis, while another strain, the Show Racer (SR), is relatively resistant and develops the disease at an older age.[2]

Many studies have been made of the cholesterol levels of the plasma of these birds, how they may change as a result of dietary treatment and how they may be modified in an attempt to prevent development of lesions or cause their regression (Table 1). A few more detailed studies[20] including those of the lipoproteins of the pigeons have also been made to provide more complete understanding of their lipid transport and metabolism.

The plasma cholesterol levels of the mature pigeon reared and maintained on standard chow diets are relatively high in comparison with the level of many mammalian species, i.e., about 250 to 350 mg/dℓ in the pigeon.[1,3,4,6] The level of serum cholesterol of both strains of mature male birds was similar.

The lipoproteins of the pigeon as in most species is distributed between a fraction with the electrophoretic mobility of α_1-globulin, i.e., α_1-lipoprotein, and one with the mobility of β-globulin, β-lipoprotein β-Lp[6] (Figure 1). The latter fraction represents only about 15 to 25% of the total lipid.

Electrophoresis of pigeon serum as is true of most species on starch gel[5] gives greater resolution of the α-lipoprotein (α-Lp) than is obtained by paper or agarose gel. By this technique three and sometimes four distinct bands with mobility in the α_1-globulin range are demonstrated. The β-Lp fraction is a relatively small component and no pre-β-Lp was found in normal male pigeon serum.

To determine the effect of growth and development the serum lipoprotein pattern and cholesterol level of WC and SR pigeons were measured when the birds were 1, 4, 9, and 12 months of age.[6] At 12 months ^{131}I uptake by the thyroid was measured. The birds were sacrificed and the thyroid glands weighed and examined. At 4 and 12 months of age the serum cholesterol level of the SR was higher than that of the WC, while at 9 months of age the levels were similar. The ratio of α/β lipoprotein of the SR was higher than ot the WC. This was true of both sexes and at all ages (Figure 2). Both groups had similar ^{131}I uptake. Despite the lower body weight of the SR their thyroid glands weighed more than those of the WC. A histologic examination of the thyroids of both groups showed occasional to moderate foci of interstitial, lymphocytic infiltration. The significance of this finding was not understood. The difference in thyroid gland weights in the two strains may suggest basic difference in metabolic activities (Figure 3).

If the beneficial effect of a high α-Lp level in delaying development of atherosclerosis is also true in the pigeon then this finding of a high level in the SR may help to explain its resistance to the disease. Langelier et al.[7] in a more recent study (1976) noted that while the overall lipid and lipoprotein distribution of the two strains of pigeons was similar, there were differences between strains in the lipid and protein content of the different fractions. The fluctuation in serum cholesterol concentration of the individual male was large, but that of the female was even greater. This emphasizes the need for study of large numbers and of frequent studies to establish any differences between strains. Consideration of the im-

Table 1
SUMMARY TABLE — LIPIDS AND LIPOPROTEINS OF PIGEONS

Author	Year	Lipids studied	Lipoprotein method of Lp studied
Clarkson, T. B. et al.	1959	Atherosclerosis, a spontaneously developing disease in the pigeon was described	No lipoprotein
Young, F.	1966	Lipid analyses, especially arachidonic acid Arachidonic acid levels were considered in relation to susceptibility of pigeons to atherosclerosis	No lipoprotein
Lofland, H. B. et al.	1966	Cholesterol The effect of modifying the fat and protein content of the diet of pigeons on development of atherosclerosis was evaluated; importance of diet composition on disease susceptibility emphasized	No lipoprotein
Lewis, L. A. and Smith, S. C.	1970	Cholesterol Changes in serum cholesterol and lipoprotein levels that occur during growth and maturation were studied in WC and SR pigeons; at all ages and in both sexes the ratio of α/β-Lp was greater in the SR than in the WC bird; this may be a factor in the delayed development of atherosclerosis in this strain of pigeon	Lipoprotein pattern by paper electrophoresis (Hatch-Lees Method)
Senterre, R. F.	1972	No lipid analyses; severity of atherosclerosis in aortic bifurcation greater in WC than SR especially evident earlier development in WC	No lipoprotein
Pattern, M. M.	1974	Serum cholesterol Development of substrains of WC and SR with levels of serum cholesterol higher than that of the original controls was achieved by the third generation; a substrain of WC with lower levels was obtained, but was unsuccessful in SR; heritability as a factor regulating serum, cholesterol is emphasized	No lipoprotein
Young	1974	Serum free, ester and total cholesterol and TG and PL Serum cholesterol, cholesterol ester, and free cholesterol levels of SR and WC birds were increased when cholesterol supplemented diet was fed; TG of SR also increased; despite high plasma lipid levels aortas of SR were normal after 12 and 24 weeks of feeding	No lipoprotein
Rymaszewski, Z. J.	1975	Skin and aorta sterol levels only Change in aorta sterol was not paralleled by change in skin sterol as index of atherogenecity; the latter remained relatively constant and therefore cannot be used	No lipoprotein
Langelier, M. et al.	1976	Lipid analysis Lipid analyses of plasma-isolated fractions from WC and SR pigeons on regular diet The major lipoproteins of both strains was the HDL; there was no VLDL fraction resolved in plasma of either strain; the LDL of WC showed a significantly lower percentage of ester cholesterol than did the SR, the TG content of WC, LDL fraction was also lower than that of SR; no significant differences in FA composition of sterol esters, PL, and triglycerides of the lipoproteins of the two strains were noted. Differences in LDL content of protein, TG, and cholesterol esters of strains may be an explanation for metabolic differences in two strains of pigeons	Lipoprotein Lipoproteins isolated by ultracentrifugation studied by paper electrophoresis by method of Hatch and Lees
Cramer, E. V. and Smith, S. C.	1976	Fat metabolism of WC and SR during embryogenesis was studied; the WC consumed significantly less of all lipid classes found in yolk of egg, during development period; major lipid class in yolk was TG; C Ester and phospholipids all studied	No lipoprotein

Table 1 (continued)
SUMMARY TABLE — LIPIDS AND LIPOPROTEINS OF PIGEONS

Author	Year	Lipids studied	Lipoprotein method of Lp studied
Flynn, K.	1976	Plasma cholesterol; fecal steroids	No lipoprotein
		Following ileal by-pass surgery pigeons failed to show significant reduction in serum cholesterol levels as occurs in most other species; however, despite failure of plasma cholesterol to decrease regression of lesion occurred	
Clarkson, T. B. et. al.	1976	Review article, no details on methodology	
Subbiah, M. T. R. et. al.	1976	Arterial sterol changes that occur between 9 and 12 months of age in WC, atherosclerotic prone and SR, resistant pigeons studied; see original for details of differences	No lipoprotein
Subbiah, M. T. R.	1977	Arterial lipid levels; correlated with age of birds; atherosclerotic prone WC and resistant SR were studied; atherosclerotic lesions developed only after 9 months of age and only in WC; sterol balance studies differed from those of SR and blood pressure of WC slightly higher after this age, see original for details	No lipoprotein
Wagner, W. D.	1977	Plasma cholesterol	No lipoprotein
		Effects of continuous or intermittent feeding of cholesterol-supplemented diets to WC pigeon for a 14-month period was evaluated; compared to controls aortic collagen content was significantly increased only in intermittently fed group. Plasma chol. of continuous cholesterol fed was 1300 to 2000 mg/dℓ, while intermittent was 300 to 2000 mg/dℓ; with return to no added cholesterol diet, both groups' plasma cholesterol fell to about 300 mg/dℓ. For details of lipid changes in vessels see text. This may approximate type of intermittent exposure of vessels of human during varying periods of stress	
Lofland, H. B. and Clarkson, T. B.	1977	Lipid studies; cholesterol	
		A review article emphasizing role of selective breeding to obtain experimental animals with desired susceptibility to atherosclerosis and appropriate serum lipid levels; section on pigeons is excellent	
Ravisubbiah, M. T.	1978	Plasma cholesterol and triglyceride; aortic lipids; cholesterol ester hydrolase of aorta	No lipoprotein
		Pigeons with atherosclerotic lesions were subjected to ileal by-pass surgery; no change in plasma cholesterol or triglyceride levels occurred but there was significant regression of lesions; it was suggested this may be due to increased activity of cholesterol ester hydrolase in lysosomal and supernatant fractions of aorta	
Wagner, W. D.	1978	Cholesterol, triglyceride blood pressure and glucose also measured	No lipoprotein
		Genetically selected White Carneau (WC-2) and randomly bred (RBWC) were studied in an attempt to explain increased susceptibility of WC-2 birds to atherosclerosis; despite similar plasma cholesterol WC-2 were more susceptible; no significant differences in risk factors studied could explain the difference in susceptibility; possible differences in arterial wall metabolism are suggested as explained	

Table 1 (continued)
SUMMARY TABLE — LIPIDS AND LIPOPROTEINS OF PIGEONS

Author	Year	Lipids studied	Lipoprotein method of Lp studied
Turner, D. M.	1979	Plasma cholesterol only	No lipoprotein
		The effect of increasing concentration of cholesterol in the diet on the development of atherosclerosis was studied in WC female pigeons was studied; 0.5% additional cholesterol was as effective as higher intake; an additional stress factor CO exposure to cause carboxyhemoglobin level of 20% was also evaluated; at 0.5% and 1% dietary cholesterol and CO-caused increased aortic and plasma cholesterol; at 1% triglyceride levels of both aorta and plasma were decreased; additional stress of CO poisoning modifies effect of high cholesterol feeding on both aortic and plasma cholesterol levels	
Hrafchak, B. B. et. al.	1980	Cholesterol	
		Atherosclerotic lesion development studied when WC birds fed an atherogenic diet from 7 weeks of age. Cholesterol increased from 307 mg/dℓ to 1112 mg after 7 weeks and 2245 after 12 weeks; aortic bifurcation a very susceptible area for lesion development and study	
Mayo, S. J. T.	1980	Complete lipid analysis, C, PL, ester, TG	Lipoproteins studied by preparative UC and by acrylamide gel electrophoresis
		Preparation of lipoprotein fractions was carried out by ultracentrifugation. HDL apo A-I was studied in detail for physical, chemical, and immunologic properties (see original for details); PAGE by method of Mao, Gotto, and Jackson used in apo A-I characterization; pigeon apo A-1 showed no significant conformational change upon lipid binding as judged by circular dichroism, under like conditions human apo A-1 showed "drastic conformational change." By ultracentrifugation most of pigeon LDL had flotation rate S_α 0 to 12 at d 1.063, with some at S_α 12 to 20. Apolipoprotein A-1 had molecular weight of approximately 28,000	
Revis, N. W.	1981	Cholesterol, protein phospholipids, triglyceride	HDL by ultracentrifugation at d 1.063 and 1.21
		The effect of altered levels of calcium, cadmium, magnesium, and lead intake on the development of atherosclerosis and on serum LDL, HDL, and lipid levels was investigated in male WC pigeons; treatment with increased calcium intake resulted in fourfold increase in LDL level and increased arteriosclerosis; importance of LDL protein in these changes was suggested; the changes induced by the other electrolytes were variable and are detailed in original article	

portance of the age factor influence on serum lipid levels of the birds must always be kept in mind. The fluctuations in serum cholesterol level of female birds are greatly influenced by hormonal factors and altered nutritional needs due to the demands of egg laying (see chapter "Lipoproteins of Chickens" by K.-J. Ho and C. B. Taylor in this volume).[8] The critical period when WC birds first show atherosclerotic lesions is when they are between 9 and 12 months old,[9,10] while SR birds develop lesions at an older age.

Possible explanation for proneness of WC to atherosclerosis and resistance of SR was sought in an investigation of lipid metabolism of the birds at prehatch level.[11] At the 1st day of development of the embryo the yolk lipid content of the two strains of pigeons was the same. Triglycerides represented 80% of the total. At pre-hatch significant differences in level of lipids had occurred. Significantly higher levels of phospholipid (PL) remained in yolk of WC. The SR birds' embryos consumed more of all varieties of lipids during

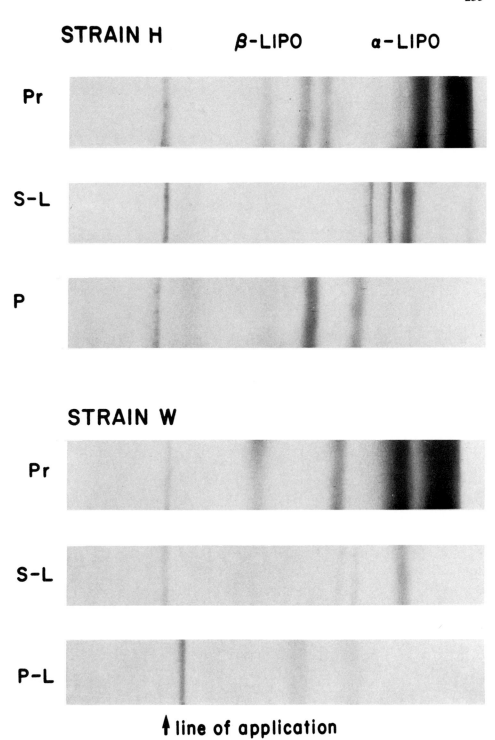

FIGURE 1. Two clearly resolved lipoprotein fractions with the electrophoretic mobility of α-globulin and β-globulin are found in pigeon serum. For comparison, a protein-stained pattern of pigeon serum proteins is included. Note the double albumin fraction.

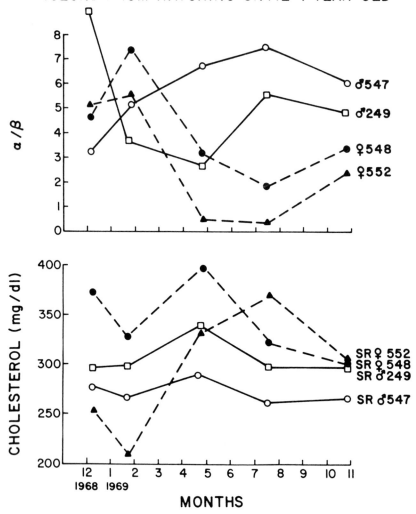

FIGURE 2.   Changes in serum cholesterol concentration during growth.

gestation than the WC embryos. These findings indicate basic difference in fat utilization during embryonic development, which differences may persist throughout life. The failure to detect differences in plasma lipid and lipoprotein levels to explain atherogenic proneness in the adult birds may be due to rapid turnover of the atherogenic components.

Wagner[12] compared serum cholesterol, uric acid, calcium, phosphorus, adrenal gland, and thyroid weights in two strains of WC pigeons, WC-2 (an atherosclerotic prone group), and RBWC (randomly bred white Carneau with only usual proneness to the disease). No significant differences in any of the parameters studied were found. It was suggested that the proneness of the WC-2 group to atherosclerosis must lie in different areas, possibly the metabolism of the artery. It is also possible that differences in plasma lipoprotein composition, which were not measured, could have existed.

It has been known since 1966 when Lewis, et al.[13] published studies on excessively obese patients that ileal by-pass operation in the human being is followed by a decrease in serum cholesterol level of approximately 40% and that the decrease is chiefly in β-Lp fraction.

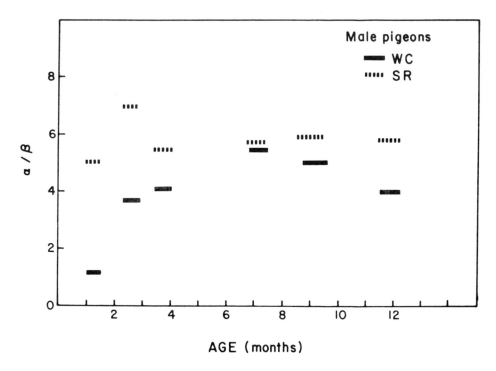

FIGURE 3. The ratio of α/β-Lp in the serum of WC and SR pigeons from 1 to 12 months of age shows consistently a greater proportion of lipid-stainable material in the α-Lp than β-Lp of the SR.

Since then with a less extensive by-pass operation the effect similar to that of the early studies on serum lipid has been found to occur without weight reduction.[29] Ileal by-pass surgery was carried out on WC pigeons which had developed early atherosclerotic lesions to determine whether this procedure would result in regression of lesions and what effect it would have on the serum lipid levels.[14] There was a 50% decrease in the cholesterol esters in the aorta of the operated birds, but no change in plasma levels of cholesterol or triglyceride. It thus appears that in this species the by-pass operation resulted in changes at the arterial wall and enzyme levels without obvious alteration at the plasma level.

That hypercholesterolemia is a necessary but not a sufficient factor in itself to enduce atherogenicity in the pigeon was suggested by studies of Young.[2,15,16] Young WC, fed an atherogenic diet, developed lesions, while SR of like age and on similar diet failed to show lesions even when the SR had serum cholesterol levels of 1000 mg/dℓ.

In the pigeon, age of puberty is not affected by a cholesterol-rich diet. Both strains of birds showed similarly elevated serum cholesterol levels when fed the cholesterol-rich diet. The level of arachidonic acid in the serum and liver cholesterol esters of WC (atherosclerotic prone) 8-month-old pigeons was lower than that of SR birds. These differences were not evident in 5-year-old birds.[2,15,16]

Other dietary and nutritional components besides basic nutritional factors, i.e., fat, carbohydrate, and protein, have been implicated in the development of atherosclerosis. Studies have been carried out on pigeons to determine the effect of varying the amount of some of the minerals in the diet of the birds.[17] Development of atherosclerosis and plasma LDL and cholesterol were measured. Calcium, magnesium, cadmium, and lead were elements selected. Lead and/or cadmium in the drinking water increased the number and size of the atherosclerotic plaques. While LDL and cholesterol were altered the changes did not seem to be directly related to plaque size. Calcium also increased LDL levels and atherosclerosis was more extensive. The study emphasizes the multifactorial nature of lipid metabolism and the etiology of atherosclerosis.

Since serum lipid levels do not serve in the pigeon as a good index of degree of atherosclerotic involvement in the arteries other measures were sought. An investigation of skin sterol concentration in pigeons was conducted and showed that the skin sterol concentration remained constant with increasing age, while the concentration in the arteries was augmented.[18] No satisfactory measure has been found yet to define the degree of arterial plaque involvement.

Diets containing 0.5% added cholesterol caused as great increase in serum cholesterol levels of WC pigeons as was observed with 1 or 2% cholesterol diets. Additional treatment of the birds with exposure to carbon monoxide supplement in air to achieve CO Hb levels of 20% resulted in increased plasma and aortic cholesterol levels, but the aortic triglyceride levels decreased. No lipoprotein studies were reported.[19] This study gives additional evidence of the suitability of this species for a wide variety of stress.

While many of the investigations using pigeons as an experimental model for atherosclerosis have limited the plasma constituents analyzed to the cholesterol and sometimes triglyceride and phospholipid, Mao et al.[20] have carefully studied the physical, chemical, and immunochemical properties of apolipoprotein A-1 of pigeon plasma high-density lipoproteins (see Table 1).

Selection utilizing a positive assortive mating plan for two strains of pigeons, WC and SR, resulted in development of third generation of birds of WC with significantly higher serum cholesterol levels (390 ± 8.3 mg/dℓ) than in initial generation (321 ± 5.6 dℓ); a low-level group was also obtained with a level of 297 ± 6.7. The high level of the racing homer of the third generation was 405 ± 9.2 mg/dℓ which was significantly higher than the initial level of 311 ± 6.0 mg/dℓ. The ability to selectively breed the pigeon to have serum lipid levels of a preselected level increases the potential usefulness of this bird in atherogenic research.[12]

The pigeon is a species which will prove valuable in permitting the researcher to obtain greater understanding of the many facets of metabolism important in development or prevention of atherosclerosis. Because of the unique nature in susceptibility of certain strains to the disease and ease of modifying this by breeding, diet, and other stress mechanisms, the pigeon will maintain the interest of future investigators.

REFERENCES

1. **Clarkson, T. B., Pritchard, R. W., Netsky, G., and Lofland, H. B.,** Atherosclerosis in pigeons. Its spontaneous occurrence and resemblance to human atherosclerosis, *Arch. Pathol.,* 68, 143, 1959.
2. **Young, F. and Middleton, C. C.,** Correlation of arachidonic acid of sterol esters with resistance and susceptibility to naturally occurring atherosclerosis in pigeons, *Proc. Soc. Exp. Biol. and Med.,* 123, 816, 1966.
3. **Lofland, H. B., Jr., Clarkson, T. B., Rhyne, A., and Goodman, H.,** Interrelated effects of dietary fats and proteins on atherosclerosis in the pigeon, *J. Atheroscler. Res.,* 6, 395, 1966.
4. **Lofland, H. B., Jr. and Clarkson, T. B.,** Interrelated effects of nutritional factors on serum lipids and atherosclerosis, in *Dairy Lipids and Metabolism,* Brink, M. F. and Kretchevsky, D., Eds., Westport, Conn., 1968, 135—148.
5. **Lewis, L. A.,** Thin-layer starch gel electrophoresis. A simple accurate method for characterization of protein components, *Clin. Chem.,* 12, 596, 1966.

6. **Lewis, L. A., Smith, S. C., and Hazard, J. B.,** Serum lipids, lipoproteins and thyroid studies in pigeons, *Circulation,* 7(Suppl. III), 18, 1970.

7. **Langelier, M., Connelly, P., and Subbiah, M. T.,** Plasma lipoprotein profile and composition in White Carneau and Show Racer breeds of pigeons, *Can. J. Biochem.,* 54(1), 27, 1976.

8. **Lewis, L. A. and Naito, H. K., Eds.,** *CRC Handbook of Electrophoresis,* Vol. 4, CRC Press, Boca Raton, Fla., 1983.

9. **Subbiah, M. T., Unni, K. K., Kottke, B. A., Carlo, I. A., and Dinh, D. M.,** Arterial and metabolic changes during the critical period of spontaneous sterol accumulation in pigeon aorta, *Exp. Mol. Pathol.,* 24(3), 287, 1976.

10. **Subbiah, M. T., Rymaszewski, Z., Carlo, I. A., and Kottke, B. A.,** Comparison of age related alterations in arterial sterol accumulation, blood pressure and cholesterol balance in spontaneously atherosclerosis-susceptible and atherosclerosis-resistant pigeons, *Adv. Exp. Med. Biol.,* 82, 249, 1977.

11. **Cramer, E. B. and Smith, S. C.,** Yolk lipids of developing atherosclerosis susceptible White Carneau and atherosclerosis resistant Show Racer pigeon embryos, *J. Nutr.,* 106(5), 627, 1976.

12. **Wagner, W. D.,** Risk factors in pigeons genetically selected for increased atherosclerosis susceptibility, *Atherosclerosis,* 31(4), 453, 1978.

13. **Lewis, L. A., Turnbull, R. B., and Page, I. H.,** Effects of jejunocholic shunts on obesity, serum lipoproteins, lipids and electrolytes, *Arch. Int. Med.,* 117, 4, 1966.

14. **Ravi Subbiah, M. T., Dicke, B. A., Kottke, B. A., Carlo, I. A., and Dinh, D. M.,** Regression of naturally occurring atherosclerotic lesions in pigeon aorta by intestinal bypass surgery. Early changes in arterial cholesteryl ester metabolism, *Atherosclerosis,* 31(2), 117, 1978.

15. **Young, F.,** Fatty acid composition of serum lipids in 5-year White Carneau and Show Racer Pigeons, *Proc. Soc. Exp. Biol. Med.,* 130, 980, 1969.

16. **Young, F.,** Serum lipids of young Show Racer and White Carneau pigeons fed a semipurified diet with or without cholesterol, *J. Nutr.,* 104(6), 719, 1974.

17. **Revis, N. W., Major, T. C., and Horton, C. Y.,** The effects of calcium, magnesium, lead, or cadmium on lipoprotein metabolism and atherosclerosis in the pigeon, *J. Environ. Pathol. Toxicol.,* 4(2—3), 293, 1980.

18. **Rymaszewski, Z. J., Subbiah, M. T., and Kottke, B. A.,** Effect of age on sterols and steryl esters in pigeon skin and aorta, *Exp. Gerontol.,* 10(6), 313, 1975.

19. **Turner, D. M., Lee, P. N., Roe, P. J., and Goush, K. J.,** Atherogenesis in the White Carneau pigeon. Further studies of role of carbon monoxide and dietary cholesterol, *Atherosclerosis,* 34(4), 407, 1934.

20. **Mao, S. J., Downing, M. R., and Kottkeba, J.,** Physical, chemical, and immunochemical studies of apolipoprotein A-I from pigeon plasma high density lipoproteins, *J. Biochem. Acta,* 620, 100, 1980.

21. **Donterre, R. F., Wight, T. N., Sumar, S. C., and Brannigan, D.,** Spontaneous atherosclerosis in pigeons. A model system for studying metabolic parameters associated with atherosclerosis, *Am. J. Pathol.,* 67, 1, 1972.

22. **Patton, N. M., Brown, R. V., and Middleton, C. C.,** Familial cholesterolemia in pigeons, *Atherosclerosis,* 19, 307, 1974.

23. **Young, F.,** Serum lipids of young show racer and white carneau pigeons fed a semipurified diet with or without cholesterol, *J. Nutr.,* 104, 719, 1974.

24. **Flynn, K. J., Schumacher, J. F., Ravisubbiah, M. T., and Kotlke, B. A.,** The effect of ileal, bypass on sterol balance and plasma cholesterol in the White Carneau pigeon, *Atherosclerosis,* 24, 75, 1976.

25. **Clarkson, T. B., Prichard, R. W., Bullock, B. C., St. Clair, R. W., Lehner, N. D., Jones, D. C., Wagner, W. D., and Rudel, L. L.,** Pathogenesis of atherosclerosis; some advances from using animal models, *Exp. Mol. Pathol.,* 24(3), 264, 1976.

26. **Wagner, W. D. and Clarkson, T. B.,** Effect on regression potential of atherosclerosis produced by intermittent or continuous hypercholesterolemia, *Atherosclerosis,* 27(3), 369, 1977.

27. **Lofland, H. B. and Clarkson, T. B.,** Genetic studies of atherosclerosis in animals, *Adv. Exp. Med. Biol.,* 82, 576, 1977.

28. **Hrafchak, B. B., Bond, M. G., Wood, L. L., and Hostetler, J. R.,** Time course of diet-exacerbated carotid artery atherogenesis in the White Carneau pigeon, *Atherosclerosis,* 39(3), 243, 1980.

TURKEY SERUM LIPOPROTEINS

P. Alaupovic and N. Dashti

INTRODUCTION

The selection of turkey as a preferred species for studying avian plasma lipoproteins was prompted by the paucity of data on the chemical composition and nature of this macromolecular system and by the large body size and availability of adequate blood volumes from individual birds.[1] Like other avian species, the turkey lipoprotein system is very sensitive to dietary and hormonal manipulations.[2-5] Turkeys fed cholesterol-supplemented diets have significantly increased levels of plasma cholesterol in comparison with control birds.[3,5] A single dose of injected diethylstilbestrol results in an altered plasma lipoprotein pattern characterized by a significant increase in triglyceride-rich lipoproteins and a virtual absence of β- and α-lipoproteins (β-Lp and α-Lp).[4] Turkeys are among the most hypertensive vertebrates with systolic pressures ranging between 190 and 250 mm Hg.[6,7] The susceptibility of Broad-Breasted white turkeys to the development of spontaneous dissecting aneurysms of the abdominal aorta has been attributed mainly to hypertension and weakening of the aortic wall.[8-10] For these reasons, the turkey also represents a very useful animal model for studying relationships between hypertension, hyperlipoproteinemia, and atherosclerosis.[3,5,9,11]

There are some important differences in the metabolism and transport of lipids between the avian and mammalian species. Birds synthesize most of their fatty acids and triglycerides in the liver rather than in adipose tissue.[12-14] Borron and Britton demonstrated that in Large White turkeys the incorporation of acetate into liver lipids was nearly 200 times greater than that for adipose tissue.[15] Another difference pertains to the absorption of exogenous fatty acids. Whereas in mammals the exogenous long-chain fatty acids are transported from the intestine into the systemic circulation through the lymphatic system as chylomicrons and large very low-density lipoproteins (VLDL), in the avian species this process occurs via the portal vein.[16,17] Bensadoun and Rothfeld have suggested that, by analogy with mammalian chylomicrons, these avian intestinal lipoproteins rich in exogenous triglycerides be referred to as portomicrons.[17]

The onset of egg production in birds exerts a more profound effect on plasma lipoproteins than ovulation in mammalian species. In response to increased levels of circulating estrogens prior to egg laying, there is an elevated hepatic production of several plasma proteins and lipoproteins which seem to be barely detectable in male or sexually immature female birds.[18-20] These include the plasma egg yolk protein vitellogenin with its subcomponents phosvitin and lipovitellin,[18] and plasma apolipoprotein VLDL-II[19-21] (apo VLDL-II) and apolipoprotein B[22] (apo B), the protein moieties of VLDL, and low-density lipoproteins (LDL). Since there are significant quantitative and, possibly, qualitative differences in the lipoprotein profiles between male and sexually mature, laying female birds, it seems appropriate to discuss lipoprotein systems of male and female turkeys separately.

PLASMA LIPOPROTEINS OF MALE TURKEYS

Plasma cholesterol levels of male turkeys seem to depend to some extent on factors such as age, diet, and heredity. The reported values for the Broad-Breasted white turkeys, 8 to 14 weeks of age, ranged between 103 and 110 mg/dℓ;[5,23] plasma cholesterol of Broad-Breasted white turkeys, 52 weeks of age, was reported to be 138 ± 28 mg/dℓ.[7] Plasma cholesterol values for White Holland turkeys, 42 to 71 weeks of age, varied from 93 to 138 mg/dℓ with an average value of 115 ± 11.5 mg/dℓ.[24] In Broad-Breasted Bronze turkeys,

Table 1
CHEMICAL COMPOSITION (%) OF MAJOR LIPOPROTEIN DENSITY CLASSES FROM MALE TURKEYS

			Cholesterol			
Density class	Protein	Phospholipid	Free	Ester	Glycerides	Ref.
VLDL[a] (n = 4)	13.7	21.7	3.6	26.6	34.2	1
VLDL[b]	12.6	12.8	3.2	15.4	56.0	11
LDL (n = 4)	26.0	26.2	5.2	38.2	5.1	1
LDL	19.3	20.7	9.0	42.3	8.6	11
HDL (n = 4)	58.5	26.4	2.9	8.7	3.5	1
HDL	49.2	29.4	3.2	16.1	2.2	11

[a]   In both studies, the VLDL were isolated at d < 1.006 g/mℓ, LDL at d 1.006 to 1.063 g/mℓ and HDL at d 1.063 to 1.21 g/mℓ.

[b]   Density classes were isolated from a serum pool of ten birds.

the plasma cholesterol levels increased from an average value of 148 mg/dℓ at 8 weeks of age to 249 mg/dℓ at 16 weeks of age; after that age the plasma cholesterol levels remained constant.[25] A similar age dependence was also reported for free cholesterol levels which increased from a value of 65 mg/dℓ at 12 weeks of age to a value of 81 mg/dℓ at 20 weeks of age.[2] Results of our previous and more recent studies showed that, similar to man and other mammalian and avian species, approximately 75 to 80% of total plasma cholesterol is present in the esterified form.[1,26] The Broad-Breasted white turkeys are characterized by relatively low levels of plasma triglycerides ranging from 38 ± 10[26] to 55 ± 10 mg/dℓ.[7] In these same studies, the plasma phospholipid levels were found to be 273 ± 48 and 283 ± 38 mg/dℓ, respectively.

The distribution of the major lipoprotein density classes in male turkeys is very similar to that of other avian species.[27-29] High density lipoproteins (HDL) account for over 75% of the total lipoprotein content, whereas LDL and VLDL comprise approximately 20% and 5% of the lipoproteins, respectively.[1] These results are very similar to those based on densitometric quantification of α_1-, β-, and pre-β-Lp bands separated by agarose gel electrophoresis.[11] In Broad-Breasted white turkeys, the concentration of VLDL was found to be minimal ranging between 10 to 50 mg/dℓ.[1] In contrast, Mills and Taylaur[29a] detected a very high concentration of VLDL (2260 mg/dℓ) in an unspecified breed of turkeys. Whether this discrepancy is due to a difference in diet, strain, or sex of turkey is not known. It should be pointed out, however, that such high concentrations of VLDL are characteristic for laying chickens and estrogenized roosters.[18,27,29a,30,31]

Results of two separate studies on the percent chemical composition of major lipoprotein density classes are shown in Table 1. In both studies, the protein and phospholipid contents of lipoprotein density classes increased and glyceride content decreased with increasing densities. Glycerides were the major lipid constituent of VLDL, while cholesterol esters and phospholipids were the predominant lipid components of LDL and HDL, respectively. In all three density classes, glycerides consisted of approximately equal amounts of triglycerides and diglycerides.[1] There were, however, some differences between these two studies. In general, all three density classes examined by Kelley and Alaupovic[1] had higher percent content of protein than those studied by Pagnan et al.[11] Another difference was in the lipid composition of VLDL. In the former study, the phospholipid and cholesterol ester contents were higher and the triglyceride content lower than those in the latter study. Whether this

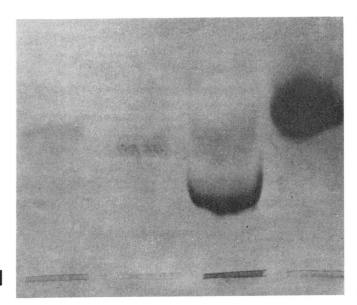

+

ORIGIN

−

VLDL   LDL₁   LDL₂   HDL

FIGURE 1.   Agarose gel electrophoresis of the major lipoprotein density classes of male turkeys. The gels were stained for lipids with oil red O.

difference could be attributed to different dietary regimens and/or age of the birds is not known. The phospholipid composition of the lipoprotein density classes was characterized by phosphatidylcholine as the major constituent and sphingomyelin, lysophosphatidylcholine, and phosphatidylethanolamine as the minor constituents.[1] The distribution of phosphatides in VLDL and LDL differed from that in HDL. The content of phosphatidylcholine (72%) was higher and the contents of lysophosphatidylcholine (6.5%) and sphingomyelin (13%) were lower in HDL than in VLDL or LDL. In these two latter density classes the percent content of phosphatidylcholine was 58 to 59%, while those of sphingomyelin and lysophosphatidylcholine were 22 to 28% and 13%, respectively. In general, the qualitative and quantitative phospholipid composition of turkey serum and lipoprotein density classes was found to be similar to those of oher avian and mammalian species including man.[32]

The relatively low flotation coefficient (S_f 33) of male turkey VLDL indicated the presence of small lipoprotein particles and reflected a chemical composition characterized by a relatively low content of glycerides (34%, Table 1) and relatively high percentages of cholesterol esters (27%), phospholipids (22%), and protein (14%). In contrast to VLDL, the mean flotation coefficient of turkey LDL (S_f 7.4) only differed slightly from that of human male LDL (S_f 6.3).[33] In accordance with these flotation rates, the chemical composition of turkey LDL was very similar to that of a human LDL subfraction of d 1.030 to 1.053.[34] The flotation rate of turkey HDL ($S_{f1.21}$ 4.2) was higher than the flotation rate of its human counterpart ($S_{f1.21}$ 2.0)[33] reflecting most probably the presence of some apo B-containing lipoprotein particles with densities greater than 1.063 g/mℓ.[1]

Electrophoretic patterns of the major density classes of male turkeys in 1% agarose are shown in Figure 1. Both VLDL and LDL₁ migrated faster (pre-β bands) than the main LDL₂ band; the migration rate of VLDL was slightly higher than that of LDL₁. The LDL₂ subfraction was resolved into two bands. The major band migrated to the β-globulin position and the minor band to the pre-β position. The immunoelectrophoretic pattern of LDL₂ showed

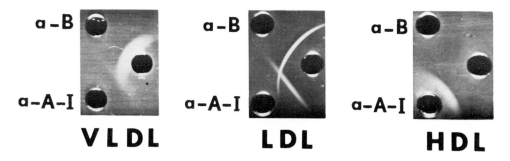

FIGURE 2.   Immunodiffusion patterns of male turkey lipoprotein density classes. The VLDL, LDL, and HDL samples were placed in the middle right well of the 1% agarose gel. The placement of antisera is indicated in each pattern; a, anti.

that the major band contained apo B and the minor band contained apolipoprotein A (apo A) as their protein moieties. In contrast, VLDL and LDL_1 only reacted with antibodies to apo B. Thus, the minor, faster migrating band of LDL_2 is actually a slower moving α-Lp. The HDL particles moved to the α_1-globulin position. It should be pointed out, however, that some turkey HDL_2 preparations contained α_1-Lp as their main constituent and β-Lp as their minor constituent.[1]

The immunochemical characterization of the major density classes was carried out with antisera to apo B and apolipoprotein A-I (apo A-I). As presented in Figure 2, the VLDL particles gave a strong, positive reaction with antibodies to apo B and a weak reaction with an antiserum to apo A-I. In agreement with the results of agarose gel electrophoresis, the double diffusion analysis of LDL showed the presence of precipitin lines with antibodies to both apo B and apo A-I. Furthermore, the nonidentity reaction between the apo B and apo A-I precipitin lines indicated that these two apolipoproteins resided on separate lipoprotein particles. The HDL particles only reacted with antibodies to apo A-I.

Studies on chicken[27-29,30,32] and turkey[7] lipoproteins have shown that both species contain apolipoproteins homologous with human apo A and apo B. It has also been established that VLDL particles from laying hens or estrogenized roosters contain another apolipoprotein referred to as apo VLDL-II.[19-21] The existence of a mammalian or human apolipoprotein homologous with the avian apo VLDL-II has not been reported.

In an attempt to characterize the major apolipoproteins of turkey lipoproteins, Kelley and Alaupovic used LDL as the starting material for the isolation of apo B and HDL_3 for the isolation of apo A.[1] To remove small amounts of apo A present in LDL, this density class was totally delipidized and the residual protein moiety was extracted with aqueous buffers. The insoluble apo B was reduced and carboxymethylated, and the carboxymethylated apo B was purified by chromatography on a Bio-Gel® A-5m column equilibrated with 6 *M* guanidine HCl. Elution of carboxymethylated apo B at the void volume indicated clearly a marked tendency of apo B towards aggregation even in a dissociating solvent. This apo B preparation contained no apo A peptides or any other proteins or apolipoproteins detectable by polyacrylamide gel electrophoresis (PAGE). Amino acid composition of apo B was characterized by aspartic acid, glutamic acid, and leucine as the major amino acids. The content of half-cystine was 15.6 mol/1000 mol. All attempts to determine the N- and C-terminal amino acids failed. The carbohydrate analysis of apo B revealed mannose, galactose, and galactosamine as the main sugar constituents. Based on the solubility properties, electrophoretic behavior, distribution in density classes, and amino acid and carbohydrate composition, the turkey apo B resembles closely the apo B preparations isolated from both the chicken[29,30] and human LDL.[34,35]

The HDL_3 subfraction gave a single precipitin line with its corresponding antiserum.

However, after total delipidization, this HDL subfraction showed the presence of two immunoprecipitin lines. Since there was no reaction with antibodies to apo B, this finding indicated that HDL$_3$ contained at least two apolipoprotein constituents. The separation of these polypeptides was achieved by column chromatography of totally delipidized HDL$_3$ on Bio-Gel® A-5m.[1] The major polypeptide migrated on SDS-PAGE to the middle part of the gel, while the minor polypeptide moved near the dye front. The major apolipoprotein with a molecular weight of approximately 27,000 daltons was designated as apo A-I and the minor apolipoprotein with a molecular weight of 10,000 was named apo A-II. The designation of these two apolipoproteins as nonidentical polypeptides of apo A was based on the same criteria that were applied to human apo A-I and apo A-II.[36] The amino acid composition of apo A-I was characterized by the absence of half-cystine. The C-terminal amino acid of apo A-I was found to be alanine and the N-terminal amino acid was identified as aspartic acid. The apo A-I contained no sugar constituents. The apo A-II was also characterized by the absence of half-cystine. The C-terminal amino acid was identified as alanine. However, the N-terminal amino acid of apo A-II was blocked. There was no evidence for the presence of a carbohydrate moiety. It is quite remarkable that, despite a distant phylogenetic relationship, several of the physical and chemical properties of turkey apo A-I and apo A-II are quite similar to those of human apo A-I and apo A-II. These similarities for apo A-I include molecular weight, electrophoretic mobility in urea-PAGE and SDS-PAGE, amino acid composition, the N-terminal amino acid, homology of the N-terminal sequence, and the absence of a carbohydrate moiety. The major difference is the substitution of alanine for glutamine as the C-terminal amino acid. While glutamine is the C-terminus of human,[37] primate,[38,39] and pig[40] apo A-I, alanine is also the C-terminal amino acid of chicken[29] and rat[41] apo A-I. The turkey apo A-II is similar to human apo A-II in the amino acid composition, blocked N-terminal amino acid and the absence of a carbohydrate moiety. It differs, however, from its human counterpart in the molecular weight and C-terminal amino acid. Turkey apo A-II lacks half-cystine, and like those of rhesus[38] and rat,[42] but unlike those of human[37] and chimpanzee[39] exists in the monomeric form. For this reason, it migrates on SDS-PAGE faster than human apo A-II. Whereas human[37] and primate[38,39] apo A-II contain glutamine as the C-terminal amino acid, turkey apo A-II contains alanine as the C-terminus. So far, there have been no reports on the isolation or identification of avian apolipoproteins homologous with human apolipoproteins C, D, E, F, and G.[36] Kelley and Alaupovic observed, however, that a turkey HDL$_3$ fraction eluted from the Bio-Gel® A-5m-column after the elution of fractions containing apo A-I and A-II activated a human plasma lipoprotein lipase preparation.[1] It was suggested on the basis of this observation that, at least in some animal species, the presence of apo C may only be detectable by one of its functional properties, i.e., activation of lipoprotein lipase.

To account for the apolipoprotein heterogeneity of ultracentrifugally isolated lipoprotein density classes of human plasma, Alaupovic and co-workers have proposed that the plasma lipoprotein system consists of distinct lipoprotein families or particles which may be recognized and identified on the basis of their apolipoprotein composition.[36,37,43] Lipoprotein families which contain a single apolipoprotein are referred to as the free forms of lipoprotein families or *simple lipoproteins*, and those which contain several apolipoproteins are designated as the associated forms of lipoprotein families or *complex lipoproteins*. Free forms of lipoprotein families constitute, thus, the simplest physicochemical entities of the plasma lipoprotein system. Human lipoproteins contain at least eight simple lipoprotein families (lipoproteins A, A-I, B, C, D, E, F, and G) and a number of complex lipoproteins.[36] In comparison, the turkey lipoproteins represent a rather simple macromolecular system of antigenically defined lipoprotein families. Results of distribution studies showed that each major density class contained both apo A and apo B. As already pointed out, the plasma levels of a polypeptide(s) analogous with human apo C and an estrogen-induced apolipoprotein, apo VLDL-II, were too small for an unequivocal chemical identification. The lipoprotein forms of turkey apo A and apo B were established by use of affinity chroma-tography of major density classes on concanavalin A-Sepharose.[44] Since concanavalin A only binds lipoproteins which contain apo B,[45] this procedure was found to be most suitable for a selective separation of these lipoprotein particles. The retained fractions of VLDL, LDL, and HDL only contained lipoprotein B (LP-B), whereas unretained fractions only consisted of lipoprotein A (LP-A).[44] In VLDL and LDL, LP-B was the major and LP-A the minor lipoprotein family. In contrast, LP-A was the major and LP-B the minor lipoprotein family of HDL. LP-B was distributed over a wider density range than LP-A. The major portion (75%) of LP-B was present in LDL, but substantial amounts of this lipoprotein family were also present in VLDL (15%) and HDL (10%). LP-B isolated from LDL particles migrated on agarose gel electrophoresis to the β-globulin position and displayed a single, symmetrical peak in the analytical ultracentrifuge with a flotation coefficient, S_f, of 10.3. The hydrated density of this LP-B preparation was 1.026 g/mℓ. The chemical composition of LP-B from LDL and HDL was very similar except for the higher protein content of the LP-B preparation from HDL. In fact, all LP-B preparations had higher protein and phos-pholipid and lower glyceride and cholesterol ester contents than the LDL preparations; this compositional discrepancy was most probably due to losses of neutral lipids during the elution of retained LP-B from concanavalin A. The LP-A preparations isolated from LDL and HDL were characterized on agarose electrophoresis by a single band with the α_1-globulin mobility. The major portion of LP-A was present in HDL (98%). A small portion of LP-A was found in LDL (2%) and only trace amounts were detected in VLDL. LP-A isolated from HDL displayed a single, symmetrical peak in the analytical ultracentrifuge with a flotation coefficient, $S_{f1.21}$, of 5.9. The hydrated density of this LP-A preparation was 1.168 g/mℓ. LP-A from HDL had more protein and less free and esterified cholesterol than LP-A isolated from DL. In contrast to LP-B, the chemical composition of LP-A was very similar to that of corresponding HDL. The LP-A preparations gave a single precipitin line with antibodies to LP-A. However, upon delipidization, the LP-A gave two precipitin lines with the same antiserum. One of the precipitin lines was due to apo A-I and the other to apo A-II. The presence of these two nonidentical polypeptides of apo A was also verified by SDS-PAGE of totally delipidized LP-A. Since the structural integrity of LP-A depends on the presence of both polypeptides, apo A-I and apo A-II represent integral constituents of apo A. There is no evidence for the occurrence in turkey plasma of lipoprotein particles with either apo A-I or apo A-II as the only protein constituent.

In summary, the male turkey plasma lipoprotein system consists of LP-A as the major and LP-B as the minor lipoprotein family. The LP-A particles occur almost exclusively at densities greater than 1.073 g/mℓ, whereas LP-B particles are present throughout a density spectrum ranging from approximately 0.94 to 1.12 g/mℓ. The presence of only trace amounts of portomicrons and low concentrations of VLDL particles are most probably due to a very efficient lipolytic system characteristic of male and nonlaying birds. The LP-A and LP-B do not form associations but exist as separate, discrete lipoprotein particles. The protein moiety of LP-A consists of two nonidentical polypeptides, apo A-I and apo A-II. These two polypeptides are integral components of apo A. The LP-B particles contain apo B as their protein moiety. Male turkeys also have trace amounts of the estrogen induced apolipoprotein, apo VLDL-II, and an apolipoprotein(s) which activates the lipoprotein lipase and, thus, may be homologous with human apo C.

PLASMA LIPOPROTEINS OF FEMALE TURKEYS

The initiation of egg production brings about profound changes in the plasma lipoprotein system of the avian species.[18,32] A highly significant increase in the levels of plasma VLDL is accompanied by a decrease in the concentration of HDL and a small increase, if any, in the levels of LDL.[27,30] These changes in the concentrations of lipoprotein density classes, especially a dramatic elevation of VLDL particles, reflect primarily an increased hepatic production of triglycerides and apo B and VLDL-II.[19-22, 46] Synthetic rates and plasma levels of these lipoprotein constituents are regulated by circulating estrogens.[18,47] Bacon et al. established that the total plasma estrogen levels of turkey hens in their prelaying, laying and postlaying periods correlated significantly with plasma neutral lipids and VLDL;[47] laying hens with their highly elevated VLDL also had the highest levels of total plasma estrogen. In addition to plasma lipids and apolipoproteins, estrogens also induce the hepatic synthesis of vitellogenin, a macromolecular, phosphorus-rich complex which includes phosvitin and lipovitellin as its subcomponents.[18] All these macromolecular compounds are transported through the systemic circulation to the ovary where they traverse the capillary and are deposited in the egg yolk.[18] It seems that the VLDL particles are transferred from plasma to egg yolk in a relatively intact form.[48] Although intended for deposition in the hen's egg yolk, these proteins and lipoproteins can also be induced by estrogen administration in male birds.[2,4,18,19,22]

A systematic study on the physical and chemical properties of laying turkeys has not yet been reported in the literature. Bacon et al. established that the laying hens had significantly higher concentrations of neutral lipids and VLDL than the sexually immature hens.[47] In a study of Eastern Wild turkeys, Lisano and Kennamer found that the plasma cholesterol levels of female birds (159.8 ± 38.1 mg/100 mℓ) were significantly higher than those of male birds (123.2 ± 26.7 mg/100 mℓ).[49] Furthermore, the plasma cholesterol levels of female turkeys increased from January (157.8 ± 66.9 mg/100 mℓ) to February (182 ± 29.8 mg/100 mℓ) coinciding with the beginning of egg-laying season. Results of our preliminary studies have shown significant differences in the levels of plasma lipids and apolipoproteins between laying and nonlaying turkeys.[26,50] The plasma levels of cholesterol (111 ± 18 mg/100 mℓ), triglyceride (68 ± 24 mg/100 mℓ), phospholipid (275 ± 25 mg/100 mℓ), apo A-I (84 ± 21 mg/100 mℓ) and apo B (24 ± 4 mg/100 mℓ) of nonlaying turkeys are similar to those of male turkeys. However, at the onset of egg production, there is a highly significant increase in the levels of triglycerides (1985 ± 825 mg/100 mℓ), cholesterol (173 ± 65 mg/100 mℓ), phospholipids (1200 ± 350 mg/100 mℓ), and apo B (185 ± 49 mg/100 mℓ) and a reduction in the levels of apo A-I (39 ± 17 mg/100 mℓ). The accumulation of triglyceride-rich lipoproteins (VLDL) in the plasma of laying turkeys is accompanied by a decrease in the levels of HDL with little or no change in the concentration of LDL. As shown in Figure 3, the 1% agarose-gel electrophoretic patterns of the major lipoprotein density classes of sexually immature female turkeys are very similar to those of male turkeys (Figure 1). The only exception is a somewhat slower mobility of the VLDL band (pre-β-Lp) of nonlaying birds in comparison with the corresponding VLDL band of male turkeys. Like the male turkey density classes, LDL$_1$ of nonlaying turkeys contain a single pre-β-Lp band, LDL$_2$ consist of a major β-Lp band and a minor α$_1$-Lp band (Figure 3). In contrast, VLDL of laying turkeys display a single, very slow migrating band (Figure 4). There is practically no difference in the mobilities of LDL$_1$, LDL$_2$, and HDL bands. The α$_1$- and β-Lp bands of LDL$_2$ from nonlaying female birds are replaced by a single band of intermediate mobility in this density class of laying birds. The α$_1$-Lp band of HDL from laying turkeys has a slower mobility than that of HDL from nonlaying turkeys.

The distribution and relationship of apolipoproteins in the major lipoprotein density classes of laying turkeys was studied by use of antisera to male turkey apo A-I and apo B. In addition, we have isolated from plasma VLDL of laying turkeys an apolipoprotein analogous to apo VLDL-II isolated from plasma VLDL of laying chicken;[21] this apolipoprotein referred to as the estrogen induced protein (EIP) was then used for the preparation of its corresponding

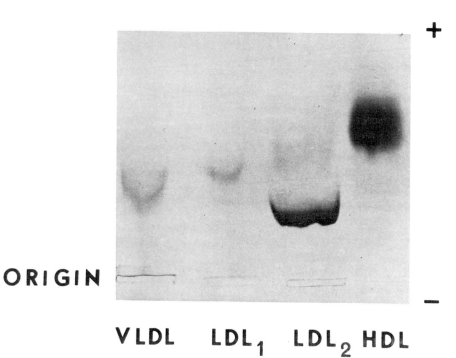

FIGURE 3.   Agarose gel electrophoresis of lipoprotein density classes isolated from plasma of nonlaying female turkeys. The gels were stained for lipids with oil red O.

FIGURE 4.   Agarose gel electrophoresis of major lipoprotein density classes isolated from plasma of laying turkey hens. The gels were stained for lipids with oil red O.

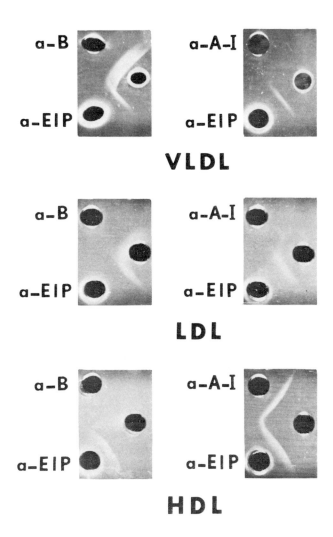

FIGURE 5.   Double diffusion analyses of lipoprotein density classes
isolated from plasma of laying turkeys. VLDL, LDL, and HDL were placed
in the middle right well of the 1% agarose gel and antisera to A-I, B, and
EIP as indicated in the patterns; a, anti.

antiserum. The double diffusion analysis of VLDL from laying turkeys (Figure 5) showed
a positive reaction with antisera to apo B and EIP. The identity reaction between the precipitin
lines of apo B and EIP indicated that these two apolipoproteins most probably reside on the
same lipoprotein molecule. The presence of a second, separate precipitin line of apo B may
be due to the existence of either lipoprotein particles which contain only apo B or lipoprotein
particles which contain apo B and another, as yet unidentified apolipoprotein(s). LDL
particles gave strong positive reactions with antisera to apo B and EIP and a very weak
reaction with an antiserum to apo A-I (Figure 5). Since EIP precipitin line showed a spur
at the junction with the apo B precipitin line, it seems that LDL contain at least two types
of lipoprotein particles. The major lipoprotein contains both apo B and EIP, while a minor
lipoprotein contains only EIP. The double diffusion pattern of HDL was characterized by
positive precipitin lines of apo A-I and EIP; the reaction of HDL with anti-apo B serum
was negative. The strong identity reaction between the precipitin lines of apo A-I and EIP
clearly indicates that HDL of laying turkeys consist of a major, complex lipoprotein which

contains both apo A and EIP as its protein constituents. The strong tendency of EIP to form associations with both apo A-I and apo B represents an unexpected finding and feature of the plasma lipoprotein system of laying turkeys. At the present time, the functional significance, if any, of these complex lipoprotein particles is not known.

In summary, the plasma lipoprotein system of sexually mature, laying turkeys differs qualitatively and quantitatively from that of sexually immature female birds and male turkeys. In addition to apo A-I, A-II, and apo B, the plasma lipoproteins also contain an EIP analogous to chicken apo VLDL-II. The main, if not exclusive, lipoprotein form of triglyceride-rich lipoproteins (portomicrons and VLDL) is a complex lipoprotein which contains both apo B and EIP (LP-BP). The main lipoprotein species of triglyceride-poor lipoproteins (HDL) is a complex lipoprotein (LP-AP) with apo A and EIP as its protein constituents. The LDL particles contain a major lipoprotein with apo B and EIP (LP-BP) and a minor lipoprotein (LP-EIP) with only EIP as its protein moiety. VLDL may contain additional, as yet unidentified, complex lipoproteins of apo B.

ACKNOWLEDGMENT

We thank Mrs. M. Farmer for typing of this manuscript. Some of the studies reviewed in this manuscript were supported in part by grant HL-23181 from the U.S. Public Health Service and by the resources of the Oklahoma Medical Research Foundation.

REFERENCES

1. **Kelley, J. L. and Alaupovic, P.,** Lipid transport in avian species. I. Isolation and characterization of apolipoproteins and major lipoprotein density classes of male turkey serum, *Atherosclerosis,* 24, 155, 1976.
2. **Heller, V. G. and Thayer, R. H.,** Chemical changes in the blood composition of chickens and turkeys fed synthetic estrogens, *Endocrinology,* 42, 161, 1948.
3. **Simpson, C. F. and Harms, R. H.,** Aortic atherosclerosis of turkeys induced by feeding of cholesterol, *J. Atheroscler. Res.,* 10, 63, 1969.
4. **Neilson, J. T. McL. and Simpson, C. F.,** Plasma lipoproteins of turkeys injected with a single dose of diethylstilbestrol, *Atherosclerosis,* 18, 445, 1973.
5. **Krista, L. M., Jackson, S., Mora, E. C., McDaniel, G. R., and Patterson, R. M.,** Blood plasma constituents in 14-week-old hypertensive and hypotensive strains of turkeys, *Poult. Sci.,* 57, 1022, 1978.
6. **Krista, L. M., Waibel, P. E., Shoffner, R. N., and Sautter, J. H.,** A study of aortic rupture and performance as influenced by selection for hypertension and hypotension in the turkey, *Poult. Sci.,* 49, 405, 1970.
7. **Pagnan, A., Pessina, A. C., Thiene, G., and Dal Palu, C.,** The natural history of hypertension in turkeys, *Clin. Sci. Mol. Med.,* 55, 213s, 1978.
8. **Pritchard, W. R., Henderson, W., and Beall, C. W.,** Experimental production of dissecting aneurysms in turkeys, *Am. J. Vet. Res.,* 19, 696, 1958.
9. **Clarkson, T. B.,** Atherosclerosis-spontaneous and induced, *Adv. Lipid Res.,* 1, 211, 1963.
10. **Kurtz, H. J.,** Aortic rupture in turkeys: lesions in tunica media and tunica adventitia associated with intimal plaques, *Am. J. Vet. Res.,* 30, 101, 1969.
11. **Pagnan, A., Thiene, G., Pessina, A. C., and Dal Palu, C.,** Serum lipoproteins and atherosclerosis in hypertensive broad breasted white turkeys, *Artery,* 6, 320, 1980.
12. **O'Hea, E. K. and Leveille, G. A.,** Lipid biosynthesis and transport in the domestic chick *(Gallus domesticus), Comp. Biochem. Physiol.,* 30, 149, 1969.
13. **Schrago, E., Glennon, J. A., and Gordon, E. S.,** Comparative aspects of lipogenesis in mammalian tissue, *Metabolism,* 20, 54, 1971.
14. **Evans, A. J.,** *In vitro* lipogenesis in the liver and adipose tissue of the female Aylesburry duck at different ages, *Brit. Poult. Sci.,* 13, 595, 1972.

15. **Borron, D. C. and Britton, W. M.,** The significance of adipose tissue and liver as sites of lipid biosynthesis in the turkey, *Poult. Sci.,* 56, 353, 1977.

16. **Noyan, A., Lossow, W. J., Brot, N., and Chaikoff, I. L.,** Pathway and form of absorption of palmitic acid in the chicken, *J. Lipid Res.,* 5, 538, 1964.

17. **Bensadoun, A. and Rothfeld, A.,** The form of absorption of lipids in the chicken, *Gallus domesticus, Proc. Soc. Exp. Biol. Med.,* 141, 814, 1972.

18. **Schjeide, O. A.,** Lipoproteins of the fowl-serum, egg and intracellular, in *Progress in the Chemistry of Fats and Other Lipids,* Vol. 6, Holman, R. T., Lundberg, W. O., and Malkin, T., Eds., The Macmillan Company, New York, 253, 1963.

19. **Chan, L., Jackson, R. L., and Means, A. R.,** Regulation of lipoprotein synthesis. Studies on the molecular mechanisms of lipoprotein synthesis and their regulation by estrogens in the cockerel, *Circ. Res.,* 43, 209, 1978.

20. **Blue, M.-L. and Williams, D. L.,** Induction of avian serum apolipoprotein II and vitellogenin by tamoxifen, *Biochem. Biophys. Res. Commun.,* 98, 785, 1981.

21. **Jackson, R. L., Lin, H.-Y., Chan, L., and Means, A. R.,** Amino acid sequence of a major apoprotein from hen plasma very low density lipoproteins, *J. Biol. Chem.,* 252, 250, 1977.

22. **Luskey, K. L., Brown, M. S., and Goldstein, J. L.,** Stimulation of the synthesis of very low density lipoproteins in rooster liver by estradiol, *J. Biol. Chem.,* 249, 5939, 1974.

23. **Kelley, J. L., Ulrich, T., Thayer, R. H., and Alaupovic, P.,** Alterations in Plasma Apolipoproteins and Lipoprotein Density Classes of Cholesterol-Fed Turkeys, Abstracts of Papers, 31st Annu. Meet., Council on Arteriosclerosis, American Heart Association, Miami Beach, Fla., November 28 to December 1, 1977, 21.

24. **Kirschner, N., Pritham, G. H., Bressler, G. O., and Goodeuk, S., Jr.,** Composition of normal turkey blood, *Poult. Sci.,* 30, 875, 1951.

25. **Speckmann, E. W. and Ringer, R. K.,** The influence of reserpine on plasma cholesterol, hemodynamics and arteriosclerotic lesions in the Broad Breasted Bronze turkey, *Poult. Sci.,* 41, 40, 1962.

26. **Dashti, N. and Alaupovic, P.,** unpublished results, 1982.

27. **Yu, J. Y.-L., Campbell, L. D., and Marquardt, R. R.,** Immunological and compositional patterns of lipoproteins in chicken (*Gallus domesticus*) plasma, *Poult. Sci.,* 55, 1626, 1976.

28. **Kruski, A. W. and Narayan, K. A.,** Effect of orotic acid and cholesterol on the synthesis and composition of chicken (*Gallus domesticus*) serum lipoproteins, *Int. J. Biochem.,* 7, 635, 1976.

29. **Hillyard, L. A., White, H. M., and Pangburn, S. A.,** Characterization of apolipoproteins in chicken serum and egg yolk, *Biochemistry,* 11, 511, 1972.

29a. **Mills, G. L. and Taylaur, C. E.,** The distribution and composition of serum lipoproteins in eighteen animals, *Comp. Biochem. Physiol.,* 40B, 489, 1971.

30. **Chapman, M. J., Goldstein, S., and Laudat, M.-H.,** Characterization and comparative aspects of the serum very low and low density lipoproteins and their apoproteins in the chicken (*Gallus domesticus*), *Biochemistry,* 16, 3006, 1977.

31. **Hillyard, L. A., Entenman, C., and Chaikoff, I. L.,** Concentration and composition of serum lipoproteins of cholesterol-fed and stilbestrol-injected birds, *J. Biol. Chem.,* 223, 359, 1956.

32. **Chapman, M. J.,** Animal lipoproteins: chemistry, structure, and comparative aspects, *J. Lipid Res.,* 21, 789, 1980.

33. **Ewing, A. M., Freeman, N. K., and Lindgren, F. T.,** The analysis of human serum lipoprotein distributions, *Adv. Lipid Res.,* 3, 25, 1965.

34. **Lee, D. M. and Alaupovic, P.,** Studies of the composition and structure of plasma lipoproteins. Isolation, composition, and immunochemical characterization of low density lipoprotein subfractions of human plasma, *Biochemistry,* 9, 2244, 1970.

35. **Lee, D. M.,** Isolation and characterization of low density lipoproteins, in *Low Density Lipoproteins,* Day, C. E. and Levy, R. S., Eds., Plenum Publishing, New York, 1976, 3.

36. **Alaupovic, P.,** The concepts, classification systems, and nomenclature of human plasma lipoproteins, in *Handbook of Electrophoresis,* Vol. 1, Lewis, L. A. and Opplt, J. J., Eds., CRC Press, Boca Raton, 1980, 27.

37. **Kostner, G. and Alaupovic, P.,** Studies of the composition and structure of plasma lipoproteins-C- and N-terminal amino acids of the two nonidentical polypeptides of human plasma apolipoprotein A, *FEBS Lett.,* 15, 320, 1971.

38. **Edelstein, C., Lim, C. T., and Scanu, A. M.,** The serum high density lipoproteins of *Macacus rhesus.* II. Isolation, purification and characterization of their two major polypeptides, *J. Biol. Chem.,* 248, 7653, 1973.

39. **Scanu, A. M., Edelstein, C., and Wolf, R. H.,** Chimpanzee (*Pan troglodytes*) serum high density lipoproteins. Isolation and properties of their two major apolipoproteins, *Biochim. Biophys. Acta,* 351, 341, 1974.

40. **Jackson, R. L., Baker, H. N., Tauton, O. D., Smith, L. C., Garner, C. W., and Gotto, A. M., Jr.,** A comparison of the major apolipoprotein from pig and human high density lipoproteins, *J. Biol. Chem.,* 248, 2639, 1973.

41. **Swaney, J. B., Reese, H., and Eder, H. A.,** Polypeptide composition of rat high density lipoprotein — characterization by SDS-gel electrophoresis, *Biochem. Biophys. Res. Commun.,* 59, 513, 1974.

42. **Herbert, P. N., Windmueller, H. G., Bersot, T. P., and Shulman, P. S.,** Characterization of the rat apolipoproteins. I. The low molecular weight proteins of rat plasma high density lipoproteins, *J. Biol. Chem.,* 249, 5718, 1974.

43. **Alaupovic, P., Lee, D. M., and McConathy, W. J.,** Studies on the composition and structure of plasma lipoproteins. Distribution of lipoprotein families in major density classes of normal human plasma lipoproteins, *Biochim. Biophys. Acta,* 260, 689, 1972.

44. **Kelley, J. L. and Alaupovic, P.,** Lipid transport in avian species. II. Isolation and characterization of lipoprotein A and lipoprotein B, two major lipoprotein families of the male turkey serum lipoprotein system, *Atherosclerosis,* 24, 177, 1976.

45. **McConathy, W. J. and Alaupovic, P.,** Studies on the interaction of concanavalin A with major density classes of human plasma lipoproteins. Evidence for the specific binding of lipoprotein B in its associated and free forms, *FEBS Lett.,* 41, 174, 1974.

46. **Kudzma, D. J., St. Claire, F., De Lalo, L., and Friedberg, S. J.,** Mechanism of avian estrogen-induced hypertriglyceridemia: evidence for overproduction of triglyceride, *J. Lipid Res.,* 16, 123, 1975.

47. **Bacon, W. L., Brown, K. I., and Musser, M. A.,** Changes in plasma calcium, phosphorus, lipids and estrogens in turkey hens with reproductive state, *Poult. Sci.,* 59, 444, 1980.

48. **Holdsworth, G., Michell, R. H., and Finean, J. B.,** Transfer of very low density lipoprotein from hen plasma into egg yolk, *FEBS Lett.,* 39, 275, 1974.

49. **Lisano, M. E. and Kennamer, J. E.,** Values for several blood parameters in Eastern Wild turkeys, *Poult. Sci.,* 56, 157, 1977.

50. **Kelley, J. L., Curry, M. D., Thayer, R. H., and Alaupovic, P.,** Quantification of the Major Serum Lipoprotein Families of Laying and Non-Laying Turkeys by Electroimmunoassay, Abstracts of Papers, 30th Annu. Meet., Council on Arteriosclerosis, American Heart Association, Miami Beach, Fla., November 15 to 17, 1976, 22.

Lipoproteins of Some Unusual Animal Species

SERUM LIPOPROTEINS OF PRAIRIE DOGS

Herbert K. Naito

The term "prairie dog" is a misnomer, even though the genus name, *Cynomys* is derived from the Greek work, *Kynos*, meaning dog. Nor are prairie dogs confined to the prairies; they have been observed at altitudes of 12,500 feet in the Rocky Mountains in Colorado.

The prairie dog is a moderately large rodent of the order Rodentia, which includes such animals as rats, mice, guinea pigs, woodchucks, beavers, and porcupines, and the family Sciuridae, to which the chipmunks, woodchucks, fox squirrels, ground squirrels, and flying squirrels also belong. Actually, the prairie dog is a large, burrowing ground squirrel, with short legs, long toenails, plump body, and short tail.

These plump, tawny rodents have proportionately shorter tails and bodies and are heavier than their ground-squirrel relatives. They vary in length from 13 to 17 in. In late summer, when fat, they weigh from 1 1/2 to more than 3 lb.

Male and female prairie dogs resemble each other. The females have eight to twelve nipples which can be observed at nursing time. As a rule, males are somewhat larger than females.

The color pattern is simple, without dark bands or other body marks. The species vary somewhat in color, but in general they are grayish, reddish brown, clear buff, or pale cinnamon. The underparts usually are paler than the backs. Depending on the species, the tail is conspicuously tipped with white or black.

Three kinds of prairie dogs have tails tipped with black: the blacktailed prairie dog of the plains, which once was the most numerous of all the species; the Arizona prairie dog, which has a short black tip on a tail that is slightly longer than the black-tailed variety of the plains (once widely distributed from southeastern Arizona, central New Mexico, southwestern Texas, and adjacent portions of Mexico, it is now rare); and the Mexican prairie dog of southeastern Coahuila and northern San Luis Potosi, Mexico, never studied as extensively as its northern relatives, which resembles the black-tailed species of the Great Plains but has a longer tail with more black on its terminal portion.

Four kinds of prairie dogs have tails tipped with white; these are mountain- and desert-dwelling species.

The white-tailed prairie dog once existed in large colonies at higher altitudes in Montana, Wyoming, Colorado, and Utah. The habitats varied from sagebrush-covered hillsides to dry mountain meadows and open ponderosa-pine timber stands. Though many have been poisoned, colonies still exist in Colorado and Utah.

Gunnison's prairie dog, another white-tailed variety, lives in mountain parks, meadows, and on the high plateaus of southern Colorado and northern New Mexico. Its communities are loosely organized, and individuals frequently live alone in the manner of ground squirrels.

The Zuni prairie dog, or *glo-un* of the Navajo Indians, is similar to the Gunnison's prairie dog but is more cinnamon in color. Its original range included northwestern New Mexico, northern Arizona, and southwestern Colorado, where they once ranged from the high mountain parks to the vast arid plains and semidesert country, but range and numbers are now greatly reduced.

The Utah prairie dog, now considered to be in danger of extinction, has always been restricted to the mountain valleys of central Utah. The terminal half of the tail is white, and a spot of black appears above the eyes.

Some prairie dogs that dig in lignite in western North Dakota are black because of the coal dust in their burrows. Deep-red prairie dogs, colored by the red soils in Oklahoma, once were described as a different species with the name *Cynomys pyrrotrichus*.

The color of prairie dogs is not greatly changed by the molt which occurs twice each year. The spring molt begins at the head and works back to the tail. The fall molt, which changes the summer fur to winter fur, progresses from tail to head. The tail molts only once each year, in summer at the conclusion of the change to summer fur. The molts produce no secondary variations between the sexes.

The prairie dog's internal structure is well adapted to its primarily vegetarian life on the prairies, the desert grasslands, and in mountain meadows. Incisors, for example, are large and molars are broad. These teeth aid in clipping and chewing grasses and weeds above ground and in removing obstructing roots during burrow construction.

The abdominal viscera are especially structured for a vegetable diet. The cecum — a reservoir for food materials — is exceptionally large, often as large or larger than the stomach.

Not all prairie dogs are dormant in winter. In higher mountain country where deep snow persists for many months the animals do go into winter sleep. But the black-tailed prairie dogs of the Great Plains may be seen on bright cold days in midwinter.

Seasonal changes in plant life cause variations in feeding and accumulation of fat. Communal life reaches its full development as the young grow into adults. When the youngsters become independent of the parents, either may move to new territory. Finally, autumn arrives and with diminishing food supplies the colony members begin their gradual retreat to the underground world they will occupy until the coming of another spring.

Recently, prairie dogs have been used as an animal model for the study of rapid induction of cholesterol cholelithiasis.[1-6] However, virtually no work on serum lipoproteins have been reported on the prairie dogs. The characterization of serum lipids and lipoproteins of prairie dogs fed a chow diet or cholesterol-supplemented diet was recently reported by our group.[6,13] These studies were conducted in the fall to prevent any influence of the possible seasonal hypothermic effects that might influence the results. Eighteen adult female prairie dogs (from Otto Martin Locke, New Braunfels, Texas) with a mean body weight of 800 g were caged individually in a photo- and thermo-regulated room. After a 2-week acclimation period and receiving a chow diet, the animals were separated into two groups: a control group (A) which remained on the standard laboratory chow, which on analysis showed <0.1% cholesterol (by weight), and a cholesterol-fed group (B) fed a 1.2% (by weight) cholesterol-supplemented diet (General Biochemicals Division, Chagrin Falls, Ohio). The formulation of the diet is shown in Table 1. After 14 days on the respective diets, the fasting animals (16 hr) were anesthetized with phenobarbitol and blood was drawn via heart puncture. The serum lipoproteins were characterized by chemical, electrophoretic, ultracentrifugal, and immunological techniques.

The serum lipid values are shown in Figure 1. Cholesterol feeding causes a rapid induction of marked hyperlipidemia in the prairie dogs within 14 days. Significant rise in lipid concentration is in the serum total cholesterol (TC), ester cholesterol (EC), free cholesterol (FC), and phospholipid (PL) fractions causing a five-, five-, seven-, and threefold increase in concentration, respectively. The statistical significance of the increases is shown in Figure 1. The percentage of EC is 80% of the total cholesterol in the chow-fed group; the cholesterol-fed group has relatively the same proportion, 70% of the TC as EC.

The lipoprotein patterns as determined by paper electrophoresis and by PAGE show two major lipoprotein fractions in both groups (Figure 2). On paper electrophoresis chow-fed animals have a β-lipoprotein (β-Lp) and an α-lipoprotein (α-Lp) band, similar to that found in man with the exception that the pre-β-Lp band is not apparent in these animals due to the low triglyceride levels. By either electrophoretic method, the α-Lp of the prairie dogs is lower in electrophoretic mobility as compared to man. The α-Lp fraction is a major fraction of the serum lipoprotein electrophoretic pattern of the female prairie dog. However, in the male prairie dogs the β-Lp appears to be the major lipoportein in the chow-fed animals. Cholesterol feeding causes the appearance of a pattern resembling a primary hyper-β-lipo-

Table 1
CHOLESTEROL-
SUPPLEMENTED PRAIRIE
DOG DIET[a,b]

Ingredients	g/kg
Sucrose	339.4
Cornstarch	140.0
Egg yolk solids	332.3
Soy protein	74.8
Nonnutritive fiber (cellulose)	58.9
Vitamin mix (Teklad)	10.0
Mineral mix (Phillips-Hart)	40.0
Cholesterol	4.7[c]

[a] Modified Brenneman's diet.[4]

[b] Made by General Biochemicals Division, Chagrin Falls, Ohio.

[c] Represents 1.2% cholesterol, by weight.

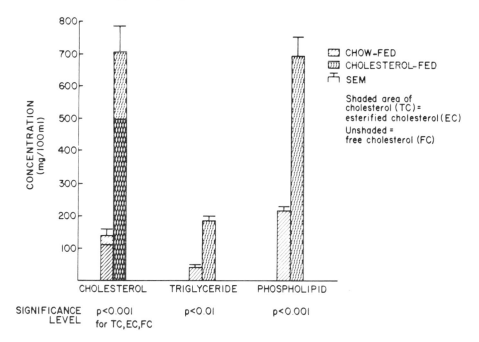

FIGURE 1.   Lipid composition (cholesterol, triglycerides, and phospholipids) of sera chow-fed and cholesterol-fed prairie dogs.

proteinemia in human serum (Figure 2). We noticed that on paper electrophoresis the α-Lp fraction is usually not sharply resolved. To obtain more discrete and definitive lipoprotein patterns we used PAGE. On a basal diet, the PAGE patterns reveal that α-Lp constitute about 70% of the lipid stainable material when scanned by a densitometer (in females), while the remaining is primarily β-Lp and trace amounts of pre-β-Lp and no evidence of chylomicra. With cholesterol feeding, the concentration of all lipoprotein fractions increases.

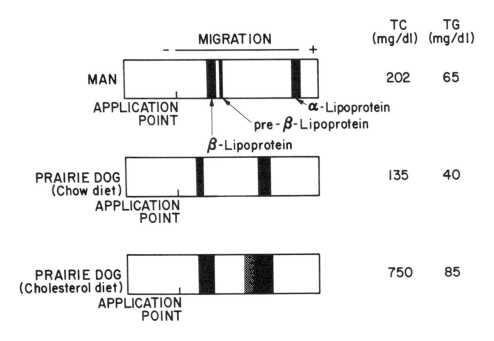

FIGURE 2.   Schematic representation of typical lipoproteins electrophoretic patterns (paper electrophoresis) of prairie dogs.

The greatest increase is seen in the β-Lp fraction with the α-Lp concentration increasing moderately (Figure 2). The PAGE patterns of the prairie dogs give us additional evidence that cholesterol feedings causes the development of a Type II-like phenotype. While most of the phenotypes were of the Type IIa, some showed a Type IIb-like pattern. The PAGE patterns of cholesterol-fed prairie dogs usually show a fast and slow component in the α-Lp area. The densitometer scans of the chow-fed animals always show a single peak (with slight trailing) in the α-Lp area, suggesting, for the most part, a homogeneity of molecules. However, the scan of the cholesterol-fed animals with the diffused α-Lp on paper electrophoretograms usually show a double peak in the α-Lp area. Whether or not these two α-Lp moieties in this group correspond to the HDL$_c$ and HDL$_2$ fractions of cholesterol-fed mongrel or purebred fox hounds and swine, described by Mahley et al.[7] is not known. They described the appearance of an HDL fraction which becomes overloaded with cholesterol and floats at a progressively lower density. We have observed this slow-moving lipoprotein (α$_2$-Lp) in beagles that become spontaneously hypercholesterolemic on a chow diet.[8] Butkus et al.[9] have also reported similar patterns in mongrel dogs fed the Malmros hypercholesterolemic diet deficient in essential fatty acids. While the physiological significance of the appearance of this new α-Lp fraction in prairie dogs is not yet determined, Mahley et al.[7] suggested that the newly formed cholesterol-rich HDL may have a pathological significance in inducing atherosclerosis in dogs. Thus, with cholesterol feeding there is a reversal of the β/α-Lp ratios which is mainly due to the great increase in β-Lp levels in the cholesterol-fed group.

Further characterization of the lipoprotein fractions shows that the lipoprotein fraction isolated by the preparative ultracentrifuge (d 1.21 to 1.063, which is the density at which human HDL is found) have an α$_1$-globulin mobility when compared to whole serum run by agarose gel (Pfizer Pol-E-Film® IV[10]) and the PAGE[11] methods. The lipoprotein fraction with a density of 1.063 to 1.006 (which is the density of human LDL) has a fast β-globulin mobility. This isolated lipoprotein fraction seems to have a slightly greater mobility than the β-Lp fraction in whole serum when run on agarose, but it is in the β-Lp area when done

RELATIVE PERCENTAGE OF COMPOSITION OF THE LIPID MOIETY IN EACH LIPOPROTEIN FRACTION

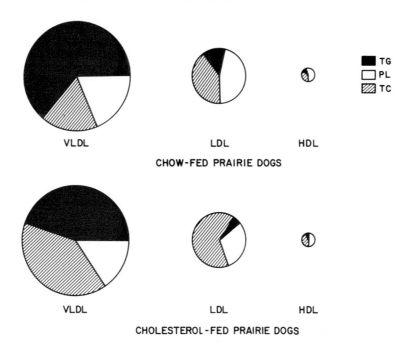

FIGURE 3.   Relative percentage composition of VLDL, LDL, and HDL of prairie dogs. VLDL = 138 ± 10, 39 ± 6, and 216 ± 12 mg/dℓ for triglycerides (TG) and phospholipids (PL), respectively, for chow-fed animals. VLDL = 705 ± 87, 85 ± 14, and 692 ± 60 mg/dℓ for TC, TG, and PL, respectively, for cholesterol-fed animals ($\overline{X}$ + SEM).

on PAGE media. The lipoprotein fractions of densities <1.006 (overnight spin in the preparative ultracentrifuge with the initial removal of chylomicra) in the control group were present in such small concentrations that they were barely visible on the agarose gel. In the cholesterol-fed group, the VLDL fraction was slightly more visible only by the PAGE method and had a mobility where the VLDL is normally present.

The chemical composition of the lipoprotein fractions of both groups of animals is shown in Figure 3. The distribution of the different lipid classes (cholesterol, triglyceride, phospholipid) of each lipoprotein fraction (VLDL, LDL, HDL) in the control animals resembles that of "normal human beings" with the exception of the LDL fraction which carries a slightly greater percentage of phospholipids and slightly lesser amounts of cholesterol and triglycerides in the prairie dogs. With cholesterol feeding, the concentration of the serum total cholesterol and β-Lp fractions increase markedly. The relative percentage of composition of the different lipoprotein fractions show that the percentage of cholesterol being carried by each lipoprotein moiety increase in the cholesterol-fed groups. Again, the chemical composition of the major lipoprotein fractions of the prairie dog fed cholesterol resembles that of the patients with hyper-β-lipoproteinemia. The main lipoprotein carrier for cholesterol was the LDL, comprising 40 and 65% of the total lipids in that fraction of the control and cholesterol-fed groups, respectively. In both groups, the triglycerides are the main components of the VLDL, and the phospholipids are the main components of the HDL (Figure 3). Thus, it appears that the lipoprotein fractions have basic physicochemical characteristics similar to those of human beings.

Studies on Ouchterlony double-immunodiffusion technique[12] show the prairie dog LDL (chow-fed) reacted against prairie dog anti-β-Lp antisera, and HDL reacted against prairie dog anti-α-Lp antisera. There is no cross-reaction between prairie dog HDL and prairie dog anti-β-Lp antiserum, nor is there a cross-reaction between prairie dog LDL and prairie dog anti- β-Lp antiserum. There is no cross-reaction between prairie dog LDL and human or dog anti-β-Lp antisera and between prairie dog HDL and human or dog anti-α-Lp antisera, suggesting complete identity of the prairie dog LDL and HDL moieties. For some reason, we consistently obtain a second, weak precipitin line of the LDL fraction when reacted against anti-prairie dog-β-Lp antiserum. This was also seen in the immunoelectrophoretic studies. It may represent protein contamination of the LDL fraction. Immunoelectrophoresis revealed that the cholesterol-fed groups always had a double precipitin line against anti-α-Lp antisera in the α-globulin area. There is only one precipitin line against anti-whole serum because that antiserum was prepared from the noncholesterol-fed animals. Therefore, it is doubtful that whole serum had the other α-Lp component seen to elicit antibody production in the cholesterol-fed animals.

In conclusion, we have characterized the serum lipoprotein of adult prairie dogs. We have demonstrated that:

1. Cholesterol feeding for 14 days causes marked hyperlipidemia.
2. The lipoprotein pattern of the animals on a low-cholesterol, low-fat diet resembles that of normal human beings.
3. Cholesterol feeding causes a change in lipoprotein pattern, resembling hyper-β-lipoproteinemia.
4. The chemical composition of the four major lipoprotein fractions of the chow-fed and cholesterol-fed prairie dogs resembles that of "normal" man and patients with hyper-β-lipoproteinemia, respectively; according to immunologic studies, the apoproteins of the prairie dog LDL and HDL are in some ways structurally different from those of dogs or of human beings.

This study demonstrates that the prairie dog may be a unique and most appropriate nonhuman model in the study of the relationship between cholesterol diets and hyper-β-lipoproteinemia. This is particularly important in view of the fact that in patients with coronary artery disease substantial evidence is now available to support the contention that plasma lipoproteins, especially the cholesterol-rich LDL fraction, play a significant role in atherosclerosis. While the precise sequence of events by which this process occurs has yet to be elucidated, studies have clearly demonstrated LDL appears to be localized in lesion areas.[14]

ACKNOWLEDGMENT

Supported in part by NIH Grant No. HL-6835 and Grant No. 3084R from the American Heart Association, N.E. Ohio Affiliate, Inc.

REFERENCES

1. **Asdel, K., Khan, B., Taylor, C. B., and Cox, G. E.,** Hypercholesteremia and vascular lesions in prairie dogs and ground squirrels, *Circulation*, 24, 1083 (Abstr.), 1961.
2. **Patton, D. E., Plotner, K., Cox, G. E., and Taylor, C. B.,** Biliary cholesterol deposits in ground squirrels and prairie dogs, *Fed. Proc.*, 20, 248 (Abstr.), 1961.
3. **Westenfelder, G., Fitzwater, J., Taylor, C. B., and Cox, G. E.,** Atherogenesis in prairie dogs and ground squirrels, *Contrast. Fed. Proc.*, 21, 477a, 1962.
4. **Brenneman, D. E., Connor, W. E., Forker, E. L., and DenBesten, L.,** The formation of abnormal bile and cholesterol gallstones from dietary cholesterol in the prairie dog, *J. Clin. Invest.*, 51, 1495—1503, 1972.
5. **Chang, S. H., Ho, K. J., and Taylor, C. B.,** Cholesterol gallstone formation and its regression in prairie dogs, *Arch. Pathol.*, 96, 417—426, 1973.
6. **Holzbach, R. T., Corbusier, C., Marsh, M., and Naito, H. K.,** The process of cholesterol cholelithiasis induced by diet in the prairie dog: a physicochemical characterization, *J. Lab. Clin. Med.*, 87, 987—998, 1976.
7. **Mahley, R. W.,** Alterations in plasma lipoproteins induced by cholesterol feeding in animals including man, in *Disturbances in Lipid and Lipoprotein Metabolism*, Dietschy, J. M., Gotto, A. M., Jr., and Ontko, J. A., Eds., American Physiological Society, Bethesda, 1978, 181—197.
8. **Wada, M., Minamisono, T., Ehrhart, L. A., Naito, H. K., and Mise, J.,** Spontaneous hyperlipoproteinemia in beagles, *Life Sci.*, 20, 999—1009, 1977.
9. **Butkus, A., Ehrhart, L. A., Robertson, A. L., and Lewis, L. A.,** Effects of diets rich in saturated fatty acids with or without cholesterol on plasma lipids and lipoproteins, *Lipids*, 5, 896—907, 1970.
10. Pfizer Pol-E-Film System® IV, Serum Protein Determination, Pfizer Diagnostics Division, New York, 1971.
11. **Naito, H. K., Wada, M., Ehrhart, L. A., and Lewis, L. A.,** Polyacrylamide-gel-disc-electrophoresis as a screening procedure for serum lipoprotein abnormalities, *Clin. Chem.*, 19, 228—234, 1973.
12. **Ouchterlony, O.,** in *Handbook of Immunodiffusion and Immunoelectrophoresis*, Ann Arbor-Humphrey Science Publishing, Ann Arbor, Mich., 1970, 21—123.
13. **Naito, H. K., Holzbach, R. T., and Corbusier, C.,** Characterization of serum lipids and lipoproteins of prairie dogs fed a chow diet or cholesterol-supplemented diet, *Exp. Mo. Pathol.*, 27, 81—92, 1977.
14. **Hoff, H. J., Heideman, C. L., Gaubatz, J. W., Titus, J. L., and Gotto, A. M., Jr.,** Quantitation of Apo B in human aortic fatty streaks: a comparison with grossly normal intima and fibrous plaques, *Atherosclerosis*, 30, 263, 1978.

SERUM LIPIDS AND LIPOPROTEINS OF A GROUND DWELLING ANIMAL: THE GROUND SQUIRREL

Herbert K. Naito

A ground squirrel may be any burrowing member of the Sciuridae, but more commonly the term is restricted to the spermophiles, genus *Citellus,* or *Spermophilus,* rodents characterized by having cheek pouches, short ears, long front claws and generally a short tail. This group of animals includes some 15 species that are well adapted for life on the ground. Active by day, they search for their food of seeds, roots, bulbs, plant stems and leaves, and insects. Numerous trips are made, with cheek pouches bulging, to their underground storage chambers. Summer is a time when ground squirrels become very fat. In the northern part of their range ground squirrels become dormant; in the southern part they may only avoid extreme weather conditions by staying in their burrows for short periods. These rodents are omnivorous, feeding upon vegetation, seeds, insects, etc. Although chiefly North American, one, known as the suslik, reaches eastern Europe and others occur in Asia. The arctic species (*C. parryi*) is about 12 in. long, the tall 5 in.; it is yellowish brown, speckled with gray. Rock squirrels (*C. variegatus* and allies) are about the same size, with a longer tail and larger ears. The mantled ground squirrels (*C. lateralis* and *C. saturatus*) are often confused with chipmunks, having black and white lateral stripes. The 13-lined ground squirrel (*C. tridecemlineatus*), is often called gopher (the true gopher, however, is of a different family), and is pale buff, marked with brownish lines and whitish spots.

Among the other ground-dwelling sciurids are the marmots (including the American woodchuck), chipmunks, and prairie dogs.

As shown in Table 1, the reports on lipids and lipoproteins of ground squirrels are very limited. One of the earliest studies was by Wilker and Musacchia,[1] They studied the genus *Citellus barrowensis*. The animals were live trapped in various parts of Arctic Alaska (one group [A] in July and another [B] in September). Their study shows that the amounts of some of the lipid fractions in the blood are relatively high, if compared with most ordinary laboratory mammals. The mean cholesterol concentration in the blood was 223.9 mg/dℓ, with a range of 197 to 266 mg/dℓ. The phospholipid and total fatty acid concentrations were 613 and 833 mg/dℓ, respectively. There were no apparent differences in the amount of blood lipids in the squirrels collected early in summer as compared to those collected in September. However, differences in liver lipids were observed. The liver contained three times as much fatty acid and almost twice as much phospholipid in Group A as compared to Group B. The hepatic cholesterol concentrations were the same in both groups. The authors suggested that the change in hepatic lipid concentrations was correlated with the preparation, by the animals, for the long-term hibernation, which begins in early fall and ends in April. During the spring and summer, fat must be synthesized and stored quickly as reserve for the winter.

According to Huang and Morton,[2] who studied two populations of Belding ground squirrels (*Spermophilus beldingi beldingi*) at high altitude throughout their season of activity, the oscillations in plasma lipid concentrations appear to be related to the seasonal rhythm of energy storage. Newly emerged adult males in one population (Tioga Pass in the Sierra Nevada Mountains) in mid-May had a mean plasma lipid concentration of 598 mg/dℓ which decreased steadily to a seasonal low in mid June to 432 mg/dℓ. This was followed by an increase that was maintained near or above 600 mg/dℓ until emergence in August.

In adult females a seasonal low in mean plasma lipid concentration of 552 mg/dℓ occurred soon after emergence. This was followed by a significant increase to 810 mg/dℓ in mid-June when most females had begun lactating. Subsequently plasma lipid decreased in females to about the same concentrations seen in males. Starting in mid-July, plasma lipid began

Table 1
STUDIES DONE ON LIPIDS AND LIPOPROTEINS OF GROUND SQUIRRELS

Authors	Lipids	Lipoproteins	Ref.
Wilber and Musacchia	A study on lipids in the blood, liver, and kidneys of the Arctic ground squirrel, *C. barrowensis,* collected at different times of the year	ND	1
Huang and Morton	A study on plasma lipids and proteins in two populations of Belding ground squirrels, *S. beldingi beldingi,* during interhibernatory season of activity and normothermia	ND	2
Galster and Morrison	The blood lipids, proteins, and hematocrit levels of squirrels, *S. tridencemlineatus,* were examined during the various months throughout the year	Lipoproteins were studied using paper electrophoretic method	4
Bragdon	A study of the blood lipids and lipoproteins of the Columbian ground squirrel, *C. Columbianus,* in captivity; several experimental diets were used	Analytical ultracentrifugal method was used to examine the serum lipoproteins	12
Naito and Gerrity	A study on blood lipids and lipoproteins of the 13-lined ground squirrel, *C. mexicanus,* fed chow diets and cholesterol-supplemented diets	Serum lipoproteins were studied by electrophoretic, ultracentrifugal, chemical, and immunological techniques	14

increasing in both sexes, maxima of 825 mg/dℓ in females and 700 mg/dℓ in males being attained in mid-August at the very end of their active season.

There was little seasonal variation in plasma lipids in the second population (Big Ben in Sierra Nevada Mountains).

At both locations plasma lipid concentrations tended to be higher in females than in males throughout the season.

According to Morton[3] the ground squirrels in the Tioga Pass area undergo extensive fattening following reproduction and enter hibernation with intracorporal lipid stores equivalent to or exceeding lean, dry body components. Upon emergence, some 8 to 9 months later, they still retain 20 to 25% of this fat, which is totally depleted in the weeks after emergence which encompass the period of reproduction.

Galster and Morrison reported that captive 13-lined ground squirrels (*S. tridencemlineatus*) also show seasonal changes in serum lipids that appear to be related directly to fat storage or utilization.[4]

During cold exposure, increased brown fat mass is characteristic of most hibernating animals. Chaffee et al.[5] described increased brown fat in cold-exposed golden-mantled squirrels (*C. lateralis*). Burlington et al.[6] analyzed brown fat from control, hibernating, and aroused 13-lined ground squirrels (*S. tridencemlineatus*) and found that the axillary brown fat mass was significantly increased in cold-exposed, hibernating animals compared to that in the control group. This increased tissue mass was primarily due to a significantly increased neutral lipid and water content. Brown fat phospholipids were also increased by approximately 2 1/2 times during hibernation. The amount of nonlipid dry residue, which includes protein, was also greater in hibernating animals; however, this accounted for only 15% of the increased tissue mass. The phospholipid and nonlipid dry residue increases were approximately proportional to the increased water. After arousal from hibernation, tissue neutral lipids were significantly decreased, but phospholipids, water, and nonlipid dry residue did not reveal comparable changes relative to tissue from hibernating squirrels.

Total DNA was unchanged during hibernation or arousal from hibernation in both tissue and isolated cells. Consequently, the ratio of lipid to DNA was significantly increased during hibernation and significantly decreased, relative to hibernation values, after arousal. In all experimental groups, the ratio of lipid to DNA in isolated cells was four to five times that observed in whole tissue.

Brown adipose tissue serves as a major source of heat during arousal from hibernation.[7-9] Spencer et al.[10] observed a decrease in brown adipose tissue glycerides during arousal of golden-mantled ground squirrels. Similar results were obtained with brown fat from 13-lined ground squirrels.[8] On the basis of morphological and histological evidence, Grodums et al.[11] described a significant depletion of lipids in cells from aroused golden-mantled squirrels. Thus, the results of Burlington et al.,[6] showing a significant reduction in the amount of lipid/DNA in isolated brown fat cells from aroused ground squirrels, are in accord with these investigators. In addition, the reduced lipid content can be ascribed to the disappearance of neutral lipid. These results support the concept that lipids from brown fat are utilized as a substrate for thermogenesis during arousal from hibernation.

In another species (*C. columbianus*) Bragdon[12] also showed seasonal variation, occasionally showing extreme hyperlipemia. Serum triglycerides comprised the major lipid fraction followed by phospholipids. This was especially true of the animals on the high-fat diet and 1% cholesterol diet as compared to a low-fat diet. His study also indicated that this species has relatively high serum cholesterol concentrations. On the low-fat diet the cholesterol values were similar to those seen in an adult human being, 215 mg/dℓ (range = 146 to 291 mg/dℓ). On the high-fat diet or 1% cholesterol diet, the ground squirrels showed concentrations that could be considered hypercholesterolemic in man. This is the first study that attempted to analyze the serum lipoproteins of ground squirrels. The lipoproteins were quantitated by analytical ultracentrifugal technique. No electrophoresis studies were done. The serum α-lipoproteins (α-Lp) were the major fraction comprising 76% of the total lipoproteins in the low-fat diet group and 68% in the high-fat diet group. Atheromatous lesions were found in the aorta of the three ground squirrels with the highest serum cholesterol concentrations. During hibernation the serum triglyceride levels were low. Bragdon suggested that the hyperlipemia was the result of dietary caloric excess.

In 1966, Galsten and Morrison[4] were the first to report on the electrophoretic patterns of serum lipoproteins of ground squirrels (*S. tridecemlineatus*). The animals were maintained in the laboratory and fed Purina Rat Chow (50% carbohydrate, 24% protein, and 4% fat). Lipoproteins were studied using the paper electrophoresis method of Jencks et al.[13] The lipoproteins exhibited yearly cycles in concentration with maxima in late fall and minima in late spring. Serum total lipid concentration rose through the summer, reaching a maxima in November. This indicated that chylomicra, α-Lp and β-lipoproteins (β-Lp) followed the general yearly cycle in total lipids; however, the lipoprotein changes exhibited individual differences. The α-Lp rose in early summer and doubled in concentration in November-December and then dropped during winter with a sharp decrease in March. Thus, the α-Lp were considered elevated during the hibernation season and the levels rose sharply at awakening and during the breeding season before declining in early June. The β-Lp also rose during summer but fell sharply in September; this was followed by a gradual decline through the winter and remained low through the spring. The chylomicra showed a brief concentration peak in November which fell to the average value by the end of December. Thus, the β-Lp and chylomicra, for the most part, declined through the early hibernating season and then remained constant until late spring.

It appears that Columbian and 13-lined ground squirrels both show concentrations of serum lipids that are about seven times those found in the laboratory rat.[15] This striking difference in the hibernators suggests greater capacity for carrying lipids in the blood.[4]

Other than the early reports of Bragdon[12] and Galston and Morrison,[4] little information concerning lipoproteins of ground squirrels is available. In 1979, Naito and Gerrity[14] did more detailed studies on lipids and lipoproteins of *C. mexicanus* (see Table 2). Five groups

Table 2
COMPOSITION OF EXPERIMENTAL DIETS
(% BY WEIGHT)

Nutrients	Low-fat/ cholesterol diet[a]	High-fat/ cholesterol diet[b]
Fat	4.5[c]	30.4[d]
Protein	23.4	25.1
Carbohydrate (NFE)[e]	58.5	38.4
Fiber	6.3	1.3
Ash	7.3	4.8

[a] Purina Rodent Laboratory Chow (Ralston Purina Co., St. Louis, Mo.).
[b] Purina Monkey Chow diet fortified with 50% egg yolk (ICN Pharmaceuticals, Inc., Cleveland, Ohio).
[c] Cholesterol content, < 0.01% (by weight). Rodent Chow fortified with cholesterol had a cholesterol content of 2.1%.
[d] Cholesterol content, 1.0% (by weight).
[e] NFE, Nitrogen-free extract.

Table 3
SERUM LIPID CONCENTRATION
(MEAN ± SEM)

	Duration on diet (weeks)	Serum lipids (mg/dℓ)		
		Chol[a]	TG[b]	PL[c]
Group A (chow-fed)	3	127.3 ±18.6	70.3 ±8.5	146.5 ±13.5
Group B (chow-fed)	26	147.9 ±7.7	92.3 ±16.8	178.0 ±19.8
Group C (chow-fed)	52	122.6 ±12.4	66.6 ±5.5	212.4 ±19.5
Group D (chol-fed)	26	156.8 ±26.4	96.5 ±21.1	196.6 ±11.0
Group E (chol-fed)	52	196.9 ±12.4	98.0 ±30.5	324.6 ±21.4

[a] Cholesterol.
[b] Triglyceride.
[c] Phospholipid.

of adult males (about 1-year old) were studied: Group A, fed Rodent Laboratory Chow (Ralston Purina Co., St. Louis, Mo.) for 3 weeks; Group B, fed the same chow diet for 26 weeks; Group C, fed the chow diet for 52 weeks; Group D, fed a semipurified diet supplemented with 2% cholesterol (ICN Pharmaceuticals, Inc., Cleveland, Ohio) for 26 weeks; Group E, fed a Purina® monkey chow diet fortified with 50% egg yolk. The serum lipids are shown in Table 3.

The feeding of 2% cholesterol-supplemented rodent chow diet to ground squirrels for 26 weeks (Group D) did not induce hypercholesterolemia or hypertriglyceridemia. However, feeding the primate diet fortified with 50% egg-yolk for 1 year (Group E) induced a modest rise (21%) in serum total cholesterol level ($p < 0.05$), but serum triglycerides were not elevated. Since no absorption and excretion studies were done, and turnover rates were not determined, we can only speculate why the primate-egg-yolk diet induced mild hypercho-

Table 4
LIVER LIPID CONCENTRATION
(MEAN ± SEM)

	Duration on diet (weeks)	Liver lipids (mg/100 g tissue)		
		Chol[a]	TG[b]	PL[c]
Group A (chow-fed)	3	412.1 ± 25.7	2,033.2 ± 477.3	3,966.0 ± 180.7
Group B (chow-fed)	26	395.4 ± 34.4	4,549.2 ± 699.2	4,046.1 ± 127.3
Group C (chow-fed)	52	333.2 ± 11.1	3,422.7 ± 1,151.4	3,306.4 ± 130.6
Group D (chol-fed)	26	500.9 ± 42.5	3,148.7 ± 693.8	3,554.5 ± 72.1
Group E (chol-fed)	52	377.7 ± 38.1	11,033.9 ± 2,249.7	2,990.4 ± 150.5

[a] Cholesterol.
[b] Triglyceride.
[c] Phospholipid.

lesterolemia in these animals. Although the cholesterol content of this diet was 50% less than the low-fat rodent chow diet fortified with 2% cholesterol (Table 2), the high fat content (30.4% vs. 4.5%) may have increased the coefficient of absorption of dietary cholesterol. It is well known that the intestinal absorption of dietary cholesterol is influenced by the amount and type of fat in the diet and the physical state of the dietary cholesterol (i.e., crystalline vs. noncrystalline) The lower proportion of polyunsaturated fat vs. saturated fats (P/S ratio) in egg yolk (P_s = 0.44) vs. that of the rodent chow (P/S = 0.81) may have had an additional effect on the observed mild hypercholesterolemia.

While cholesterol feeding did not induce hypercholesterolemia or hypertriglyceridemia, elevation in serum phospholipid concentrations was observed. This increase was apparent only when Group D was compared against Group A, while no significant differences were found between Groups B and D. Thus, this increase in phospholipid concentration is apparently not due to feeding the 2% cholesterol-supplemented chow diet, but rather to aging, particularly since Group C (52-week control) had even higher serum phospholipid levels as compared to Groups A, B, or D. The animals fed the monkey diet fortified with 50% egg yolk (Group E) had significantly elevated serum phospholipid levels compared to Group C, suggesting that the egg-yolk diet had an effect in addition to the aging effect. This finding may not be surprising in view of the high phospholipid content of egg yolk. The results suggest that like the dog and rat, but unlike the prairie dog, rabbit, cynomolgus monkey, and guinea pig, the ground squirrel does not readily develop hypercholesterolemia in a low-fat, cholesterol-supplemented diet.

Liver cholesterol concentration (Table 4) increased slightly in cholesterol-fed animals in Group D, but not in Group E. Liver triglyceride and phospholipid levels did not change, except in Group E which had a marked accumulation of triglyceride due to the high-fat, egg-yolk diet. The liver triglycerides in the younger ground squirrels (Group A) were lower in concentration than in Groups B to E. One animal in Group D developed gallstones. This is in contrast to the *C. tridencilineatus* which readily develops gallstones when given a similar egg-yolk diet.[16]

The distribution of serum lipoprotein as determined by paper and by polyacrylamide-gel (Figure 1) electrophoresis (PAGE) (Table 5) did not markedly change after cholesterol feeding. The predominant lipoprotein fraction (about 75% of the polyacrylamide-gel lipid

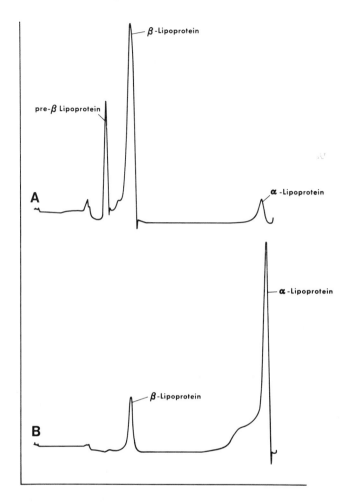

FIGURE 1. Densitometric scans of PAGE disc lipoprotein patterns of man(A) and of ground squirrel (B).

Table 5
SERUM LIPOPROTEIN DISTRIBUTION AND CONCENTRATION

| | Serum lipoprotein distribution[a] | | | Serum lipoprotein concentration[b] | | | | |
| | | | | | | LDL | | |
Group	pre-β + β	α-Lp	β/α ratio	VLDL −S 70 to 400	40 to 70	25 to 40	HDL 0 to 10
A	22.2 ±4.7	77.8 ±4.7	0.30 ±0.09	0	Trace	Trace	60
B	26.9 ±5.2	73.1 ±5.2	0.41 ±0.11	0	Trace	Trace	81
C	43.4 ±11.9	64.3 ±6.2	0.80 ±0.25	0	Trace	Trace	98
D	26.7 ±5.0	73.3 ±5.0	0.38 ±0.09	0	42	84	308[c]
E	31.9 ±6.9	72.7 ±4.8	0.47 ±0.11	16	12	80	442

[a] PAGE (relative percent distribution).
[b] Flotation rate, −S, at density 1.21; expressed as mg/dℓ.
[c] Double peak.

stainable fractions when quantitated by densitometry) was α-Lp which floated at a density of 1.063 to 1.21. Analytical ultracentrifugal data correlated with the electrophoretic studies (Table 5). The main lipoprotein fraction in all groups was HDL. Groups A to C had only traces of LDL with the remainder of the lipoproteins in the HDL fraction.

Group D had 71% of the serum lipoproteins as HDL, while 29% was LDL. Group E had a slight increase in the distribution of HDL (80% of the lipoproteins as HDL) while the LDL was slightly lower (17%) with trace amounts of VLDL (3%). Despite the apparent minimal change in serum lipoprotein distribution, dietary-induced hypercholesterolemia in Group E caused primarily an increase in HDL concentration, followed by elevation in LDL and VLDL levels (Table 4). Chylomicra were observed in some of the fasting ground squirrels fed the egg-yolk diet. According to Bragdon,[12] when the lipoprotein fractions of ground squirrels were isolated according to the conventional densities of the major lipoproteins found in man, only one band for each of the major lipoprotein classes was observed. Furthermore, the VLDL fraction had the same mobility as the pre-β-Lp in man. When the serum of ground squirrels is electrophoresed on paper, the mobility of the β-Lp fraction is slightly faster than that of the β-Lp band of man, and the α-Lp fraction is present in greater concentration than that of man. This is in agreement with the analytical ultracentrifuge data (Table 4). The isolated LDL fraction had the mobility of human β-Lp, while the HDL fraction had the mobility of human α-Lp.

The lipid composition of the various lipoprotein fractions of each of the three groups in the first study is shown in Table 5. The VLDL fraction (d < 1.006) was composed primarily of triglyceride in Groups A, B, and D (61, 65, and 72%, respectively). The remainder of the VLDL fraction was made up of cholesterol with only trace (<1%) amounts of phospholipids. This extremely low phospholipid content makes this VLDL moiety unusual when compared to human VLDL.

The LDL fraction (d 1.006 to 1.063) was composed mainly of cholesterol in all three groups (61, 62, and 57% in Groups A, B, and D, respectively), while phospholipid was the second most abundant lipid moiety (36, 34, and 37%, respectively). In this respect, the LDL moiety has some chemical similarities with human LDL. The HDL fraction (1.063 to 1.21) was made up chiefly of cholesterol (50%) and phospholipids (25 to 45%) with variable amounts of triglycerides (2 to 20%). Human HDL is usually comprised primarily of phospholipid (52%), with lesser amounts of cholesterol (32%) and triglycerides (12%). This unusual ability of ground squirrel HDL to carry such large amounts of cholesterol may partially explain why the corresponding increase in LDL concentration was not as great as the rise in serum total cholesterol seen in Group E.

We have been unable to induce hypercholesterolemia in ground squirrels with a rodent chow diet supplemented with 2% cholesterol. This is in contrast to others.[12,16] Feeding a monkey chow diet fortified with 50% egg yolk produced only mild elevations in serum cholesterol concentrations, not the marked hypercholesterolemia reported by Patton et al.,[16] Asdel et al.,[17] and Westenfelder et al.[18] The reasons for this discrepancy are unclear, but the age, sex, and in particular, the species of ground squirrels used in other studies differ from those in the present study. It is interesting that "control" animals on the vegetable monkey chow without egg-yolk supplementation[16-18] had a mean serum cholesterol concentration of 340 mg/dℓ (range 177 to 513) while our control animals on a rodent chow had a mean concentration of 148 mg/dℓ (range 137 to 153). These data indicate a 50% increase in baseline levels of serum cholesterol over those in our study.

Bragdon[12] reported that *C. columbianus* fed a low-fat vegetable diet had a mean serum cholesterol value of 215 mg/dℓ, but had a mean cholesterol level of 324 mg/dℓ when fed a high-fat vegetable diet. Wilber and Musacchia[1] analyzed blood of Alaskan ground squirrels *(C. barrowensis)* immediately after trapping and found that the mean value for total cholesterol was 224 mg/dℓ and for phospholipid was 6.5 mg/dℓ. Thus, the observed differences in the lipid values in our study compared to previous ones[1,12] is probably due to large species differences in baseline values and their response to diets. This apparent species effect on blood lipid levels may be one of the more important factors associated with the resistance of the species of animals that we studied to become markedly hypercholesterolemic.

This review on ground squirrels should emphasize the importance of concentration and distribution differences in lipid and lipoprotein of different classes of animals, including different species within the same genus. It should also be emphasized that different colonies of animals from the same species can differ in their lipid and lipoprotein concentations and their response to different diets. This review also points out the effect of seasons on the changes in blood lipids and lipoprotein concentrations. Thus, before selecting an animal model for lipid and lipoprotein studies, including that of induction of atherogenesis, background information on the animal species should be examined in detail.

ACKNOWLEDGMENT

Supported in part by NIH Grant No. HL-6835 and Grant No. 3084R from the American Heart Association, N.E. Ohio Affiliate, Inc. I would like to express my gratitude to Debra Maciejewski for typing this manuscript.

REFERENCES

1. **Wilber, C. B. and Musacchia, X. J.,** Fat metabolism in the arctic ground squirrel, *J. Mamml.,* 31, 304, 1950.
2. **Huang, S. and Morton, M. L.,** Seasonal changes in plasma proteins and lipids in the Bolding ground squirrel *(Spermophilus Beldingi Beldingi), Comp. Biochem. Physiol.,* 54A, 239, 1976.
3. **Morton, M. L.,** Seasonal cycles of body weights and lipids in Belding ground squirrels, *Bull. South. Calif. Acad. Sci.,* (1976).
4. **Galster, W. A. and Morrison, P.,** Seasonal changes in serum lipids and proteins in the 13-lined ground squirrel, *Comp. Biochem. Physiol.,* 18, 489, 1966.
5. **Chaffee, R. R. J., Pengelley, E. T., Allen, J. R., and Smith, R. E.,** Biochemistry of brown fat and liver of hibernating golden-mantled ground squirrels *(Citellus lateralis), Can. J. Physiol. Pharm.,* 44, 217—223, 1966.
6. **Burlington, R. F., Therriault, D. G., and Hubbard, R. W.,** Lipid changes in isolated brown fat cells from hibernating and aroused thirteen-lined ground squirrels *(Citellus tridencemilineatus), Comp. Biochem. Physiol.,* 29, 431, 1969.
7. **Hayward, J. S. and Ball, E. G.,** Quantitative aspects of brown adipose tissue thermogenesis during arousal from hibernation, *Biol. Bull. Woods Hole,* 131, 94—103, 1966.
8. **Joel, C. D.,** The physiological role of brown adipose tissue, in *Handbook of Physiology,* Sect. 5, Renold, A. E. and Cahill, G. F., Eds., Waverly Press, Baltimore, 1965, 59—85.
9. **Smith, R. E. and Hock, R. J.,** Brown fat: thermogenic effector of arousal in hibernators, *Science,* 140, 199—200, 1963.
10. **Spencer, W. A., Grodums, E. I., and Dempster, G.,** The glyceride fatty acid composition and lipid content of brown and white adipose tissue of the hibernator, *Citellus lateralis, J. Cell Physiol.,* 67, 431—441, 1966.
11. **Grodums, E. I., Spencer, W. A., and Dempster, G.,** The hibernation cycle and related changes in the brown fat tissue of *Citellus lateralis, J. Cell. Physiol.,* 67, 421—430,1966.
12. **Bragdon, J. H.,** Hyperlipemia and atheromatosis in a hibernator, *Citellus columbianus, Circ. Res.,* 11, 520, 1954.
13. **Jencks, W. R., Durrum, E. L., and Jetton, M. R.,** Paper electrophoresis as a quantitative method for lipoproteins, *J. Clin. Invest.,* 34, 1437, 1955.

14. **Naito, H. K. and Gerrity, R. G.,** Unusual resistance of the ground squirrel to the development of dietary-induced hypercholesterolemia and atherosclerosis, *Exp. Mol. Pathol.,* 31, 452, 1979.

15. **Lewis, L. A. and Naito, H. K.,** Relation of hypertension, lipids, and lipoproteins to atherosclerosis, *Clin. Chem.,* 24(12), 2081—2098, 1978.

16. **Asdel, K., Khan, B., Taylor, C. B., and Cox, G. E.,** Hypercholesterolemia and vascular lesions in prairie dogs and ground squirrels, *Circulation,* 24, 1083, 1961.

17. **Patton, D. E., Plotner, K., Cox, G. E., and Taylor, C. B.,** Biliary cholesterol deposits in ground squirrels and prairie dogs, *Fed. Proc.,* 20, 248, 1961.

18. **Westenfelder, G., Fitzwater, J., Taylor, C. B., and Cox, G. E.,** Atherogenesis in prairie dogs and ground squirrels: contrast, *Fed. Proc.,* 21, 477, 1962.

SERUM LIPOPROTEINS OF HAITIAN BOA CONSTRICTORS

Herbert K. Naito

INTRODUCTON

The diversity of opinions currently held concerning atherogenesis suggests that a comparative study of spontaneous arterial lesion formation could prove of value in emphasizing any factor(s) common to this pathogenic condition in animals and in man. There have been few reports on the prevalence and nature of spontaneous arterial disease in reptiles. Finlayson et al.[11] examined the hearts and aortas of 98 captive members of this class and found only minimal fatty streaking in 3% of the 98 animals. Their findings suggest that reptiles are distinct from mammals and birds, which are less immune to the development of both intimal fatty streaking and atheromatous lesions.

Snakes are poikilothermic animals that belong to class Reptilia, which includes turtles, crocodiles, alligators, and lizards. Reptiles flourished during the Mesozoic time (150,000,000 years ago) and were the first group among the vertebrates adapted for life in dry places on land.

Apparently, the reptiles are strikingly resistant to the development of atheroma, in spite of elevated plasma lipids in some species.[1] This difference in susceptibility to the development of atheroma could reflect differences in lipid and lipoprotein composition, endothelial structure and permeability to macromolecules, a low systemic blood pressure, or a difference in metabolism, i.e., being a poikilothermic animal. Snakes are also different from most animals because of their ability not to eat for long periods of time. Benedict[2] reported a boa which lived 1 year and 5 months without food. It is possible that hibernating reptiles have reduced body temperature and consequently have lower metabolic rates which enable them to survive the fast. Martin and Bagby[4] indicated that blood cholesterol concentrations showed a tendency to drop in male rattlesnakes when fasted for 20 weeks. This phenomenon was also reported in the Indian house lizard in which case the total body cholesterol remained fairly stable for 9 months out of the year but increased during the summer, presumably due to marked increase in dietary intake.[4a] Lance[5] in 1975 studied female cobras (*Naja naja*) and found considerable individual variation in plasma cholesterol concentrations (155 to 550 mg/dℓ) with significant seasonal differences with a sharp drop in concentration in May and increase in September, followed by a second decrease in November. He suggested that the decrease in plasma cholesterol concentration was strongly correlated with maximum ovarian development, and the decrease observed in November appeared to be associated with body fat deposition. Whether these differences in metabolism and biochemical changes during different seasons have any important physiological bearing on the snakes' resistance to atherosclerosis is not known.

This chapter will review the lipids and lipoproteins characteristic of snakes with particular emphasis on the Haitian boa constrictor. It becomes readily apparent that the amount of data on snakes are very limited and the reported data on serum lipoproteins are even more sparce. Some comparison will be made to other nonsnake species that belong to the reptilian family for comparison purposes only.

Snakes, in general, appear to have a relatively high concentration of cholesterol but low triglycerides. Table 1 summarizes the findings of various investigators on lipids and lipoproteins of snakes.

Ardlie and Schwartz[1] studied 148 noncaptive reptiles obtained from many parts of Australia of which 39 were snakes and 109 were lizards. One of the striking findings was observance of a wide range of serum total cholesterol values, 21 to 759 mg/dℓ. The lipoprotein pattern

Table 1
SERUM LIPIDS AND LIPOPROTEINS OF VARIOUS SNAKES

Species	Lipids (total cholesterol)	Lipoproteins	Ref.
Common garter snake	238 mg/dℓ	Two electrophoretic bands	7
Red-barred garter snake	179 mg/dℓ	One major electrophoretic band	7
Rattlesnake	158 ± 45.7	ND[a]	4
Cobra snake	155—550 mg/dℓ but were generally between 200—300 mg/dℓ; concentration dropped sharply in May and November and rose in September	ND	5
Pythons	50—270 mg/dℓ (TC)	15—20 mg/dℓ for β- cholesterol and 16—35 mg/dℓ for α-cholesterol	8
Green snake	ND	LDL = 1168 mg/dℓ, IDL = 92 mg/dℓ, VLDL = 64 mg/dℓ	6
Grass snake	ND	LDL = 359 mg/dℓ, IDL = 48 mg/dℓ, VLDL = 207 mg/dℓ	6
Water snake	ND	LDL = 267 mg/dℓ, IDL = 211 mg/dℓ, VLDL = 297 mg/dℓ	6
Rat snake	ND	LDL = 320 mg/dℓ, IDL = 111 mg/dℓ, VLDL = 43 mg/dℓ	6
Scrub python	205, 82—303	ND	1
Children's python	438, 250—759	ND	1
Olive python	265, 195—319	ND	1
Diamond python	107	ND	1
Carpet python	168, 290—715	ND	1
Brown tree snake	477, 290—715	ND	1

[a]   ND, not determined.

appears to be complex with a predominant concentration of low-density lipoprotein (LDL). Mills and Taylaur[6] studied the distribution and composition of serum lipoproteins in 18 animals and found that the green snake (*Liopeltis vernalis*) had a very high concentration of LDL (1168 mg/dℓ). All of the snakes (grass snake, *Natrix natrix;* water snake, *N. piscator;* rat snake, *Ptyas mucosus*) had LDL, while others had, in addition, intermediate density lipoprotein (IDL) and very low-density lipoprotein (VLDL). No high-density lipoproteins (HDL) were measured in their study.

Martin and Bagby[4] like Ardlie and Schwartz[1] showed large variation in concentration of serum total cholesterol in snakes (72 to 715 mg/dℓ) which could reflect species differences. A review by Chapman[9] suggests that other animals are not immune to this phenomenon.

Alexander and Day[7] studied 36 mammals, amphibians, fish, and birds. Using agarose-gel electrophoretic methods they found that the common garter snake (*Thamnophis sirtalis sirtalis*) and red-barred garter snake (*T.s. parietalis*) have two major bands, one of which they suggested was LDL.

Dangerfield et al.[8] examined more than 164 different species which included mammals, birds, and reptiles. The snakes comprised five species which included Indian pythons (*Python molurus*) and African pythons (*Python sebae*). As a group, the total cholesterol concentration in the blood was 50 to 270 mg/dℓ, of which most was in the β-cholesterol fraction. Lipo-

Table 2
PHYSICAL PARAMETERS OF *EPICRATUS S. STRIATUS*

Sex	Length (cm)	Body wt (g)	Liver wt (g)	Liver wt/ body wt
M	155	515	7.8	0.015
M	157	703	11.5	0.016
M	137	560	11.4	0.020
Mean ± SE	149 ± 6	593 ± 57	10 ± 1.2	0.017 ± 0.0015
F (gravid)	229	1845	18.1	0.010
F (gravid)	183	1770	33.5	0.019
F	163	961	14.5	0.015
Mean ± SE	191 ± 20	1525 ± 283	22 ± 5.8	0.0147 ± 0.0026
Overall mean	170 ± 13	1059 ± 245	16 ± 3.7	0.0158 ± 0.0014

protein electrophoresis studies suggested that the α-lipoprotein (α-Lp) had a very fast mobility compared to that of man. However, because 49% of the analyses were carried out on post-mortem blood samples, they suggested that the study be regarded with some circumspection.

Recently, Chapman[9] did a comprehensive review of animal lipoproteins of both invertebrates and vertebrates and concluded that snakes were unique in the animal kingdom because of the very high concentration of S_f 0 to 12 LDL, exceeding those of man. The tortoise differed from other reptiles, in that although it possessed a moderate quantity of HDL, it lacked VLDL and its concentration resembled that of amphibians.

Because no lipid and lipoprotein studies were done on the Haitian boa constrictor (*Epicrates s. striatus*) we examined six adult (three male, three female) snakes. The lipid, lipoprotein, protein, and histological methods are described in detail elsewhere.[10,14] The aims of the study were:

1. To examine whether another species of snakes from another geographical region is also resistant to atherogenesis
2. To study, in detail, the structure of the arterial wall by light and electron microscopic techniques
3. To characterize the serum lipid and lipoprotein by ultracentrifugal, electrophoretic, chemical, and gas chromatographic techniques

After 2 weeks' acclimation to laboratory conditions the snakes were placed in the cold room for about 1 hr and then anesthesized with amytol. Fasting blood was obtained from the aorta. Serum was collected and processed for lipid and lipoprotein studies. Preselected sites of the aorta were sectioned for light and electron microscopic studies. For the electron microscopic studies, two snakes were pressure perfused at 45 mm pressure with 2% glutaraldehyde fixative in 0.1 M sodium cacodylate buffer at pH 7.35. For light microscopic work, plastic embedded sections were stained with methylene blue-azure II and counterstained with basic fuchsin.

Table 2 provides the physical parameters of the individual animals. The serum lipid and lipoprotein concentrations are shown in Tables 3 to 6. Serum cholesterol was the major lipid comprising about 60 to 70% of the total lipid (Table 3). The range of total cholesterol was 137 to 405 mg/dℓ, of which ester cholesterol was the predominant fraction (60 to 70%). The serum phospholipids were the next major fraction, comprising 35% of the total lipids (Table 4). It was surprising to see only trace amounts of triglycerides (<10 mg/dℓ). The cholesterol ester fatty acid composition is shown in Table 6. Oleic acid (35.7%), arachidonic acid (27.2%), and linoleic acid (23.3%) were the major fatty acids which, in all probability, was a reflection of the animals' dietary intake. In comparison to the rat it is not considered

Table 3
SERUM LIPIDS IN *EPICRATUS S. STRIATUS*
(mg/100 mℓ)

Sex	Total chol	Triglycerides	Phospholipids	Total lipid
M	405	3	192	600
M	157	5	117	297
M	137	6	130	392
F	285	0	126	411
F	303	0	156	459
F	163	7	162	532
Mean ± SE	298 ± 33	3.5 ± 1	147 ± 11	449 ± 44

Table 4
SERUM LIPIDS IN *EPICRATUS S.*
STRIATUS
(% DISTRIBUTION)

Sex	% Chol	% TG	% PL
M	68	0.5	32
M	59	1.7	39
M	65	1.5	33
F	69	0	31
F	66	0	38
F	68	1.3	31
Mean ± SE	66 ± 1.5	0.8 ± 0.3	34 ± 1.5

Table 5
SERUM CHOLESTEROL OF *EPICRATUS S. STRIATUS*
(mg/dℓ)

Sex	Total cholesterol	Esterified cholesterol	Free cholesterol	% Esters
M	405	310	95	77
M	256	145	111	57
F	285	205	80	72
F	303	223	80	74
F	363	240	123	66
Mean ± SE	322 ± 27	225 ± 27	98 ± 9	69 ± 4

high. Dogs, on the other hand, do not have as much arachidonic acid, resembling the distributions found in man.

The preparative and analytical ultracentrifugal studies show that four of the six snakes had no VLDL (d < 1.006) while the LDL (d 1.006 to 1.063) was clearly the major serum lipoprotein (Table 7, Figure 1). The analytical ultracentrifugal pattern suggest that the S_f 10 to 20 fraction is a major fraction of d 1.006 to 1.063. This lipoprotein distribution is expected insofar as the low serum triglycerides and elevated cholesterol concentrations that were observed (Table 3). In two of the animals, the HDL (d 1.063 to 1.21) was surprisingly elevated, especially in view of the fact that they were males (Figure 2). The chemical compositional studies on the lipoprotein fractions indicate that about 85% of the serum lipids are carried in the d 1.006 to 1.063 fraction which consisted primarily of cholesterol and phospholipids with trace amounts of free fatty acids (FFA). The HDL fraction consisted of mainly phospholipids with trace amounts of cholesterol.

Table 6
COMPARISON OF PLASMA
CHOLESTERYL ESTER FATTY ACID
COMPOSITION OF SNAKE, RAT,
DOG, AND MAN
(% DISTRIBUTION)[a]

Fatty acid	Snake	Rat	Dog	Man
C14:0	0.5 ± 0.4	0.2	0.5	0.4
C16:0	3.6 ± 0.4	9.0	10.7	10.9
C16:1	2.6 ± 0.4	5.1	3.1	1.5
C18:0	1.3 ± 0.2	0.9	1.0	8.8
C18:1	35.7 ± 0.6	14.7	19.4	23.7
C18:2	23.2 ± 0.7	21.9	56.2	36.2
C18:3	1.0 ± 0.8	0.7	0.4	1.8
C20:0	0.8 ± 0.9	0.8	0.3	1.8
C20:4	27.2 ± 0.2	45.0	7.6	10.5
C22:0	0.5 ± 0.1	0.5	0.2	1.5
C24:0	0.6 ± 0.2	0.8	0.3	1.9
C26:0	3.0 ± 0.1	0.5	0.3	1.0

[a]   Snake (n = 6), rat (n = 10), dog (n = 8), man (n = 24).

Table 7
LIPOPROTEIN DISTRIBUTION IN
EPICRATUS S. STRIATUS
DETERMINED BY ANALYTICAL
ULTRACENTRIFUGATION
(mg/100 mℓ)

Sex	VLDL	LDL	HDL
M	26	260	247
M	0	216	54
M	24	349	162
F	0	311	78
F	0	333	54
F	0	332	78
Mean ± SE	8.3 ± 5.3	300 ± 21	112 ± 31

Serum electrophoretic studies on paper medium (Figure 3) clearly illustrated that the major lipid-stainable band migrates in the β-globulin area. This β-Lp band of snakes migrates faster than the corresponding β-Lp band of human beings, perhaps, because of the FFA content. The α-Lp is found in trace amounts on paper electrophoretic medium (except for the two snakes) with another lipid-stainable band in the albumin area. This band (d > 1.21), which is found ahead of the α-Lp, is the albumin-FFA complex and can be found in appreciable amounts in some snakes. Figure 4 shows the PAG densitometric tracing of the same sera used for the paper electrophoresis medium (Figure 3) where the α-Lp is barely visible. Also, the pre-β-Lp band is either absent or hardly visible, which is in keeping with the low serum triglyceride concentrations.

Histological studies of the descending aortas revealed no lesions whatsoever in any of the six snakes. This is in agreement with the study of Finlayson et al.[11] and Ardlie and Schwartz.[1]

The aortas of these snakes exhibit a common overall morphology regardless of sampling site from the triple arch, through the double aorta to the distal part of the single aorta just above the level of the cloaca.

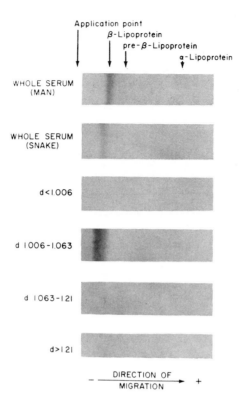

FIGURE 1.   Paper electrophoresis of whole serum and isolated lipoprotein fractions (VLDL, LDL, HDL, albumin-FFA complex) of snake.

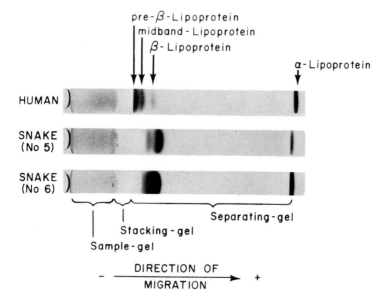

FIGURE 2.   PAGE disc patterns of snake serum lipoproteins, one with low α-Lp, and one with high α-Lp.

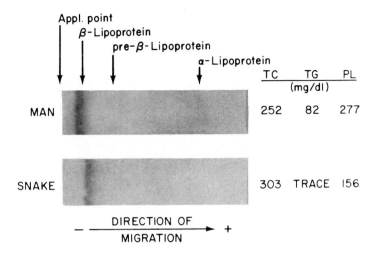

FIGURE 3. Comparison of serum lipoproteins of snake on human beings on paper electrophoretograms.

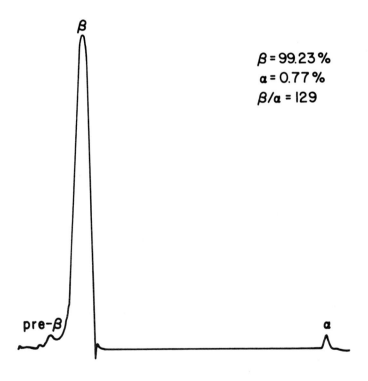

FIGURE 4. Densitometric tracing of paper electrophoretic pattern of snake serum in Figure 3.

Elastic laminae are well defined but thin, ranging in number from 8 to 12, regardless of the site (Figure 5). They are frequently branched both between and within laminal tracts. Elastic laminae do not in general fill all of the interstitial space between smooth muscle cell layers, a condition differing from that in the sexually mature mammalian aorta. This space is largely filled with copious amounts of collagen, primarily orientated along the long axis of the vessel.

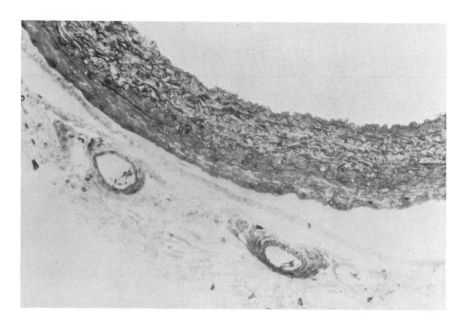

FIGURE 5. Cross-section of aorta of snake. (Magnification × 10)

Medial smooth muscle cells are elongate and extremely large compared to mammalian cells. They contain very few cellular organelles, their cytoplasm being mostly occupied with myofilaments. Dense bodies (attachment sites for myofilaments) are conspicuous both within the cytoplasm, and particularly along the plasma membranes. The basement membrane is very prominent.

The endothelium of these vessels generally rests immediately on the internal elastic lamina, precluding the existence of a subendothelial space (SES). Where present, the SES contains collagen fibrils, elastic tissue, and an occasional smooth muscle cell. The endothelium is thin and tenuous, but unremarkable and similar to other aortic endothelia in other animals except for one feature. There are numerous electron-dense, membrane-bound granules both scattered along the length of the cell and clustered in groups in the cytoplasm. At low magnification they have the appearance of Weibel-Palade bodies, but at higher magnification, they do not demonstrate the prominent microtubular substructure known to exist in the latter with the fixation technique used (Figure 6). Rather, their substructure appears granular or crystalline, depending on the plane of section. The endothelial glycocalyx, as observed by ruthenium red staining, is thin and consistent in thickness along the length of the vessel. The endothelial basement membrane is discontinuous and variable in thickness.

The adventitia of these vessels presents a very interesting structure, and is very different from that found in the mammals. Lumenally, there is a thick layer of dense collagenous tissue in which numerous fibrocytes are embedded. Exterior to this, numerous vasa vasorum and nerves can be seen. These do not enter the media. In this area also are frequently seen large tissue macrophages and other unidentified granulated cells. Other cells common to this area have a cytoplasm totally filled with ovoid profiles of dilated rough endoplasmic reticulum containing granular material. Of greatest interest is that the entire length of the aorta is enclosed by a monolayer of thin epithelioid cells located, in section, at the exterior boundary of the adventitia. The vessel is, in essence, contained in its own visceral cavity, and separated from the other organs by this cellular layer. It is perhaps an extension of the pericardium. To my knowledge, this has never been described in any other animal, and its significance is unknown.

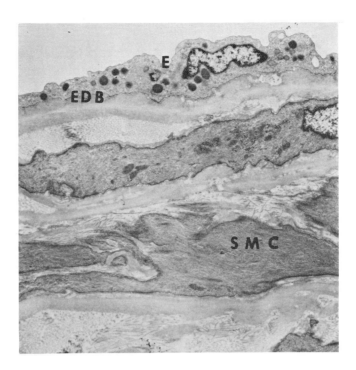

FIGURE 6.   Electron micrograph of snake aorta showing endothelium (E)
with electron-dense bodies (EDB), smooth muscle cell (SMC).

I would term these vessels normal in structure. There is no lipid accumulation in any cell type or extracellularly, and no lesion of any kind was observed. Aside from the differences from the normal mammalian aorta described above, which are probably the result of biological specialization, there is nothing to suggest these vessels are abnormal in any way.

The serum protein pattern of snakes as demonstrated by protein electrophoresis (using cellulose acetate medium) shows that the distribution of proteins is slightly different from that of man (Table 8). The relative percent of albumin is lower than that of man and the β-globulins are higher with a significant amount of another protein-stainable material in the post-γ-globulin area. The total protein content of the boa constrictors' sera was 6.06 ± 0.06 mg/dℓ (mean ± SEM), slightly lower than that of human beings. The significance of this is not apparent at this time.

It would appear that high serum cholesterol, hence LDL, alone is not a sufficient stimulus for atherogenesis to occur in Haitian boa constrictors. Snake aortic blood pressure is reported to be 50 to 60 mm Hg.[12] Fantl[13] reported that snakes have prolonged clotting times. In addition to the snakes being poikilothermic animals, the architecture of the aorta of snakes is slightly different from that of other species. These factors may all play a vital role in the snake's resistance to the formation of atheromatous lesions despite their high serum cholesterol and LDL concentrations. In summary:

1. Despite the elevated serum total cholesterol, no sponanteous arterial lesions were observed in the Haitian boa constrictor.
2. No intra- or extracellular lipids were observed in the arterial wall.
3. There were some unique arterial morphologic features: i.e., electron-dense bodies in the endothelium, frequently branched elastic laminae which do not fill the interstitial space, very large smooth muscle cells, and the enclosure of the adventitia with a monolayer of thin epitheloid cells.

Table 8
PROTEINS OF *EPICRATUS S. STRIATUS*
(MEAN ± SEM)

Animal	Total protein (mg/dℓ)	Relative percent distribution					
		Post γ	γ	β	α_2	α_1	Albumin
Man	7.5	—	13.6	9.2	8.9	3.8	64.5
	± 1.0		± 3.5	± 1.6	± 1.2	± 1.2	± 4.7
Snake	6.06	9.5	15.5	25.2	8.4	2.8	38.5
	± 0.6	± 2.5	± 2.8	± 6.1	± 2.0	± 1.2	± 5.1

4. Serum triglycerides and pre-β-Lp concentrations were in low or trace amounts.
5. The serum lipoprotein lipid composition was also very unusual.
6. These unique biochemical and morphologic features of the serum and arterial blood vessels warrant further studies to understand the unusual resistance of this reptile to atherogenesis.

ACKNOWLEDGMENTS

This study was supported by Grant No. HL-6835 from the National Heart, Lung, and Blood Institute, and Grant No. 8557 from the Bleeksma Fund. I am grateful to Dr. Ross Gerrity for providing the electron micrographs, and I would like to express my gratitude to Debra Maciejewski for typing this manuscript.

REFERENCES

1. **Ardlie, N. G. and Schwartz, C. J.**, Arterial pathology in the Australian reptile — a comparative study, *J. Pathol. Bacteriol.*, 90, 487—494, 1965.
2. **Benedict, F. G.**, The physiology of large reptiles with special reference to the heat production of snakes, tortoises and alligators, Lubrecht and Cramer, Monticello, New York, 1973 (reprint from 1932 edition).
3. **Klauber, L. M.**, Rattlesnakes. *Their Habits, Life Histories and Influence of Mankind,* University of California Press, Berkeley, 1956.
4. **Martin, J. H. and Bagby, R. M.**, Effects of fasting on the blood chemistry of the rattlesnake, *Crotalus atrox, Comp. Biochem. Physiol.*, 44A, 813—820, 1973.
4a. **Sanyal, M. K. and Prasad, M. R.**, Seasonal variations in testis and total body cholesterol in the Indian house lizard, *Hemidactylus nauiridus* Ruppell, *Steroids*, 6, 313—322, 1965.
5. **Lance, V.**, Studies on the annual reproductive cycle of the female cobra, *Naja Naja.* I. Seasonal variation in plasma cholesterol, *Comp. Biochem. Physiol.*, 52A, 519—525, 1975.
6. **Mills, G. L. and Taylaur, C. E.**, The distribution and composition of serum lipoproteins in eighteen animals, *Comp. Biochem. Physiol.*, 4013, 489—501, 1971.
7. **Alexander, C. and Day, C. E.**, Distribution of serum lipoproteins of selected vertebrates, *Comp. Biochem. Physiol.*, 4613, 295—312, 1973.
8. **Dangerfield, W. G., Finlayson, R., Myatt, G., and Mead, M. G.**, Serum lipoproteins and atherosclerosis in animals, *Atherosclerosis*, 25, 95—106, 1976.
9. **Chapman, M. J.**, Animal lipoproteins: chemistry, structure and comparative aspects, *J. Lipid Res.*, 21, 789—853, 1980.
10. **Naito, H. K. and Gerrity, R. G.**, Unusual resistance of the ground squirrel to the development of dietary-induced hypercholesterolemia and atherosclerosis, *Exp. Mol. Pathol.*, 31, 452—467, 1979.
11. **Finlayson, R., Symons, C., and T-W-Finennes, R. N.**, Atherosclerosis: a comparative study, *Br. Med. J.*, 5227, 501—507, 1962.

12. **Johansen, K.,** Circulation in the three-chambered snake heart, *Circ. Res.,* 7, 828—832, 1959.
13. **Fantl, P.,** A comparative study of blood coagulation in vertebrates, *Aust. J. Exp. Biol. Med. Sci.,* 39, 403—412, 1961.
14. **Naito, H. K., Holzbach, R. J., and Corbusier, C.,** Characterization of serum lipids and lipoproteins of prairie dogs fed a chow diet or cholesterol-supplemented diet, *Exp. Mol. Pathol.,* 27, 81—92, 1977.

A BROAD PERSPECTIVE ON INVERTEBRATE AND LOWER VERTEBRATE APOLIPOPROTEINS: CRUSTACEANS, INSECTS, FISH, AMPHIBIANS, AND REPTILES*

M. John Chapman

INTRODUCTION

Although the importance of the protein components, i.e., the apolipoproteins, of the serum lipoproteins has been recognized for more than 50 years, it is only within the past decade that we have come to understand the complexities of their structure and of their metabolic roles. Furthermore, knowledge of the mechanisms and sites of their biosynthesis and degradation is now available, to the extent that on the one hand, the structure of a gene coding for an avian apolipoprotein has recently been established,[1-4] while on the other, a membrane receptor mediating the cellular degradation of an apolipoprotein has been isolated from bovine tissue and characterized.[5,6]

Such advances have been made in species which appeared relatively late in evolution, and thus we remain largely ignorant of the molecular biochemistry of lipid transport systems in many animals, and more especially in the invertebrates and certain of the lower vertebrates. Nonetheless, the rudimentary data available suggest that even in invertebrates such as the Arthropoda (specifically the crustaceans and insects), lipids are primarily transported in intimate association with certain specialized proteins; these lipid-protein complexes take the form of soluble hemolymph lipoproteins whose precise physicochemical properties vary according to species.[7,8]

The aim of this treatise is, then, to review present knowledge of the apolipoproteins of crustaceans and insects, together with those of certain lower vertebrates, namely of fish, amphibians, and reptiles, and whenever possible, to compare them with their counterparts in man. In order to place these observations in perspective, discussion of the apolipoproteins of each group of invertebrates or vertebrates will be preceded by consideration of the nature of the lipid transport system in which they occur. In this regard, particular emphasis will be given to the chemical and physical properties of the various classes of lipoprotein particle constituting each system. Wherever possible, animal counterparts to human serum apolipoproteins will be designated according to the nomenclature propounded for the latter by Alaupovic.[9]

INVERTEBRATES

The invertebrates constitute a large and highly diversified section of the animal kingdom, ranging from the bacteria and protozoa, through the coelenterates (jellyfish and corals) to the chordate ancestors (lancelets and sea squirts) on the one hand, and on the other, to the platyhelminthes (flatworms), nematodes (roundworms), annelids (segmented worms), molluscs (snails), and arthropods. Of the invertebrate phyla, only that of the arthropods has received attention from students of lipid transport, and this extends to two of its four classes, i.e., the crustaceans and insects.

Crustaceans

A considerable amount of information is presently forthcoming on the absorption, biosynthesis, and tissue distribution of free and esterified cholesterol and plant sterols in crus-

---

* A list of abbreviations follows the text on page 307.

taceans.[10-12] In contrast, we remain ill informed of the manner by which these species effect the transport and exchange of sterols and other lipids between their sites of absorption and formation and those of storage and utilization. Nonetheless, a variety of forms of lipoprotein have been superficially documented in the hemolymph of certain marine decapod crustaceans, namely crabs and lobsters;[13-19] regrettably, however, the procedures typically employed to detect and characterize them have been limited to electrophoresis,[16-19] and further analyses have been lacking. Most of these lipoproteins appear to be female specific and closely linked to egg formation. In their physical and chemical properties, they approximate the high-density substances of mammalian plasma. For example, the major lipoprotein could be isolated ultracentrifugally from the spiny lobster (*Panulirus interruptus*) in the density range corresponding to an HDL_3, and contained 50 to 55% protein.[13] Similarly the hemolymph lipoprotein of the blue crab, *Callinectes sapidus*, isolated by paper curtain electrophoresis, contained some 72.5% protein, thereby resembling the VHDL of man.[14] Despite the paucity of data on the lipid composition of crustacean lipoproteins, the predominant lipid is seemingly phospholipid,[13,14] with lesser amounts of cholesterol, triacylglycerol, and partial glycerides; minor amounts of hydrocarbons, wax esters, cholesteryl esters, and free fatty acids (FFA) have also been detected, either in the lipoproteins themselves or in the unfractionated hemolymph.[13-15] Carotenoids are generally present too, an indication that crustacean lipoproteins not only serve to transport lipids between a variety of tissues, but also to supply nutrients to the maturing oocyte, itself rich in β-carotene.[14]

Data on the protein moieties of such lipoproteins is cursory, this aspect having been addressed in only two studies, that of Lee and Puppione in the spiny lobster (*P. interruptus*),[13] and that of Kerr in the blue crab, *C. sapidus*.[14] In the former, the apoprotein of the hemolymph HDL_3 became insoluble upon delipidation by TMU.[13] In this respect, the lobster protein resembled apolipoprotein B (apo B) in man, which if not modified chemically prior to delipidation, is only soluble in detergents or solutions of strong denaturing agents.[20] Similar behavior was described by Kerr for the apoprotein of the hemolymph lipoprotein of the blue crab,[14] whose dissolution upon delipidation could only be assured in SDS solution. Such observations imply that an apo B-like protein may have been the fundamental vector for lipid transport in the primitive circulatory systems of lower animals and raise a number of fascinating questions. For example, what is the structural relationship between this apolipoprotein and human apo B, and do regions of homology exist? Moreover, if indeed the crustacean protein is a counterpart to apo B, one may ask whether the delivery of cholesterol to organs and tissues is assured by the same cell membrane receptor as that discovered and documented by Brown and Goldstein in mammalian tissues.[21]

Insects

Whereas the hemolymph lipoproteins of the crustacea appear to be relatively undifferentiated, simultaneously supplying lipids both for egg formation as well as for more general metabolic requirements, the lipid transport systems of insects are generally more complex. Thus, insects typically possess a distinct group of lipoproteins which are associated with vitellogenesis, i.e., the vitellogenins, and a separate group which transport lipids for use in such diverse processes as locomotion, embryogenesis, metamorphosis, and hormone biosynthesis. In this respect, they appear to be the earliest members of the animal kingdom in which separate classes of lipoproteins have evolved to fulfill specific transport roles. This latter group have recently been termed "lipophorins" by Chino and colleagues,[22] a designation which should facilitate some clarification of nomenclature in this confused area. These authors have on previous occasions referred to the lipophorins as "diacylglycerol-carrying lipoproteins".[22]

Lipophorins

Characteristics which may be used to differentiate between the lipophorins and the vitellogenins are the predominance of diacylglycerol as the major lipid (in the range of 10 to 35%) and the superior proportion of total lipid in the former (representing from some 35 to 75% of total mass), endowing them with hydrated densities comparable to LDL and HDL in man. Moreover, while the vitellogenins function exclusively in egg formation (ovogenesis), the lipophorins serve to transport lipids (mainly diacylglycerol) from both sites of absorption (intestine) as well as storage (in the fat body) and synthesis to sites of deposition or utilization. Among the lipids transported by the lipophorins are diverse sterols, which due to the inability of both crustaceans and insects to synthesize cholesterol must be obtained from dietary sources.[10-12]

As is evident in Table 1, lipophorins have often been isolated by ultracentrifugal flotation,[23,29] so taking advantage of their low densities; alternatively, precipitation procedures, followed by ion-exchange chromatography, have been used.[26,27] The major lipid carrier in pupal and adult insect hemolymph is almost invariably a lipoprotein of high density (i.e., of d 1.063 to 1.21 g/mℓ). To date, the only insect lipoprotein isolated in a density interval (d 1.046 to 1.063 g/mℓ) corresponding to LDL in man is that of the saturniid silkmoth, *Hyalophora cecropia*.[23] In addition to diacylglycerol (DG), it contained significant amounts of other lipids, such as sterols, sterol esters, hydrocarbons, and phospholipids. Indeed, the contents of these lipids vary markedly between the HDL fractions of the various species appearing in Table 1, to the extent that hydrocarbons are a major component of the lipophorins of one silkmoth (*H. cecropia*), but are absent from that of another (*P. cynthia*).[23,24] The high contents of hydrocarbons in the HDL of the American cockroach and of the locust are noteworthy, especially since these are important constituents of their cuticular lipids.[26,27] Triacylglycerols, monoacylglycerols, and nonesterified fatty acids are consistently minor components. Only the contents of phospholipid and protein in the insect substances are akin to those seen in human HDL, a certain reflection of the contrasting metabolic roles of these macromolecules, diacylglycerol functioning as the essential form in which fat is transported in insects.[32] The principle phospholipid of insect lipoproteins is phosphalidyl choline,[23,24,28,29] in which the major fatty acids in locust HDL are palmitate (16:0) and oleate (18:1);[25] these same fatty acids also predominate in the diacylglycerols of *L. migratoria*.[25] Finally, it should be mentioned that for the most part, the lipophorins appear to be glycolipoproteins, containing as much as 2% by weight of carbohydrate.[26-29]

Vitellogenins

By contrast with the lipophorins, the vitellogenins form a more homogeneous group, typically containing 90% or more protein; this characteristic affords them hydrated densities (>1.21 g/mℓ) well in excess of those of the lipophorins and has occasionally been employed for their isolation in purified form (Table 1).[23,29,31] Diacylglycerol is again preponderant among their lipid components, although representing only 3% or less of the total weight of the particle. As the DG content of the egg vitellin is still lower, at least in the locust,[30] it has been suggested that this lipid is progressively released from vitellogenin once within the oocyte.[30]

In their physical properties, and specifically particle diameter and molecular weight, the insect lipophorins and vitellogenins both exhibit some resemblance to the HDL$_2$ of man, but are quite distinct from human HDL$_3$ and VHDL.[20,23-33] In protein contents however, they are clearly more akin to the latter substances.

It would be misleading to leave the impression that the only insect species in which lipoproteins have been detected are those listed in Table 1. In fact, electrophoresis has served to identify lipid-protein complexes in numerous insects and particularly in Lepidopterans; such lipoproteins have been detected on the basis of their specific staining with lipophilic dyes after hemolymph electrophoresis in a variety of supporting media. Examples are the lepidopterans *Hyalophora gloveri*, *Antheraea polyphemus*, *A. mylitta*, and *Callosamia pro-*

Table 1
CHEMICAL COMPOSITION (MEAN WEIGHT %) OF INSECT HEMOLYMPH LIPOPROTEINS: LIPOPHORINS AND VITELLOGENINS

Species	Lipophorins									Vitellogenins				
	Hyalophora cecropia (silkmoth)	*H. cecropia*	*Philosamia cynthia* (silkmoth)	*Locusta migratoria* (locust)	*L. migratoria*	*Periplaneta americana* (cockroach)	*Manduca sexta* [a] (tobacco hornworm)	*Galleria mellonella* [b] (waxmoth)	*G. mellonella* [b]	*G. mellonella* [b]	*L. migratoria*	*H. cecropia*	*M. sexta* [c]	*P. cynthia*
Density (g/mℓ)	1.046—1.063	1.158—1.170	—	—	—	—	1.15	1.05—1.10 (Ia)	1.152—1.160 (Ib)	1.265	—	< 1.26	1.29	—
Human counterpart	LDL	HDL	HDL	HDL	HDL	HDL	HDL	HDL	HDL	VHDL	VHDL	VHDL	VHDL	VHDL
Component														
Triacylglycerols	8.0	2.9	0.5	0.6	0.7	1.0	2.3	2.0	1.3	0.3	TR	0.5		
Diacylglycerols	33.4	27.2	24.8	13.1	13.4	7.6	12.6	20.4	10.8	2.9	1.9	2.0		3.3
Monoacylglycerols	2.1	1.0	ND	1.8	ND	ND	0.2	0.1	0.1	TR	—	0.2	ND	ND
Sterols	8.9	2.6	5.8	2.3	3.2	2.5	2.0	2.9	1.3	0.4	0.6	0.9		1.2
Sterol esters	4.4	2.0	ND	ND	0.1	ND	[5.0]	1.1	0.6	0.1	—	0.2	[5.8]	
Freefatty acids	1.6	1.3	—	1.5	—	—	1.0	6.1	3.2	0.5	—	0.4		
Hydrocarbons	9.3	2.4	11.4	12.1	8.6	14.2	[5.0]	—	—	—	—	0.3		ND
Phospholipids	8.4	8.6	—	—	14.8	21.4	14.0	7.3	4.0	1.2	5.7	1.7	5.6	ND
Protein	24.0	52.0	56.0	68.5	59.0	50.0	61.0	58.0	77.7	94.3	91.8	94.0	86.0	90.3
Ref.	23	23	24	25	26	27	28	29	29	29	30	23	31	24

Note: ND, not detectable; TR, trace.

[a] Reference 28: [], sterol esters and hydrocarbons determined together.
[b] Reference 29: 20% lipid unidentified in HDL subfraction Ia, 0.9% in Ib, 0.3% in VHDL.
[c] Reference 31: [] neutral lipids quantitated together, sterol esters and hydrocarbons absent.

methea,[34] the grasshopper (*Melanoplus differentialis*),[35] the fruit fly (*Drosophila hydei*),[36] the housefly (*Musca domestica*),[37] the Colorado potato beetle (*Leptinotarsa decemlineata*),[38] and the silkworm (*Bombyx mori*).[39]

Data on the circulating concentrations of the lipophorins and vitellogenins in the various insect species is notably lacking, possibly as a result of difficulties involved in assessing the total volume of the "open" hemolymph compartment in these Arthropods. The studies of Chino and colleagues do, however, indicate that hemolymph lipoprotein protein may in certain cases, such as the cockroach, account for up to half of the total protein present.[26,27]

Apolipoproteins

In considering the apolipoproteins of insect hemolymph, the same distinction between lipophorins and vitellogenins as that denoted above will be respected.

Our present knowledge of insect apolipoproteins is restricted to four species, the American cockroach (*P. americanus*),[27] the locust (*L. migratoria*),[26] the tobacco hornworm (*M. sexta*),[28] and the silkworm (*P. cynthia*).[24] Moreover, analyses are limited to amino acid compositions of the delipidated, unfractionated apoproteins and to polyacrylamide gel electrophoresis (PAGE) in SDS, with the exception of *M. Sexta* apo HDL in which the constitutive polypeptides have been separated and purified.[28] Nonetheless, the resemblance between the amino acid compositions of the protein moieties of the HDL from the respective species is remarkable (Table 2). Thus, all display elevated levels of aspartic and glutamic acids, lysine, valine, and leucine. Closer inspection reveals, as might be predicted, that these apo HDLs uniformly exhibit markedly similar polypeptide band patterns upon SDS-gel electrophoresis. This pattern is characterized by two distinct components, one of high and the other of low molecular weight (2.5 to 2.8 $\times$ 10^5 and 8.1 to 8.5 $\times$ 10^4, respectively). Only the larger subunit appears to contain carbohydrate, consisting mainly of mannose.[26,27] When the two subunits were separated from tobacco hornworm HDL by gel filtration chromatography in SDS, they possessed amino acid profiles which were essentially identical (Table 2).[28] Only their location within the native particle differed, the larger polypeptide being entirely accessible to tryptic cleavage while the smaller subunit remained intact. This observation led Pattnaik et al. to propose that a large portion of the smaller component might lie inside the particle.[28]

Like its HDL, the apoprotein of *M. sexta* vitellogenin (VHDL) is constituted of two discrete polypeptide chains, although of lower molecular weight (1.8 $\times$ 10^5 and 5 $\times$ 10^4, respectively) than their counterparts in apo HDL. Two subunits have also been identified in the vitellogenins of *P. cynthia*,[40] *Blattella germanica*,[41] *H. cecropia*,[41] and *Culex pipiens*.[42] In contrast, multiple components have been described in the apoproteins of the locust, which appear to exist in an oligomeric series.[43]

In an attempt to relate the structural features of the various polypeptide subunits of insect vitellogenins, Mundall and Law have proposed that they are derived from some simple form, probably corresponding to that of *M. sexta*,[31] *P. cynthia*,[40] or *H. cecropia*;[41] subsequently the vitellogenin might be subject to posttranslational modification. Indeed, this occurs in *L. migratoria* in which the apoprotein is first synthesized in the fat body as a precursor of 26,000 mol wt, and is later modified by proteolysis and aggregation.[43] The general validity of this hypothesis however awaits further data on vitellogenins in a wider range of insects.

Three comments are pertinent in concluding this section: first, that the insect lipophorins must be considered as multifaceted lipid carriers, being distinct from the plasma lipoproteins of the higher vertebrates;[26] second, that their molecular organization differs markedly from that of the mammalian HDL particle with some protein occupying the core region;[28] and third, that the apolipoprotein systems of insects seem infinitely less developed than those of higher vertebrates, displaying only two polypeptides which in their solubility are more akin to human apo B than to apo AI or apo AII.

Table 2
AMINO ACID COMPOSITION OF THE PROTEIN MOIETIES AND PURIFIED APOLIPOPROTEINS OF INSECT HEMOLYMPH HIGH-DENSITY LIPOPROTEINS (LIPOPHORINS) (mol/100 mol)

| | Apo HDL | | | | Apolipoproteins (*M. sexta*)[a] | |
Species	*Philosamia cynthia* (silkworm)	*Periplaneta americana* (cockroach)	*Locusta migratoria* (locust)	*Manduca sexta* (tobacco hornworm)	Apo (L)	Apo (S)
Amino acid						
Lysine	10.7	9.3	9.2	8.5	7.9	7.6
Histidine	2.8	3.9	3.7	2.9	3.3	3.2
Arginine	3.7	2.7	2.6	4.1	4.1	5.4
Aspartic acid	12.6	11.0	11.4	12.5	10.3	10.9
Threonine	4.9	6.6	6.0	3.9	4.5	3.4
Serine	6.9	6.9	7.1	7.7	6.6	8.0
Glutamic acid	10.4	10.8	11.2	9.7	9.9	9.8
Proline	4.7	3.8	3.8	5.0	4.6	4.6
Glycine	6.7	6.4	6.3	6.6	7.2	8.1
Alanine	6.3	6.8	7.2	7.5	7.7	8.6
Valine	7.4	8.4	8.4	7.1	7.8	7.6
Methionine	5.0	3.0	2.0	1.3	1.5	1.8
Isoleucine	5.8	4.1	4.2	5.4	5.9	4.8
Leucine	9.0	10.7	10.6	8.7	9.4	9.3
Tyrosine	2.8	3.0	2.7	3.8	3.8	3.2
Phenylalanine	4.8	4.7	4.7	5.2	5.0	3.4
Cysteine	NM[b]	6.0	7.0	NM	NM	NM
Ref.	24	27	26	28	28	28

Note: Results expressed as mol/100 mol. of amino acid residues.

[a] The large (L) and small (S) subunit apolipoproteins of *M. sexta* HDL were isolated by gel-filtration chromatography using an SDS-containing elution buffer.

[b] Not measured.

VERTEBRATES

The Fishes

The earliest recognized representatives of the vertebrates are the fishes, and more specifically, the Agnates, i.e., cyclostomes or jawless fish. This class is one of three into which the fishes are grouped taxonomically, and one of seven into which the vertebrates are subdivided as a whole. As shown in Table 3, the remaining two groups of fishes are the Elasmobranchs or cartilaginous fish, and the Osteochthyes or bony fish. The latter group encompasses the Crossopterygii and Actinopterygii, subclasses which comprise the ancient and modern bony fishes, respectively; as such they account for by far the greatest number of living fish species.

Lipoproteins

It is in the fish that the type of lipoprotein profile typical of man and other mammals,[44,45] i.e., an essentially continuous spectrum of particles with densities spanning the range ~0.98 to 1.21 g/mℓ and with concentration minima in the regions of 1.006 and 1.063 g/mℓ, makes its earliest appearance. Indeed, such a profile is evident in the hagfish, *Myxine glutinosa*, the only Cyclostome studied to date,[46-48] with the exception that *Myxine* possesses large

Table 3
THE THREE MAIN GROUPS OF LIVING FISH

Class	Subclass	Order	Members
Cyclostomata (Agnates)			Hagfish, lampreys
Elasmobranchii			Sharks, dogfish, skates,
(Chondrichthyes)			rays
Osteochthyes	Crossopterygii		Lungfish, coelocanth
	Actinopterygii	Chondrostei	Sturgeons
		Teleostii	Modern bony fish

amounts of substances of intermediate density (i.e., IDL) which overlap the d 1.006 g/mℓ region.

These data bear further scrutiny however. Thus, in almost all analytical ultracentrifugal analyses of piscine serum lioproteins conducted to date, the same criteria as those used to define the density limits of VLDL, LDL, and HDL (i.e., d >1.006 , 1.006 to 1.063 and 1.063 to 1.21 g/mℓ, respectively) in man have been tacitly applied.[44] Indeed, this comment is not restricted to fish, but rather is relevant to many reported studies of lipoprotein profile in a large variety of animals. The first limitation of this approach arises from the fact that the densities at which minima occur in the species under study may be quite distinct from that of man, as is the case in the adult rainbow trout, *Salmo gairdnerii*, for example, which lacks a minimum at d 1.063 g/mℓ.[49] Secondly, in man the LDL region is characterized by the presence of particles containing primarily apo B, and that of HDL by apo AI- and apo A-II-containing substances; in fish, apo B-containing particles may contribute considerably to quantitation of material in the HDL region,[49,50] while conversely, an appreciable amount of apo A may be present in particles floating in the LDL region. This latter situation occurs to a marked degree in the dog (*Canis familiaris*)[51] and European badger (*Meles meles L.*),[52] but seemingly to a minor degree in fish, at least it seems so in Teleosts.[49,50,53] A further qualification of the analytical ultracentrifugal quantitation of fish lipoproteins concerns the application of the same constants (such as refractive increment, partial specific volume, and concentration dependence of flotation) which are only available for human lipoproteins.

The published data on lipoprotein levels in the three classes of fishes are presented in Table 4; most were obtained by an analytical ultracentrifugal methodology. The marked variations in absolute concentrations are immediately evident; for example, the prespawning pink salmon, *Oncorhynchus gorbuscha*, lacked detectable amounts of both VLDL and LDL, but displayed one of the highest HDL levels (>3000 mg/100 mℓ serum) seen to date. Elevated HDL concentrations were also a feature of the rainbow trout and pacific sardine. By contrast, HDL was uniformly found at extremely low levels in Elasmobranchs, and in addition, was associated with high concentrations of VLDL (with the exception of the dogfish, *Scyliorhinus canicula*)[48] and moderate amounts of LDL.

Again, a cautionary note is in order. Thus, in several instances, measurements have been made on pooled plasmas or sera from fish whose sex, age, and nutritional status were not established. Moreover, to the knowledge of this author, studies have not been performed as a function of the growth and development of the respective species; such investigations would be invaluable, particularly in Teleosts which frequently display well-defined life cycles. In the light of the above considerations, the data in Table 4 should therefore be regarded within the context of the physiological state of the donor animal(s), if indeed this was determined.

Clearly then, the fishes display complex, multicomponent lipoprotein profiles whose characteristics are more akin to those of the mammalia than to the relatively simple lipid transport systems of the invertebrates. Such dissimilarities are entirely compatible with the

Table 4
THE CONCENTRATIONS OF SERUM LIPOPROTEINS IN CYCLOSTOMES AND FISH[a,b]

	VLDL		LDL		HDL	Ref.
	S_f 100—400	S_f 20—100	S_f 12—20	S_f 0—12		
Cyclostomata						
Myxine glutinosa (hagfish)	35—593	515—1074	139—221	189—489	553—753	46—48
Elasmobranchii						
Centrophorus squamosus (shark)	159	256	45	185	40	47,48
Centrophorus granulosus (shark)	133	132	36	98	23	47
Centroscymnus coelolepsis (shark)	3	193	93	52	38	47
Scyliorhinus canicula (dogfish)	0	28	18	136	23	48
Conger vulgaris (conger eel)	228	228	36	189	ND[f]	46,48
Squalus acanthias[c] (spiny dogfish)		307	120		33	54
Actinopterygii						
Chondrostei						
Acipenser stellatus (sturgeon)	201	147	9	14	204	47
Acipenser guldenstadtii (sturgeon)	334	162	76	88	672	47
Huso huso (sturgeon)	500	226	155	355	649	47
Teleostii						
Pleuronectes platessa (plaice)	0	9	4	29	ND	46
Oncorhynchus nerka[d] (sockeye salmon)		167	246		238	55
Oncorhynchus gorbuscha (pink salmon)	0	0	0	0	3300	53
Salmo gairdnerii (rainbow trout)	0—1	8—46	11—44	105—220	1500	49
Salmo gairdnerii R.[d] (rainbow trout)	201—212		193—392		1062—2216	50
Sardinops caerulea G[d] (Pacific sardine)	~115		145		1120	56
Crossopterygii						
Latimeria chalumnae (coelocanth)	230	875	120	74	127	48,57
Man (urban)[e]	49	83	53	321	230	45

a Concentrations are expressed as mg lipoprotein/100 mℓ of serum.

b Distributions were normally determined by analytical ultracentrifugation (48) on pools of serum; S_f is defined as the sedimentation flotation rate in Svedberg units, at a solvent density of 1.063 g/mℓ at 26°C. HDL is 1.063—1.21 g/mℓ.

c VLDL is of d < 1.019 g/mℓ and LDL, 1.019—1.063 g/mℓ; values include FFA content and cholesterol as total sterol.

d Data based on lipoprotein recoveries after centrifugal flotation; values in *S. caerulea* G were calculated assuming protein contents of 10% in VLDL, 20% in LDL, and 50% in HDL.[56]

e Data from fasting males, age 16—29, and excluding HDL_1.

f ND, not determined.

From Chapman, M. J., *J. Lipid Res.*, 21, 789, 1980. With permission.

more intricate, "closed" vascular systems of the fishes, and concomitantly, with the diversity of their tissues and organs.

At the same time however, at least one feature of the chemical properties of piscine lipoproteins is indicative of the intermediary and transitional position of these poikilotherms between the invertebrates and chordate ancestors on the one hand, and the amphibians, reptiles, birds, and higher vertebrates on the other. Thus, while the chemical compositions of the major classes of fish lipoproteins are comparable to those of their human counterparts, i.e., VLDL, LDL, and HDL (Table 5), they could nonetheless be distinguished by their contents of certain "primitive" lipids, i.e., hydrocarbons and monoalkyldiacylglycerols, the former being a frequent constituent of insect lipoproteins (Table 1). The presence of such lipids was however restricted to the lower piscine species, i.e., the cyclostome *Myxine*, the Elasmobranchs (the sharks, *Centrophorus squamosus*, *C. granulosus*, and *Centroscymnus coelolepsis*) and the Chondrosts (the sturgeons, *Acipenser stellatus*, *A. guldenstadtii*, and *Huso huso*). In this regard, it is noteworthy that in the exhaustive lipid analyses of human serum lipoproteins of Skipski et al.,[58] trace amounts of hydrocarbons were detectable.

The lipoproteins of both *Myxine* and *Latimeria* were remarkable in their low contents of cholesteryl ester (<5%), endowing them with ratios of esterified:free cholesterol less than 1. Conversely, the minor proportions of free cholesterol in the LDL of sturgeons results in particularly elevated ester:free ratios (>3:1).

The phospholipid constituents of fish lipoproteins have been examined in only eight species, although these are representative of the three major classes, i.e., the hagfish,[59] sharks,[47] sturgeons,[47] and salmonids.[53,59] Despite such limitations, it can be said that the major phospholipid in VLDL, LDL, and HDL is uniformly phosphatidyl choline (Table 6), a finding comparable to that in man.[58]

Fish lipoproteins are further distinguished by the remarkable fatty acid compositions of their lipid esters (Table 7). Thus the markedly elevated levels of highly unsaturated fatty acids such as eicosapentaenoate (22:5) and docosahexanoate (22:6) surely represent a specific adaptation to their lives as nonheat-sustaining animals. This adaptation permits the fishes then to maintain lipoprotein lipids in a liquid, disordered state at low temperatures, thereby facilitating their metabolism.

Finally, despite the differing proportions of neutral lipids (i.e., cholesteryl esters, triacylglycerols, hydrocarbons, and monoalkyldiacylglycerols) in the VLDL, LDL, and HDL of the respective species, consideration of their physicochemical properties (including particle size)[7,48-50,59] tends to support the contention that their overall particle structure conforms largely to that proposed for the human substances,[60] i.e., a hydrophobic core comprising the neutral lipids, sterol esters, and some free cholesterol (and probably hydrocarbons and monoalkyldiacylglycerol when present), covered by a hydrophilic outer surface coat constituted of the polar components, i.e., phospholipids and apoproteins.

Vitellogenins

Like the invertebrates, the fish also possess circulating lipoproteins specifically linked to egg formation. These vitellogenins again present as VHDL and have been identified in two teleosts, the cod (*Gadus morhua L.*)[61,62] and rainbow trout.[63] Their protein components could be differentiated from the remaining serum apolipoproteins in both species, suggesting the association of specific transport proteins with egg formation in Teleosts.

Apolipoproteins

Studies of the protein components of fish lipoproteins have not progressed nearly so far as those of their lipids. Indeed, information to date relates only to the cyclostome *Myxine*,[64] the shark *Centrophorus squamosus*,[48] and the salmonids *Salmo gairdnerii* (rainbow trout)[49,50] and *Oncorhynchus gorbuscha* (pink salmon).[53] Nonetheless, it has become apparent that the

Table 5
CHEMICAL COMPOSITION OF FISH AND HUMAN SERUM LIPOPROTEINS (% BY WEIGHT)

VLDL

Species	Myxine glutinosa (hagfish)[48]	Centrophorus squamosus (shark)[48]	Scyliorhinus canicula (dogfish)[48]	Salmo gairdneri[a] (rainbow trout)[49,50]		Latimeria chalumnae (coelocanth)[57]	Man[46]
Cholesteryl ester	2.5	21.4	8.3	14.9	26.7	3.1	14.9
Free cholesterol	5.1	6.7	7.4	6.4	11.5	7.0	6.4
Triglyceride	48.2	23.0	43.5	49.9	38.5	64.0	49.9
Monoalkyldiacylglycerol	10.6	8.2					
Hydrocarbon	4.4	18.8					
Phospholipid	17.3	15.3	12.4	18.6	16.1	11.5	18.6
Protein	12.1	3.1	28.4	7.7	7.2	14.4	7.7

LDL

Species	M. glutinosa (hagfish)[48]	C. squamosus (shark)[48]	Centrophorus granulosus (shark)[47]	Centroscymnus coelolepsis (shark)[47]	S. canicula (dogfish)[48]	Acipenser stellatus (sturgeon)[47]	Acipenser guldenstadtii (sturgeon)[47]	Huso huso (sturgeon)[47]	S. gairdnerii[a] (trout)[49,50]		L. chalumnae (coelocanth)[57]	Man[46]
Cholesteryl ester	3.0	25.9	20.8	11.5	22.8	13.5	28.7	30.1	15.6	27.9	3.1	38.0
Free cholesterol	9.6	8.4	8.1	7.2	12.5	3.1	3.8	2.5	6.7	9.5	5.8	9.0
Triglyceride	29.5	8.7	5.4	8.6	20.1	30.4	20.6	22.1	26.9	12.5	49.7	11.2
Monoalkyldiacylglycerol	8.7	10.9	12.1	23.9								
Hydrocarbon	5.7	5.9	10.7	5.3		16.9	1.9	6.7				
Phospholipid	22.4	20.5	20.8	22.0	14.8	15.2	22.5	18.9	27.1	14.9	13.6	22.1
Protein	21.1	17.6	22.2	21.5	29.8	21.0	22.5	19.8	24.7	35.2	27.9	20.9

HDL

Species	M. glutinosa (hagfish)[48]	C. squamosus (shark)[48]	S. canicula (dogfish)[48]	S. gairdnerii[a] (trout)[49,50]		Oncorhynchus gorbuscha[b] (salmon)[53]	L. chalumnae (coelocanth)[57]	Man[46]
Cholesteryl ester	0.7	13.9	15.6	7.7	20.1	18.3	1.3	15.0
Free cholesterol	9.6	3.8	9.4	3.4	4.1	3.2	2.2	2.9
Triglyceride	10.5	3.4	12.2	15.5	5.7	6.8	12.5	8.0
Monoalkyldiacylglycerol	1.5							
Hydrocarbon	6.1	6.9						
Phospholipid	29.6	14.5	9.8	26.5	27.9	30.0	7.0	22.7
Protein	42.1	47.7	53.1	46.9	42.2	39.9	77.1	51.9

[a] VLDL isolated at d < 1.020 g/ml, LDL 1.020—1.085 and HDL 1.096—1.21 g/ml.

[b] Protein content includes 2.4% carbohydrate; FFA content of 1.7%.

From Chapman, M. J., *J. Lipid Res.*, 21, 789, 1980. With permission.

TABLE 6
PHOSPHOLIPID DISTRIBUTION IN FISH LIPOPROTEINS (% BY WEIGHT)

Species	VLDL		LDL							HDL		
	M. glutinosa (hagfish)[59]	S. gairdnerii (Trout)[59]	M. glutinosa (hagfish)[47]	C. squamosus (shark)[47]	C. granulosus (shark)[47]	C. coelolepsis (shark)[47]	A. guldenstadtii (sturgeon)[47]	H. huso (sturgeon)[47]	S. gairdnerii (trout)[59]	M. glutinosa (hagfish)[47]	S. gairdnerii (trout)[59]	O. gorbuscha (salmon)[53]
Phosphatidylcholine	73.4	81.3	61.4	63.5	54.6	72.3	74.5	71.2	74.2	63.3	84.0	83.1
Sphingomyelin	6.3	11.7	12.9	28.0	23.8	20.8	5.6	7.5	11.6	15.6	7.8	6.7
Phosphatidylethanolamine	11.9	2.6	7.4		1.3	4.7	5.6	2.9	4.0	10.3	1.4	2.3
Phosphatidylserine } Phosphatidylinositol	2.0	1.5	2.8		1.3		1.9	2.9	3.7	3.1	1.1	3.9
Lysolecithin	4.3	2.9	12.0	6.9	17.2	2.1	9.9	11.7	4.5	5.3	4.2	3.1

From Chapman, M. J., *J. Lipid Res.*, 21, 789, 1980. With permission.

Table 7
FATTY ACID COMPOSITIONS OF THE LIPID ESTERS OF FISH LIPOPROTEINS (%)

Species	Lipoprotein fraction	Lipid	Fatty acid																	Uniden.	Ref.
			14:0	15:0	16:0	16:1	18:0	18:1	18:2	18:3	20:1	20:2	20:4	20:5	22:0	22:1	22:4	22:5	22:6		
Latimeria chalumnae (coelocanth)	VLDL	CE	2.0		17.0	11.5	6.2	57.6	1.7	2.8			1.6								57
Centrophorus squamosus (shark)	VLDL	CE(1)	2.8		11.4	8.8	2.8	62.9	2.6		5.6		1.4							1.8	48
Centrophorus squamosus (shark)	VLDL	CE(2)	0.2		2.0	0.7	0.9	1.9	0.6	0.6			9.7	37.5				8.6	33.4	4.0	48
Salmo gairdnerii (trout)	VLDL	CE(1)	0.6		25.7	4.3	6.5	28.5	9.4		12.7	5.9	4.1								49
Salmo gairdnerii (trout)	VLDL	CE(2)	0.8		7.0	2.4	3.1	3.3	0.8				5.5	6.7					65.5	5.0	49
Latimeria chalumnae (coelocanth)	LDL	CE	3.2		19.1	12.7	7.5	46.0	2.5	3.6			5.5								57
Centrophorus squamosus (shark)	LDL	CE(1)	2.2		12.1	8.0	3.3	58.2	3.6		6.3		3.7							2.6	48
Centrophorus squamosus (shark)	LDL	CE(2)	0.3		1.9	0.6	1.1	1.1	0.6	0.7			11.9	38.1			1.2	5.4	32.8	4.3	48
Myxine glutinosa (hagfish)	LDL	CE			6.1	3.7	5.8	29.6	3.0	21.5			30.3								57
Scyllium canicula (dogfish)	LDL	CE			2.8	3.8		12.7	1.9	0.4			15.7	21.4				0.9	38.8		57
Salmo gairdnerii (trout)	LDL	CE(1)	0.6		30.5	4.0	8.5	25.2	8.2	11.4		5.3	4.0		1.3						49
Salmo gairdnerii (trout)	LDL	CE(2)	0.7		4.8	2.3	2.5	2.5	1.3	0.8		1.0	5.9	6.9	1.2			64.5		5.8	49
Sardinops caerulea G. (sardine)	LDL	CE	1.3		15.9	0.9	1.2	6.5	0.3		0.1		0.8	19.7					52.1		56
Latimeria chalumnae (coelocanth)	HDL	CE	1.4		20.7	8.8	15.8	40.2	2.4	6.7			4.2								57
Centrophorus squamosus (shark)	HDL	CE(1)	2.0		14.6	7.5	3.9	55.3	3.5		5.7		11.8	36.2				2.4			48
Centrophorus squamosus (shark)	HDL	CE(2)	0.3		2.9	1.0	1.6	1.4	0.6	0.6	5.7		2.3			5.2	1.8	4.8	32.8	4.0	48
Salmo gairdnerii (trout)	HDL	CE(1)	0.5		4.2	2.1	1.8	2.8	3.2	0.5		4.9	2.3					66.9			49
Salmo gairdnerii (trout)	HDL	CE(2)	1.5		17.8	6.0	3.8	32.8	1.2	5.7			1.7	4.8		4.5			17.3	3.1	49
Sardinops caerulea G. (sardine)	HDL	CE	2.6		14.5	10.1	1.3	6.1	0.5	6.6				15.5				0.2	35.1		56
Latimeria chalumnae (coelocanth)	VLDL	TG	3.3		22.7	6.8	1.8	41.9	3.2	0.5			6.0	6.5	2.0						49
Centrophorus squamosus (shark)	VLDL	TG	2.6		14.5	10.1	1.3	62.1	0.5	6.6	5.8		1.7								48
Salmo gairdnerii (trout)	VLDL	TG	2.0		18.9	6.8	3.5	39.5	0.8	6.6	5.8		1.7	4.3		4.5		12.4		1.5	48
Latimeria chalumnae (coelocanth)	LDL	TG	1.8		16.6	9.0	3.5	41.9	10.1				1.9	4.3					6.9		49
Centrophorus squamosus (shark)	LDL	TG	1.0		14.4	8.3	3.2	59.2	1.1	8.5			4.3								57
Myxine glutinosa (hagfish)	LDL	TG	1.5		17.8	6.0	3.8	32.8	1.2		5.7		1.7	4.8		4.5			17.3	3.1	48
Scyliorhinus canicula (dogfish)	LDL	TG	1.0		8.8	3.7	3.4	19.2	0.9	2.3			4.3					1.0			57
Salmo gairdnerii (trout)	LDL	TG	1.5		17.2	8.9	3.6	42.8	10.4	5.2			3.1	20.5				8.2		35.3	57
Sardinops caerulea G. (sardine)	LDL	TG	3.9	0.6	24.7	7.9	2.1	17.0	1.5		1.1	3.2	3.3	16.4			0.2	1.9	8.2		49
Latimeria chalumnae (coelocanth)	HDL	TG	1.8		14.4	11.7	5.4	44.2	1.4	8.5			1.5					1.9	10.9		56
Centrophorus squamosus (shark)	HDL	TG	1.5		17.8	6.0	3.8	32.8	1.2		5.7		1.7	4.8		4.5			17.3	3.1	54
Salmo gairdnerii (trout)	HDL	TG	1.6		17.2	8.9	3.5	40.7	10.3	4.8			1.4	4.8				17.3		3.1	48
Sardinops caerulea G. (sardine)	HDL	TG	5.6	0.7	24.2	7.4	3.8	15.7	1.2		2.0		1.4	20.5	0.4			9.7			49
Centrophorus squamosus (shark)	VLDL	MAGE	1.7		20.7	4.9	3.4	36.2	3.6	7.5	1.5	3.7	2.5			8.2		11.4	9.5	2.5	48
Centrophorus squamosus (shark)	LDL	MAGE	1.6		16.2	3.7	3.3	26.7	2.0	8.7		3.7	2.5	2.5		11.9	2.3	11.4	14.2	7.0	48

Table 7 (continued)
FATTY ACID COMPOSITIONS OF THE LIPID ESTERS OF FISH LIPOPROTEINS (%)

Species	Lipoprotein fraction	Lipid	Fatty acid																		
			14:0	15:0	16:0	16:1	18:0	18:1	18:2	18:3	20:1	20:2	20:4	20:5	22:0	22:1	22:4	22:5	22:6	Uniden.	Ref.
Centrophorus squamosus (shark)	HDL	MAGE	1.8		18.6	4.4	5.9	27.0	10.1	1.2	6.6			1.9		7.6			11.8	3.1	48
Latimeria chalumnae (coelocanth)	VLDL	PL	2.3		20.6	8.2	13.6	49.0	2.7	9.2			15.1								57
Centrophorus squamosus (shark)	VLDL	PL	1.0		42.6	3.7	2.8	23.4	1.7	0.8			4.7	7.1					11.0	1.2	48
Salmo gairdnerii (trout)	VLDL	PL	1.2		25.0	3.5	11.5	14.1	8.3	2.3		2.1	1.6		1.8				23.7		49
Latimeria chalumnae (coelocanth)	LDL	PL	2.1		18.1	8.1	11.3	41.0	2.9	5.4			11.1								57
Centrophorus squamosus (shark)	LDL	PL	1.9		35.6	3.6	3.6	19.0	1.9		1.4		4.5	7.5					18.5	2.6	48
Myxine glutinosa (hagfish)	LDL	PL			7.7	4.6	9.0	35.7	1.7	6.5		8.1						14.2			57
Scyliorhinus canicula (dogfish)	LDL	PL			19.7	2.3	2.9	11.7	0.1	2.0				7.0						42.5	57
Salmo gairdnerii (trout)	LDL	PL	1.2		25.9	3.3	11.6	14.6	8.7	2.0									25.6		49
Sardinops caerulea G. (sardine)	LDL	PL	1.6	0.2	1.3	35.0	3.3	7.7	0.7	2.5	0.4		1.8	11.8			0.6	0.6	25.5		56
Latimeria chalumnae (coelocanth)	HDL	PL	0.4		11.3	4.5	11.7	42.1	2.6	0.9			1.8	1.8							57
Centrophorus squamosus (shark)	HDL	PL	1.3		32.1	3.4	4.9	21.0	1.4		1.3		5.4	7.2			3.0		16.6	2.3	48
Salmo gairdnerii (trout)	HDL	PL	1.0		24.8	2.9	10.5	13.3	8.7	3.0		2.3	2.4	2.7					28.6		49
Sardinops caerulea G. (sardine)	HDL	PL	0.8		28.6	3.5	0.3	6.1	0.6			0.7	4.5	17.3					37.4		56

Note: Values are weight % of fatty acids recovered. CE, cholesteryl ester; TG, triglyceride; MAGE, monoalkyldiacylglycerol; PL, phospholipid; CE(1) and CE(2) represent two forms of CE separated by thin-layer chromatography.[48] Uniden., unidentified fatty acids. Fatty acids present in trace amounts (<0.1%) omitted. Contents of the 14:1, 15:1, 16:2, 17:0, 17:1, 19:0, 20:0 fatty acids in sardine lipid esters[56] less than 1%; the 14:1 represented 8.9% of LDL phospholipid fatty acids and the 22:2 form, 2.8% of LDL triglyceride fatty acids.

From Chapman, M. J., *J. Lipid Res.*, 21, 789, 1980. With permission.

major apolipoprotein of the low-density lipoproteins of cyclostomes,[64] elasmobranchs,[48] and teleosts[49] is a counterpart to human apo B, whereas one of the principle polypeptides of salmonid HDL appears homologous with the human A-I protein.[50,53] On this basis, and considering the data in Table 4, it seems likely that homologs of the predominant apolipoproteins in normal human serum, i.e., B and A-I,[65] may be similarly prominent in fish.

Isolation of the apo B- and apo A-I-like polypeptides from fish serum has been performed by methods originally applied to their human counterparts, involving delipidation of the native lipoprotein with subsequent fractionation of the protein moiety by gel filtration or ion-exchange chromatography in suitable solvents.

The apo B-like proteins have been prepared in detergent (SDS)-containing buffer solutions,[48,49,64] since despite the enormous evolutionary distance between fish and primates, the piscine apo Bs exhibit the intractable properties (limited solubility in aqueous buffers in the absence of detergents or strong denaturing agents) typical of their human counterpart. These early indications of the presence of an apo B-like protein in fish lipoproteins were confirmed by SDS-PAGE, when use of low monomer concentration (4% or less) permitted migration of the protein into the gel and assessment of its molecular weight.[48,49,64] Under such conditions, the principle component behaved as a protein with molecular weight equal to or larger than 250,000: the human protein (i.e., apo-B_{100}) behaves primarily as a dimer of 500,000 in such gels.[66,67] This measurement should be regarded strictly as empirical however, since application of the SDS-gel method is subject to considerable limitations in the case of apo B, the reasons for which are beyond the scope of the present treatise.

Despite the lack of evidence from immunological investigations (using polyvalent antisera to human LDL and apo B) for similarity in the antigenic structure of the human and fish B proteins,[64] further data to support some degree of homology was provided by chemical analysis. Thus, the amino acid compositions of the chromatographically purified piscine proteins markedly resembled that of human apo B prepared in a similar manner (Table 8); indeed, all of these proteins were enriched in aspartic and glutamic acids, leucine, lysine, and serine.

Knowledge of the apolipoproteins of HDL is restricted to the salmonids,[49,50,53] although the electrophoretic pattern of urea-soluble polypeptides in the HDL of the shark *C. squamosus* has been reported.[48] In fact, the latter was dominated by rapidly migrating peptides, in contrast to the equivalent patterns in the rainbow trout and pink salmon in which two (or more) components with mobilities comparable to those of human apo A-I and apo A-II predominated.[49,50,53] Further studies in sharks will thus be required to ascertain whether analogs to human apo A-I or apo A-II are components of their HDL. Additional evidence for the presence of apo A-I in salmonid HDL was derived from electrophoresis in SDS-polyacrylamide gel, in which case a major band of trout apo HDL displayed a molecular weight (approximately 26,500)[20,49] similar to that of human apo A-I. Moreover, the amino acid composition of the purified salmonid apo A-Is exhibited a close resemblance to the human protein (Table 8). Some dissimilarities were evident, including the presence of isoleucine in the salmon and trout polypeptides, as well as increased proportions of alanine and methionine and a diminished leucine content.[50,53] While the second major apolipoprotein of trout HDL could not be purified by gel filtration chromatography,[50] the application of an ion-exchange procedure facilitated isolation from salmon apo HDL of a second polypeptide (denoted ''DEAE fraction 2''),[53] which was more akin to human apo A-II than to apo A-I in its amino acid composition (Table 8). This apolipoprotein did, however, differ considerably from human apo A-II in lacking cystine and containing less lysine, threonine, serine, glutamic acid, and phenylalanine and rather more histidine, arginine, aspartic acid, glycine, alanine, methionine, and isoleucine.

In addition to the dominant apolipoproteins of fish low- and high-density lipoproteins, a number of minor peptides (about ten in the rainbow trout and about five in the pink salmon

Table 8
AMINO ACID COMPOSITIONS OF FISH APOLIPOPROTEINS RESEMBLING THE HUMAN B, A-I, AND A-II PROTEINS

Species Parent lipoprotein Apolipoprotein	Myxine glutinosa (hagfish) LDL Apo-B[64]	Centrophorus squamosus (shark) LDL Apo B[43]	Salmo gairdnerii (trout) VLDL Apo B[49]	Salmo gairdnerii (trout) LDL-2[a] Apo B[49]	Man LDL Apo B[64]	S. gairdnerii (trout) HDL A-I[50]	Oncorhynchus gorbuscha (salmon) HDL A-I[b53] (3)	(4)	Man HDL A-I[c68]	O. gorbuscha (salmon) HDL A-II[d53]	Man HDL A-II[c,e68]
Amino acid											
Lysine	6.8	9.2	5.7	7.3	8.0	9.9	9.7	9.9	8.7	8.6	11.7
Histidine	2.0	2.4	1.6	1.2	2.2	1.2	1.6	1.6	2.1	2.4	0
Arginine	4.0	4.2	2.8	2.8	3.3	6.5	4.7	5.4	6.6	2.3	0
Aspartic acid	13.2	11.5	12.4	11.6	10.8	5.4	5.8	5.5	8.7	8.2	3.9
Threonine	5.5	7.3	8.0	8.4	6.5	4.3	5.3	5.5	4.2	2.3	7.8
Serine	9.9	9.5	8.5	6.5	8.6	4.1	4.9	4.8	5.8	3.1	7.8
Glutamic acid	11.9	11.5	12.6	13.6	12.5	18.1	19.9	20.6	19.5	13.4	20.8
Proline	4.0	3.0	1.0	0.9	4.0	4.8	3.8	3.8	4.1	5.8	5.2
Glycine	6.6	7.0	5.5	4.7	5.0	2.6	2.1	2.2	4.1	9.4	3.9
Alanine	6.9	6.6	9.2	10.8	6.5	13.2	12.0	11.5	7.9	11.0	6.5
Valine	4.1	5.3	9.4	9.5	5.0	6.7	6.5	8.0	5.4	9.2	7.8
Methionine	3.0	0.9	2.7	1.4	1.4	2.4	3.3	3.2	1.3	3.4	1.3
Isoleucine	3.9	5.3	4.9	5.0	5.2	4.5	3.7	3.2	0	3.1	1.3
Leucine	10.3	9.7	9.2	11.2	12.2	10.3	11.0	10.2	16.2	9.5	10.4
Tyrosine	3.1	2.8	2.6	2.3	3.3	3.7	4.0	3.8	2.9	6.7	5.2
Phenylalanine	4.8	3.8	2.8	2.9	5.2	2.3	1.6	1.1	2.5	1.7	5.2

Note: Results are expressed as moles/100 mol. amino acid residues.

a   Trout LDL-2 was of d 1.020—1.078 g/mℓ.[49]
b   DEAE fractions (3) and (4).
c   Calculated from the primary structure.
d   DEAE fraction (2).
e   Human apo A-II contains 1.3 mol cystine/100 mol amino acid residues.

Table 9
CLASSIFICATION OF THE MAJOR
GROUPS OF LIVING AMPHIBIANS AND
REPTILES

Class	Subclass	Order	Members
Amphibians	Apoda		Caecilians
	Urodela		Newts, salamanders
	Anura		Frogs, toads
Reptiles	Anapsids	Chelonia	Tortoises, turtles
	Diapsids	Squamata	Lizards, snakes
		Crocodilia	Alligators, crocodiles

and shark *C. squamosus*) have been detected in alkaline-urea and SDS-polyacrylamide gels.[48,49,50,53] To date, none of these have been purified to homogeneity, and it is a matter of conjecture as to whether any of them may correspond, structurally or metabolically, to the C peptides of man. A further possibility concerns the polypeptide of ~43,000 mol wt in trout HDL which may correspond to the human and rat A-IV polypeptide (mol wt 46,000),[49,69] although this apoprotein is essentially located in chylomicra and VLDL in the latter species.

It seems clear from the foregoing that the major apolipoproteins of fish — and particularly of Teleosts — may be functionally similar to their counterparts in man. Thus, despite differences in amino acid composition which probably reflect single point mutations, the fish proteins are able to bind lipids in such a manner as to form lipoprotein particles whose physicochemical characteristics are comparable to those of man. Finally, observations on the lipid transport systems of fish indicate that complex lipoprotein profiles appeared early in evolution, at the level of the chordate ancestors or cyclostomes, and that these in turn resulted from the appearance of a series of genes coding for specific apolipoproteins. Subsequently, these genes were apparently highly conserved through the course of evolution, a proposal consistent with the common ancestry hypothesis of Barker and Dayhoff.[70]

Amphibians and Reptiles

The amphibians and reptiles made their appearance some 100 to 150 million years ago, the former appearing during the late Devonian and the latter during the Carboniferous period. Each of these groups of vertebrates is subdivided into two or more subclasses, as indicated in Table 9. Since the evolution of these animals occurred in parallel with an important change in habitat from a predominantly aquatic to a largely terrestrial environment, some signficant modifications of their lipid transport systems might have been anticipated. As noted elsewhere, however,[7] there is a regrettable paucity of data on the lipoproteins and apolipoproteins of amphibians and reptiles, a situation which effectively limits assessment of alterations which may have represented adaptations to environmental change.

Lipoproteins

Reports in the literature describing the lipid transport systems of amphibians and reptiles are restricted to serum lipid and lipoprotein concentrations, electrophoretic profiles, and some chemical analyses. The latter include the chemical compositions of the major lipoprotein classes (as VLDL, LDL, and HDL), the distribution of phospholipid subclasses and the fatty acid profiles of their lipid esters (i.e., cholesteryl esters, triacylglycerols, and phospholipids).

Table 10
SERUM LIPOPROTEIN CONCENTRATION IN AMPHIBIANS AND REPTILES[a,b]

Species	VLDL		LDL		HDL	Ref.
	S_f 100—400	S_f 20—100	S_f 12—20	S_f 0—12		
Amphibians						
Rana temporaria (frog)	0	99	38	55	29	46
Rana catesbeiana (bullfrog)						
Adult	10—30		55—65		5—10	71
Tadpole	ND[d]		ND		200—300	71
Pleurolides waltii (salamander)	0	2	6	6	17	—[c]
Reptiles						
Natrix natrix (grass snake)	149	58	48	359	382	46
Liopeltis vernalis (green snake)	0	64	92	1168	N.D.	46
Natrix piscator (water snake)	84	213	211	267	222	46
Ptyas mucosus (rat snake)	0	43	111	320	N.D.	46
Varanus salvator (water monitor)	0	0	0	130	58	46
Testudo graeca (tortoise)	0	0	0	45	117	46

[a]   Concentrations are expressed as mg lipoprotein/100 mℓ serum.

[b]   Distributions determined by analytical ultracentrifugation.[48]

[c]   Values in the salamander *P. waltii* are from the unpublished data of Chapman, M. J., M. Flavin, and G. L. Mills.

[d]   ND, not determined.

From Chapman, M. J., *J. Lipid Res.*, 21, 789, 1980. With permission.

Furthermore, and as noted in the fishes, blood samplings have rarely been taken in relation to nutritional status, nor to sex, age, or stage of the reproductive cycle (especially in females); such is the case in the most extensive study of amphibian and reptilian lipoproteins to date, involving one amphibian (frog, *Rana temporaria*) and six reptilian (grass snake, *Natrix natrix;* green snake, *Liopeltis vernalis;* water snake, *Natrix piscator;* rat snake, *Ptyas mucosus;* water monitor, *Varanus salvator;* and, tortoise, *Testudo graeca*) species.[46] In addition, VLDL, LDL, and HDL were isolated ultracentrifugally by application of the density limits originally devised for the human lipoproteins.[46] The limitations of this approach are well illustrated by the almost continuous distribution of low-density lipoproteins seen in the water monitor and which was spread across the "boundary" between LDL and HDL at d 1.063 g/mℓ.[46]

The circulating levels of lipoproteins in three amphibians and six reptiles are shown in Table 10. Values in the amphibians are almost uniformly lower, with the possible exception of the bullfrog tadpole whose HDL attained concentrations of 300 mg/100 mℓ serum. Our early indications in studies of a pooled plasma from the Urodele *Pleurolides waltii* were of low concentrations in all classes (Table 10); however, in more recent studies of plasmas from a number of these salamanders, (adult males and females), we have observed rather higher levels (VLDL, 25 to 51; LDL, 52 to 101; and, HDL, 72 to 114 mg/100 mℓ plasma, respectively).[79] Such data highlight the problems implicit to studies of small numbers of animals, which in turn possess extremely small blood volumes (1 mℓ or less).

It is noteworthy that a marked reduction (some 30-fold) in HDL levels occurred consequent to the development of the adult bullfrog from the tadpole.[71] Such a finding is not inconsistent with the observation of Gillett and Lima that plasma cholesterol concentrations fall with age in male lizards,[72] although further studies will be required to determine whether such phenomena are widespread among these two groups of vertebrates.

The lipoprotein profiles of the reptiles (except for the tortoise and water monitor), examined by Mills and Taylaur, are remarkable in displaying S_f 0 to 20 LDL levels which were uniformly in excess of 400 mg/dℓ (range 407 to 1260 mg/dℓ),[46] making them distinct from the higher vertebrates which almost exclusively display HDL as the predominant class.[7] In contrast, the majority of the reptiles lacked measurable amounts of the lighter (S_f 100 to 400) subclass of VLDL. The overall nature of the lipoprotein distributions in amphibians and reptiles documented in electrophoretic studies tends to confirm those described above and determined primarily by analytical ultracentrifugation.[46,73,74]

The chemical compositions of VLDL in the bullfrog, grass, and water snakes were comparable to that typically seen in man (Table 11), although water snake VLDL possessed a high content of free cholesterol, such that its ester:free cholesterol ratio was less than one. The reduced proportion of esterified sterol in grass snake VLDL had the same effect. Like human LDL (d 1.006 to 1.063 g/mℓ), the amphibian and reptilian fractions were enriched in cholesteryl ester. In the tortoise however, triacylglycerol content surpassed that of cholesteryl ester, but the total neutral lipid content in the LDL of all six species (range 40.8 to 53.6%) resembled that in man (49.2%). A range of 10% in protein content was noted, presumably arising from differences in the overall particle distributions between the species within the density interval 1.006 to 1.063 g/mℓ.

An even greater variation in protein occurred in the HDL, ranging from only 28.9% in the water monitor to 63.5% in the tortoise. A possible explanation for the above finding in the monitor relates to the nature of the density distribution of its LDL as discussed previously; whether this snake possesses a true HDL is yet to be established. The remaining HDLs shared similar proportions of cholesteryl ester and triacylglycerol; considerable variability in free cholesterol contents did however result in rather distinct esterified:free cholesterol ratios, the lowest (1:1) occurring in the amphibia. A twofold range in phospholipid content is noteworthy (13.1% and 25.2% in frog and bullfrog, respectively); an explanation of this finding must await subfractionation and structural characterization of the HDL subclasses in these species.

As in the mammals,[7] fish, and some insects, the predominant phospholipid of amphibian lipoproteins is phosphatidyl choline,[71] (46.8 and 75.9% of total in adult bullfrog LDL and HDL, respectively), with lesser amounts of sphingomyelin (17 to 41% in the VLDL and LDL, respectively). No data are available in reptiles.

The fatty acid profiles of the lipid esters of bullfrog and water snake lipoproteins (Table 12) are distinct from those of the fishes (Table 7) in lacking large amounts of the polyunsaturated long-chain components. It is notable that the fatty acid distribution of the various lipids in each lipoprotein class were hardly distinguishable, palmitic (16:0), oleic (18:1), and linoleic (18:2) acids predominating, with the exception of arachidonate (20:4) in water snake LDL lipids (range 12.4 to 50.4%), and of the long-chain unsaturated 24:1 acid in the sphingomyelin of bullfrog lipoproteins (range 8.2 to 25.8%). The degree to which the respective fatty acid patterns result from adaptation to habitat (occurring at the level of their endogenous biosynthesis), from the nature of the dietary lipid, from the action of lipid transfer proteins, or from the activity of LCAT will be of interest to determine.

Vitellogenins

Although both the amphibians and reptiles are oviparous vertebrates, our knowledge of their plasma vitellogenins is scant at most. In the toad *Xenopus laevis* at least, the circulating vitellogenin takes the form of a VHDL, with a protein content of some 88% by weight and a hydrated density of 1.35 g/mℓ.[76] Whether the VLDL of amphibians and reptiles can contribute directly to egg formation, as is the case in certain avians and notably chicken,[77] is open to question.

Apolipoproteins

Knowledge of the apoproteins in these vertebrates is confined to rat snake LDL,[46] and to

Table 11

MEAN WEIGHT PERCENT CHEMICAL COMPOSITION OF AMPHIBIAN AND REPTILIAN SERUM LIPOPROTEINS

Species	VLDL							LDL							HDL				
	Bull-frog[71]	Grass snake[46]	Water snake[46]	Man[46]	Bullfrog[75]		Frog[46]	Grass snake[46]	Water snake[46]	Water monitor[46]	Tor-toise[46]	Man[46]	Bull-frog[71]	Frog[46]	Grass snake[46]	Water snake[46]	Water monitor[46]	Water Tor-toise[46]	Man[46]
					Adult	Tadpole													
Cholesteryl ester	17.3[a]	4.6	9.2	14.9	30.3	37.6	42.5	35.0	33.2	42.4	20.2	38.0	19.3[a]	22.4	24.4	21.8	40.4	14.3	15.0
Free cholesterol	8.8	7.9	13.0	6.7	5.6	3.2	13.9	13.0	14.6	13.5	9.4	9.0	11.9	10.3	7.3	14.8	12.1	3.4	2.9
Triglyceride	48.1	61.7	52.9	49.9	12.5	16.0	1.4	5.8	11.0	5.8	23.9	11.2	3.7	3.1	3.5	4.7	4.5	5.9	8.0
Phospholipid	11.6	14.4	14.4	18.6	23.3	22.4	16.8	17.7	18.2	11.2	16.1	22.1	25.2	13.1	24.7	21.6	14.2	12.9	22.7
Protein	12.0	11.4	10.6	7.7	22.4	20.0	25.4	28.5	23.0	27.1	30.5	20.9	36.8	51.0	40.0	37.1	28.9	63.5	51.9

[a] The FFA and partial glyceride contents (total 2—3%) of bullfrog VLDL and HDL are omitted.[71]

From Chapman, M. J., J. Lipid Res., 21, 789, 1980. With permission.

Table 12
FATTY ACID COMPOSITIONS OF THE LIPID ESTERS OF AMPHIBIAN AND REPTILIAN LIPOPROTEINS (%)

Species	Lipoprotein fraction	Lipid	Fatty acid												Ref.
			14:0	16:0	16:1	18:0	18:1	18:2	18:3	20:0	20:1	20:4	22:0	24:1	
Bullfrog	VLDL	Cholesteryl ester	Trace	7.7	20.7	1.4	57.2	8.8			4.1				71
Bullfrog	LDL	Cholesteryl ester	Trace	8.3	17.8	Trace	45.2	16.1			8.1	8.0			75
Water snake	LDL	Cholesteryl ester					19.7	27.1				50.4			46
Bullfrog	HDL	Cholesteryl ester	Trace	8.5	15.8	1.7	54.8	12.3			4.8	1.8			75
Bullfrog	VLDL	Triglyceride	3.8	28.9	20.4	3.4	37.1	5.0			1.5				75
Bullfrog	LDL	Triglyceride	3.5	26.1	18.5	3.2	32.3	11.8			4.7				75
Water snake	LDL	Triglyceride		11.1	2.4	5.4	36.1	18.5				12.4			46
Bullfrog	HDL	Triglyceride	3.6	30.3	21.2	2.5	33.2	7.7			1.2				75
Water snake	LDL	Phospholipid		5.0		17.0	20.0	18.0				34.2			46
Bullfrog	VLDL	Phosphatidylcholine	0.9	60.6	1.0	4.2	28.9	4.3							71
	LDL	Phosphatidylcholine	0.9	44.1	7.8	4.9	19.1	9.4			2.6	1.4			75
	HDL	Phosphatidylcholine	3.0	46.4	9.5	4.9	28.9	4.9			2.0				71
	VLDL	Sphingomyelin	Trace	61.2	Trace	4.1	2.1			2.8			4.1	25.8	71
	LDL	Sphingomyelin	1.0	68.6		2.9	5.3	0.9		0.9			2.2	14.5	75
	HDL	Sphingomyelin	1.6	54.9	4.7	4.8	21.4	2.9		Trace			1.0	8.2	71

From Chapman, M. J., *J. Lipid Res.*, 21, 789, 1980.

bullfrog VLDL, LDL, and HDL;[75] more recently, the protein moieties of salamander (*Pleurolides waltii*) lipoproteins have been investigated in our laboratory.[79]

Evidence suggestive of the presence of an apo B-like protein in the LDL of rat snake and in both VLDL and LDL of the salamander was obtained from electrophoresis in SDS-polyacrylamide gel, in which case the major component(s) displayed high molecular weights (>250,000 in the case of rat snake).[64] These observations were confirmed in the case of the rat snake by isolation upon gel-filtration chromatography of the protein and subsequent amino acid analysis. Thus, its amino acid composition exhibited an overall resemblance to that of human apo B and to that of the fraction isolated in a similar manner from bullfrog LDL.[64,75] Both the amphibian and reptilian proteins required the presence of a detergent (SDS) for complete solubility. The extremely small amounts of salamander LDL available to us precluded fractionation of its protein moiety. Nonetheless, the immunological cross-reactivity of *P. waltii* LDL with antisera to LDL from the chicken, and that between salamander VLDL and antisera to pig, to guinea pig, and to trout LDL, provides strong support for the presence of an apo B-like protein in this amphibian.[79] In addition to this protein, other polypeptides of lower molecular weight were detected in salamander VLDL and LDL by electrophoresis in both SDS- and alkaline urea-polyacrylamide gel systems; the size and mobility of certain of these was compatible with the proposal that they may be counterparts to the apo Cs in man. Similar peptides were also identified in bullfrog VLDL.[75]

The major apolipoproteins of salamander HDL exhibited molecular weights of 27,000 and 31,000; in the bullfrog, the principle polypeptide was of M_r 28,000. Since further characterization of these proteins was not performed, and since the salamander apoproteins migrate substantially further into alkaline-urea gels than human apo A-I, it would be premature to propose that they might be analogous to the latter. Indeed the lack of immunological cross-reactivity between the HDL of salamander and antisera to HDL from a number of higher vertebrates (including avians and man), clearly suggests that the structure of their antigenic determinants — and thus primary sequence — may differ substantially.

CONCLUSION

Through the course of this review, it has become apparent that counterparts to human apo B and A-I appeared as early as the fishes; indeed putative forms of apo B may have appeared in the invertebrates. Furthermore, while apo A-I has undergone considerable evolutionary modification, at least in that portion of the sequence bearing its major antigenic determinants, the primary structure of apo B seems to have been more highly conserved. Explanations for such a high degree of conservation are not immediately apparent, unless apo B plays a similar role in lower vertebrates as in the mammals and man, i.e., of delivering cholesterol to cells via the LDL receptor, a highly specific mechanism. Resolution of this question must await sequence data on this intractable and complex protein, as well as further understanding of its physiological role(s) in fish, amphibians, and reptiles. Finally the presence of a putative, monomeric form of apo A-II in fish HDL is of considerable interest, since the role of this apolipoprotein in mammalian HDL is poorly understood. Thus, sequencing studies, together with investigation of the recombination of the piscine protein, and of peptides derived from it, with phospholipids, could provide further data on the regions of its sequence essential to lipid binding.[78]

ACKNOWLEDGMENTS

This manuscript was completed during the author's tenure as a Visiting Scientist at the Gladstone Foundation Laboratories for Cardiovascular Disease, San Francisco, Calif., whose Director, Dr. R. W. Mahley, most generously provided facilities for its preparation. Thanks are also expressed to Ms. M. Prator for assistance in preparation of the typescript.

ABBREVIATIONS USED

Apo	Apolipoprotein
VLDL	Very low density lipoproteins of d <1.006 g/mℓ, unless otherwise defined
IDL	Intermediate density lipoproteins, density 1.006—1.019 g/mℓ
LDL	Low density lipoproteins, density as defined
HDL	High density lipoproteins, density as defined
HDL subclasses are:	HDL$_2$, d $= 1.063$—1.125 g/mℓ, and HDL$_3$, d $= 1.125$—1.21 g/mℓ, unless defined otherwise
VHDL	Very high density lipoproteins of d >1.21 g/mℓ, unless otherwise defined
LCAT	Lecithin:cholesterol acyltransferase (EC. 2.3.1.43)
SDS	Sodium dodecyl sulfate
TMU	Tetramethyl urea
TG	Triacylglycerol
DG	Diacylglycerol
CE	Cholesteryl ester
FC	Free cholesterol
PL	Phospholipid
PRN	Protein

REFERENCES

1. **Wiskocil, R., Goldman, P., and Deeley, R. G.,** Cloning and structural characterization of an estrogen-dependent apolipoprotein gene, *J. Biol. Chem.,* 256, 9662, 1981.
2. **Meijlink, F. C. P. W., van het Schip, A. D., Arnberg, A. C., Wieringa, B., AB., G., and Gruber, M.,** Structure of the chicken apo very low density lipoprotein II gene, *J. Biol. Chem.,* 256, 9668, 1981.
3. **Chan, L., Dugaiczyk, A., and Means, A. R.,** Molecular cloning of the gene sequences of a major apoprotein in avian very low density lipoprotein, *Biochemistry,* 19, 5631, 1980.
4. **Wieringa, B., AB., G. and Gruber, M.,** The nucleotide sequence of the very low density lipoprotein II in RNA from chicken, *Nucl. Acids Res.,* 9, 489, 1981.
5. **Schneider, W. J., Basu, S. V., McPhaul, M. J., Goldstein, J. L., and Brown, M. S.,** Solubilization of the low density lipoprotein receptor, *Proc. Natl. Acad. Sci., U.S.A.,* 76, 5577, 1979.
6. **Schneider, W. J., Goldstein, J. L., and Brown, M. S.,** Partial purification and characterization of the low density lipoprotein receptor from bovine adrenal cortex, *J. Biol. Chem.,* 255, 11442, 1980.
7. **Chapman, M. J.,** Animal lipoproteins: chemistry, structure and comparative aspects, *J. Lipid Res.,* 21, 789, 1980.
8. **Gilbert, L. I. and Chino, H.,** Transport of lipids in insects, *J. Lipid Res.,* 15, 439, 1974.
9. **Alaupovic, P.,** The concepts, classification systems and nomenclatures of human plasma lipoproteins, in *Handbook of Electrophoresis,* Vol. 1, Lewis, L. A. and Opplt, J. J., Eds., CRC Press, Boca Raton, Fla., 1979, 1.
10. **Douglass, T. S., Connor, W. E., and Lin, D. S.,** The biosynthesis, absorption and origin of cholesterol and plant sterols in the Florida land crab, *J. Lipid Res.,* 22, 961, 1981.
11. **Goad, L. J.,** The sterol of marine invertebrates, in *Marine Natural Products, Chemical and Biological Perspectives,* Scheuer, P. J., Ed., Academic Press, New York, 1978, 101.
12. **Teshima, S., Kanazawa, A., and Haruhito, O.,** Absorption of sterols and cholesterol esters in a prawn, *Peanaeus japonicus, Bull. Jpn. Soc. Sci. Fish.,* 40, 1015, 1974.
13. **Lee, R. F. and Puppione, D. L.,** Serum lipoproteins in the spiny lobster, *Panulirus interruptus, Comp. Biochem. Physiol.,* 59B, 239, 1978.
14. **Kerr, M. S.,** The hemolymph proteins of the blue crab, *Callinectes sapidus.* II. A lipoprotein serologically identical to oocyte lipovitellin, *Dev. Biol.,* 20, 1, 1969.
15. **Allen, W. V.,** Lipid transport in the Dungeness crab, *Cancer magister* Dana, *Comp. Biochem. Physiol.,* 43B, 193, 1972.

16. **Fielder, D. R., Rao, K. R., and Fingerman, M.,** A female-limited protein and the diversity of hemocyanin components in the dimorphic variants of the fiddler crab, *Uca pugilator*, as revealed by disc electrophoresis, *Comp. Biochem. Physiol.,* 39B, 291, 1971.

17. **Ceccaldi, H. J. and Martin, J-L. M.,** Evolution des proteines de l'hemolymph chez *Carcinus maenus* L. devant l'ovogenese, *C.R. Soc. Biol.,* 163, 2638, 1969.

18. **Adiyodi, R. G.,** Protein metabolism in relation to reproduction and moulting in the crab, *Paratelphusa hydrodomous.* II. Fate of conjugated protein during vitellogenesis, *Indian J. Exp. Biol.,* 6, 200, 1968.

19. **Barlow, J. and Ridgway, G. J.,** Changes in serum proteins during the molt and reproductive cycles of the American lobster, *Homarus americanus, J. Fish Res. Bd. Can.,* 26, 2101, 1969.

20. **Kane, J. P.,** Plasma lipoproteins: structure and metabolism, in *Lipid Metabolism in Mammals,* Vol. 1, Snyder, F., Ed., Plenum Press, New York, 1978, 209.

21. **Goldstein, J. L. and Brown, M. S.,** The low density lipoprotein and its relation to atherosclerosis, *Ann. Rev. Biochem.,* 1977, 46, 897.

22. **Chino, H., Downer, R. G. H., Wyatt, G. R., and Gilbert, L. I.,** Lipophorins, a major class of lipoproteins of insect hemolymph, *Insect Biochem.,* 7, 491, 1981.

23. **Thomas, K. K. and Gilbert, L. I.,** Isolation and characterization of the hemolymph lipoproteins of the American silkmoth, *Hyalophora cecropia, Archiv. Biochem. Biophys.,* 127, 512, 1968.

24. **Chino, H., Murakami, S., and Harashima, K.,** Diglyceride-carrying lipoproteins in insect hemolymph: isolation, purification and properties, *Biochim. Biophys. Acta,* 176, 1, 1969.

25. **Peled, Y. and Tietz, A.,** Isolation and properties of a lipoprotein from the hemolymph of the locust, *Locusta migratoria, Insect Biochem.,* 5, 61, 1975.

26. **Chino, H. and Kitazawa, K.,** Diacylglycerol-carrying lipoprotein of hemolymph of the locust and some insects, *J. Lipid Res.,* 22, 1042, 1981.

27. **Chino, H., Katase, H., Downer, R. G. H., and Takahasi, K.,** Diacylglycerol-carrying lipoprotein of hemolymph of the American cockroach: purification, characterization, and function, *J. Lipid Res.,* 22, 7, 1981.

28. **Pattnaik, N. M., Mundall, E. C., Trambusti, B. G., Law, J. H., and Kézdy, F. J.,** Isolation and characterization of a larval lipoprotein from the hemolymph of *Manduca sexta, Comp. Biochem. Physiol.,* 63B, 469, 1979.

29. **Thomas, K. K.,** Isolation and partial characterization of the hemolymph lipoproteins of the wax moth, *Galleria mellonella, Insect Biochem.,* 9, 211, 1979.

30. **Chinzei, Y., Chino, H., and Wyatt, G. R.,** Purification and properties of vitellogenin and vitellin from *Locusta migratoria* migratorioides, *Insect Biochem.,* 11, 1, 1981.

31. **Mundall, E. C. and Law, J. H.,** Physical and chemical characterization of vitellogenin from the hemolyph and eggs of the tobacco hornworm, *Manduca sexta, Comp. Biochem. Physiol.,* 63B, 459, 1979.

32. **Gilbert, L. I. and Chino, H.,** Transport of lipids in insects, *J. Lipid Res.,* 15, 439, 1974.

33. **Alaupovic, P., Sanbar, S. S., Furman, R. H., Sullivan, M. L., and Walraven, S. L.,** Studies of the composition and structure of serum lipoproteins. Isolation and characterization of very high density lipoproteins of human serum, *Biochemistry,* 5, 4044, 1966.

34. **Whitmore, E. and Gilbert, L. I.,** Hemolymph proteins and lipoproteins in Lepidoptera: a comparative electrophoretic study, *Comp. Biochem. Physiol.,* 47B, 63, 1974.

35. **Chino, H. and Gilbert, L. I.,** Lipid release and transport in insects, *Biochem. Biophys. Acta,* 98, 94, 1965.

36. **Klages, G. and Emmerich, H.,** Juvenile hormone binding proteins in the hemolymph of third instar larvae of *Drosophila hydei, Insect Biochem.,* 9, 23, 1979.

37. **Dwivedy, A. K. and Bridges, R. G.,** The effect of dietary changes on the phospholipid composition of the hemolymph lipoproteins of larvae of the housefly, *Musca domestica, J. Insect Physiol.,* 19, 559, 1973.

38. **Kramer, S. J. and deKort, C. A. D.,** Juvenile hormone carrier lipoproteins in the haemolymph of the Colorado potato beetle, *Leptinotarsa decemlineata, Insect Biochem.,* 8, 87, 1978.

39. **Gamo, T.,** Low molecular weight lipoproteins in the haemolymph of the silkworm, *Bombyx mori:* inheritance, isolation and some properties, *Insect Biochem.,* 8, 457, 1978.

40. **Chino, H., Yamagata, M., and Sato, S.,** Further characterization of lepidopteran vitellogenin from hemolymph and mature eggs, *Insect Biochem.,* 7, 125, 1977.

41. **Kunkel, J. G. and Pan, M. L.,** Selectivity of yolk protein uptake: comparison of vitellogenins of two insects, *J. Insect Physiol.,* 22, 809, 1976.

42. **Atlas, S. J., Roth, J. F., and Falcone, A. J.,** Purification and partial characterization of *Culex pipiens fatigans* yolk protein, *Insect Biochem.,* 8, 111, 1978.

43. **Chen, T. T., Strahlendorf, P. W., and Wyatt, G. R.,** Vitellin and vitellogenin from locusts (*Locusta migratoria*). Properties and posttranslational modification in the fat body, *J. Biol. Chem.,* 253, 5325, 1978.

44. **DeLalla, O. F. and Gofman, J. W.,** Ultracentrifugal analysis of serum lipoproteins, *Methods Biochem. Anal.,* 1, 459, 1954.

45. **Nichols, A. V.,** Human serum lipoproteins and their interrelationships, *Adv. Biol. Med. Phys.,* 11, 109, 1967.
46. **Mills, G. L. and Taylaur, C. E.,** The distribution and compositions of serum lipoproteins in eighteen animals, *Comp. Biochem. Physiol.,* 40B, 489, 1971.
47. **Mills, G. L. and Taylaur, C. E.,** Composition studies of fish low density lipoproteins, *Protides Biol. Fluids Proc. Colloq.,* 25, 477, 1978.
48. **Mills, G. L., Taylaur, C. E., Chapman, M. J., and Forster, G. R.,** Characterization of serum lipoproteins of the shark *Centrophorus squamosus, Biochem. J.,* 163, 455, 1977.
49. **Chapman, M. J., Goldstein, S., Mills, G. L., and Leger, C.,** Distribution and characterization of the serum lipoproteins and their apoproteins in the rainbow trout *(Salmo gairdnerii), Biochemistry,* 17, 4455, 1978.
50. **Skinner, E. R. and Rogie, A.,** The isolation and partial characterization of the serum lipoproteins and apolipoproteins of the rainbow trout, *Biochem. J.,* 173, 507, 1978.
51. **Mahley, R. W. and Weisgraber, K. H.,** Canine lipoproteins and atherosclerosis. I. Isolation and characterization of plasma lipoproteins from control dogs, *Circ. Res.,* 35, 713, 1974.
52. **Laplaud, P. M., Beaubatie, L., and Maurel, D.,** A spontaneously seasonal hypercholesterolemic animal: plasma lipids and lipoproteins in the European badger *(Meles meles* L.), *J. Lipid Res.,* 21, 724, 1980.
53. **Nelson, G. J. and Shore, V. G.,** Characterization of the serum high density lipoproteins and apolipoproteins of pink salmon, *J. Biol. Chem.,* 249, 536, 1974.
54. **Lauter, C. J., Brown, E. A. B., and Trams, E. G.,** Composition of plasma lipoproteins of the spiny dogfish, *Squalus acanthias, Comp. Biochem. Physiol.,* 24, 243, 1968.
55. **Reichert, W. L. and Malins, S. C.,** Interaction of mercurials with salmon serum lipoproteins, *Nature (London),* 247, 569, 1974.
56. **Lee, R. F. and Puppione, D. L.,** Serum lipoproteins of the Pacific sardine *(Sardinops caerulae* Girard), *Biochim. Biophys. Acta,* 270, 272, 1972.
57. **Mills, G. L. and Taylaur, C. E.,** The distribution and composition of serum lipoproteins in the coelocanth *(Latimeria), Comp. Biochem. Physiol.,* 44B, 1235, 1973.
58. **Skipski, V. P., Barclay, M., Barclay, R. K., Fetzer, V. A., Good, J. J., and Archibald, F. M.,** Lipid composition of human serum lipoproteins, *Biochem. J.,* 104, 340, 1967.
59. **Taylaur, C. E.,** Comparative Study of the Composition of Serum Lipoproteins in Fish, M. Phil. thesis, University of London, 1977.
60. **Shen, B. W., Scanu, A. M., and Kezdy, F. J.,** Structure of human serum lipoproteins inferred from compositional analysis, *Proc. Natl. Acad. Sci., U.S.A.,* 74, 837, 1979.
61. **Plack, P. A., Pritchard, D. J., and Fraser, N. W.,** Egg proteins in cod serum, *Biochem. J.,* 121, 847, 1971.
62. **Skinner, E. R. and Rogie, A.,** The relationship between cod egg protein and its precursor, *Biochem. Soc. Trans.,* 4, 1118, 1976.
63. **Skinner, E. R. and Rogie, A.,** Trout egg lipoprotein and its relationship to normal serum lipoproteins, *Protides Biol. Fluids Proc. Colloq.,* 25, 491, 1978.
64. **Goldstein, S., Chapman, M. J., and Mills, G. L.,** Biochemical and immunological evidence for the presence of an apolipoprotein B-like component in the serum low-density lipoproteins of several animal species, *Atherosclerosis,* 28, 93, 1977.
65. **Alaupovic, P.,** Structure and function of plasma lipoproteins with particular regard to hyperlipoproteinemias and atherosclerosis, *Ann. Biol. Clin.,* 38, 83, 1980.
66. **Watt, R. M. and Reynolds, J. A.,** Solubilization and characterization of apolipoprotein B from human serum low-density lipoprotein in *n*-Dodecyl octaethylene glycol monoether, *Biochemistry,* 19, 1593, 1980.
67. **Kane, J. P., Hardman, D. A., and Paulus, H. E.,** Heterogenity of apolipoprotein B: isolation of a new species from human chylomicrons, *Proc. Natl. Acad. Sci. Wash.,* 77, 2465, 1980.
68. **Osborne, J. C., Jr. and Brewer, H. B., Jr.,** The plasma lipoproteins, *Adv. Protein Chem.,* 31, 253, 1977.
69. **Beisiegel, U. and Utermann, G.,** An apolipoprotein homolog of rat apolipoprotein A-IV in human plasma, *Eur. J. Biochem.,* 93, 601, 1979.
70. **Barker, W. C. and Dayhoff, M. O.,** Evolution of lipoproteins deduced from protein sequence data, *Comp. Biochem. Physiol.,* 57B, 309, 1977.
71. **Suzuki, N., Degachi, K., Ueta, N., Nagano, H., and Shukuya, R.,** Chemical characterization of the serum very low density and high density lipoproteins from bullfrog, *Rana catesbeiana, J. Biochem.,* 80, 1241, 1976.
72. **Gillett, M. P. T. and Lima, V. L. M.,** Marked decrease in plasma cholesterol concentration with increasing age in male lizards, *Atherosclerosis,* 32, 461, 1979.
73. **Dangerfield, W. G., Finlayson, R., Myatt, G., and Mead, M. G.,** Serum lipoproteins and atherosclerosis in animals, *Atherosclerosis,* 25, 95, 1976.

74. **Alexander, C. and Day, C. E.,** Distribution of serum lipoproteins of selected vertebrates, *Comp. Biochem. Physiol.,* 46B, 295, 1973.
75. **Suzuki, N., Kawashima, S., Fukushima, S., Ueta, N., Nagano, H., and Shukuya, R.,** Isolation and partial characterization of the serum low-density lipoprotein from bullfrog, *Rana catesbeiana, J. Biochem.,* 81, 1231, 1977.
76. **Wallace, R. A.,** Studies on amphibian yolk. IX. Xenopus vitellogenin, *Biochim. Biophys. Acta,* 215, 176, 1970.
77. **Holdsworth, G., Michell, R. H., and Finean, J. B.,** Transfer of very-low density lipoprotein from hen plasma into egg yolk, *FEBS Lett.,* 39, 275, 1974.
78. **Mao, S. J. T., Jackson, R. L., Gotto, A. M., Jr., and Sparrow, J. T.,** Mechanism of lipid-protein interaction in the plasma lipoproteins: identification of a lipid-binding site in apolipoprotein A-II, *Biochemistry,* 20, 1676, 1981.
79. **Takahashi, Y. I., Slavin, M., Goldstein, S., Weech, P. K., and Chapman, M. J.,** Density, distribution, characterization, and comparative aspects of plasma lipoproteins in the salamander, *Pleurodiles waltii, Comp. Biochem. Physiol. Pt. B,* in press.

Future Studies

OVERVIEW: IT TAKES A PAST TO MAKE A FUTURE

Irvine H. Page

Dr. Lewis and Dr. Naito flattered me by asking me to write on the future of lipoproteins in their relationship to atherosclerosis. I happily accepted without the slightest idea of what I had to say. At my age I have become proficient at obituaries and funeral orations. But the future — I wish I knew my own!

To have a future you must have a past. Much of the long gone past related to studies of atherogenesis, I know, simply because I lived with it. Of the future, your guess is as good as mine. I can only say that atherogenesis is "essential", in the sense that hypertension is "essential".

Atherosclerosis as a subject has at last come of age. The era of trying to convince the biomedical community and the public that atherosclerosis was not simply a result of age, and inevitable, is over. To convince them that it was a disease took a lot of doing. The discovery of penicillin was probably the most single persuasive force.

The sulfonamides and antibiotics did away with the diseases of microbial origin on which as a young physician I was financially sustained. Livelihood depended on lobar pneumonia, typhoid fever, syphilis, tuberculosis, and other infections. When they were no longer prevalent, atherosclerosis, hypertension, and cancer came to the forefront of medical practice. Our future, and the future of lipoproteins, depended as much on the change in medical practice as on all other factors together.

Oddly, the chemistry of the fats and sterols was not considered of much importance to biochemists in my youth. The "oil chemists" and a few agricultural chemists were interested from the practical aspects but "pure" organic chemists were rated almost the same as automobile salesmen and politicians in the public polls.

It is little wonder that few saw any need for lipoproteins either in theory or practice. Exceptional people such as Theorell, Oncley, and Macheboef pleaded for their recognition, but Gofman sold the idea to medical research along with millions of federal dollars' worth of ultracentrifuges and electrophoretic equipment. Then began the great age for committees, and committees to oversee committees, and best of all, travel to exotic places with often exotic people.

It took about a decade for the dust to settle and investigation in depth to begin with Lindgren, Lewis, Scanu, Aloupovic, and others. Now it is becoming evident that the lipoproteins are following the usual scenario in research, building a solid body of knowledge on which to base useful application.

A "Diet-Heart Study" was missing to convince the world that we researchers were down-to-earth people who aimed to prevent coronary disease and stroke rather than feeding rabbits to produce atherosclerosis. Diet was, of course, a happy choice, because one thing history teaches is that diet and sex rate higher than any other interests for most of the public!

Advances in diagnostic techniques such as coronary and cerebral angiography have brought the problem of atherogenesis and the lipoproteins to the sick bed along with the surgeons because they are always quick to see a need for their services. They, as always, proved to be powerful allies and helped validate the basic science of atherogenesis, not to speak of its financial support. So we have such desirable twinning as DeBakey-Gotto.

As with most science, it takes organizations as glue to keep a field unified. The new journal *Atherosclerosis* has signaled the present age of receptor theory, conformational measurements, and a host of new techniques now being used in the hope of understanding the mechanism and treatment of atherosclerosis. It is an age in which gifted youngsters such as Brown and Goldstein, Ross, and Gimbrone are pressing on the current authorities of the Steinberg-Benditt-Havel-Goodman-Zilbersmit-Mustard generation. These are, of course, only "for instances".

In 1947 some of us defied the powers and started the American Society for Atherosclerosis which later became the Council of the American Heart Association. Now everything is in place for a great future study of atherogenesis in which lipoproteins constitute an integral part.

The thickets are full of phonies and frauds ready to capitalize on so widespread and subtle a disease as atherosclerosis. I know of none so far who have invaded the territory of the legitimate investigations. But this situation will not last. So beware, and be courageous.

The ground has been prepared, the skilled young people are in place, and the challenge is clear. Genetics and molecular biology will be much more evident, but the older concepts need to be assimilated, not rejected and forgotten. New criteria need to be introduced to rejuvenate epidemiology. Chemists in industry should be stimulated to contribute their very great talent, and clusters of scientists and physicians established who are devoted to work in the laboratory and clinic rather than the committee rooms and meetings "elsewhere". More work and less talk is a requisite for greatness and this applies to me! The past must not be forgotten in creating the future.

Index

INDEX